Partial Student's Solutions Manual
for use with

Linear Algebra
Fourth Edition

with Applications

W. Keith Nicholson
University of Calgary

McGraw-Hill Ryerson

Toronto   Montréal   Boston   Burr Ridge, IL   Dubuque, IA   Madison, WI   New York
San Francisco   St. Louis   Bangkok   Bogotá   Caracas   Kuala Lumpur   Lisbon   London
Madrid   Mexico City   Milan   New Delhi   Santiago   Seoul   Singapore   Sydney   Taipei

McGraw-Hill
Ryerson Limited
A Subsidiary of The **McGraw·Hill** Companies

Partial Solutions Manual for use with
Linear Algebra with Applications
Fourth Edition

ISBN: 0-07-089232-6

1 2 3 4 5 6 7 8 9 10 CP 0 9 8 7 6 5 4 3

Printed and bound in Canada

Care has been taken to trace ownership of copyright material contained in this text; however, the publisher will welcome any information that enables them to rectify any reference or credit for subsequent editions.

Senior Sponsoring Editor: Cathy Koop
Developmental Editor: Darren Hick
Marketing Manager: David Groth
Production Coordinator: Jennifer Wilkie
Cover Design: Greg Devitt
Printer: Canadian Printco

TABLE OF CONTENTS

Chapter 1: Systems of Linear Equations

Exercises 1.1   Solutions and Elementary Operations

1(b) Substitute these values of x_1, x_2, x_3 and x_4 in the equation

$$2x_1 + 5x_2 + 9x_3 + 3x_4 = 2(2s + 12t + 13) + 5(s) + 9(-s - 3t - 3) + 3(t) = -1$$
$$x_1 + 2x_2 + 4x_3 = (2s + 12t + 13) + 2(s) + 4(-s - 3t - 3) = 1$$

Hence this is a solution for every value of s and t.

2(b) The equation is $2x + 3y = 1$. If $x = s$ then $y = \frac{1}{3}(1 - 2s)$ so this is one form of the general solution. Also, if $y = t$ then $x = \frac{1}{2}(1 - 3t)$ gives another form.

4. Given the equation $4x - 2y + 0z = 1$, take $y = s$ and $z = t$ and solve for x: $x = \frac{1}{4}(1 + 2s)$. This is the general solution.

5. $0x = 1$ has no solution; $2x = 3$ has one solution; $0x = 0$ has infinitely many solutions.

7(b) The augmented matrix is $\begin{bmatrix} 1 & 2 & | & 0 \\ 0 & 1 & | & 1 \end{bmatrix}$

(d) The augmented matrix is $\begin{bmatrix} 1 & 1 & 0 & | & 1 \\ 0 & 1 & 1 & | & 0 \\ -1 & 0 & 1 & | & 2 \end{bmatrix}$

8(b) A system with this augmented matrix is

$$\begin{array}{rcrcrcr} 2x & - & y & & & = & -1 \\ -3x & + & 2y & + & z & = & 0 \\ & & y & + & z & = & 3 \end{array}$$

9(b) $\begin{bmatrix} 1 & 2 & | & 1 \\ 3 & 4 & | & -1 \end{bmatrix} \rightarrow \begin{bmatrix} 1 & 2 & | & 1 \\ 0 & -2 & | & -4 \end{bmatrix} \rightarrow \begin{bmatrix} 1 & 2 & | & 1 \\ 0 & 1 & | & 2 \end{bmatrix} \rightarrow \begin{bmatrix} 1 & 0 & | & -3 \\ 0 & 1 & | & 2 \end{bmatrix}.$

Hence $x = -3, y = 2$.

(d) $\begin{bmatrix} 3 & 4 & | & 1 \\ 4 & 5 & | & -3 \end{bmatrix} \rightarrow \begin{bmatrix} 4 & 5 & | & -3 \\ 3 & 4 & | & 1 \end{bmatrix} \rightarrow \begin{bmatrix} 1 & 1 & | & -4 \\ 3 & 4 & | & 1 \end{bmatrix} \rightarrow \begin{bmatrix} 1 & 1 & | & -4 \\ 0 & 1 & | & 13 \end{bmatrix} \rightarrow \begin{bmatrix} 1 & 0 & | & -17 \\ 0 & 1 & | & 13 \end{bmatrix}.$

Hence $x = -17, y = 13$.

10(b) $\begin{bmatrix} 2 & 1 & 1 & | & -1 \\ 1 & 2 & 1 & | & 0 \\ 3 & 0 & -2 & | & 5 \end{bmatrix} \rightarrow \begin{bmatrix} 1 & 2 & 1 & | & 0 \\ 2 & 1 & 1 & | & -1 \\ 3 & 0 & -2 & | & 5 \end{bmatrix} \rightarrow \begin{bmatrix} 1 & 2 & 1 & | & 0 \\ 0 & -3 & -1 & | & -1 \\ 0 & -6 & -5 & | & 5 \end{bmatrix} \rightarrow \begin{bmatrix} 1 & 2 & 1 & | & 0 \\ 0 & 1 & \frac{1}{3} & | & \frac{1}{3} \\ 0 & 0 & -3 & | & 7 \end{bmatrix}$

$\rightarrow \begin{bmatrix} 1 & 0 & \frac{1}{3} & | & -\frac{2}{3} \\ 0 & 1 & \frac{1}{3} & | & \frac{1}{3} \\ 0 & 0 & 1 & | & -\frac{7}{3} \end{bmatrix} \rightarrow \begin{bmatrix} 1 & 0 & 0 & | & \frac{1}{9} \\ 0 & 1 & 0 & | & \frac{10}{9} \\ 0 & 0 & 1 & | & -\frac{7}{3} \end{bmatrix}$. Hence $x = \frac{1}{9}$, $y = \frac{10}{9}$, $z = \frac{-7}{3}$.

11(b) $\begin{bmatrix} 3 & -2 & | & 5 \\ -12 & 8 & | & 16 \end{bmatrix} \rightarrow \begin{bmatrix} 3 & -2 & | & 5 \\ 0 & 0 & | & 36 \end{bmatrix}$. The last equation is $0x + 0y = 36$, which has no solution.

14. The substitution gives $\begin{aligned} 3(5x' - 2y') + 2(-7x' + 3y') &= 5 \\ 7(5x' - 2y') + 5(-7x' + 3y') &= 1 \end{aligned}$; this simplifies to $x' = 5$, $y' = 1$. Hence $x = 5x' - 2y' = 23$ and $y = -7x' + 3y' = -32$.

17. If John gets \$$x$ per hour and Joe gets \$$y$ per hour, the two situations give $2x + 3y = 24.6$ and $3x + 2y = 23.9$. The solution is $x = \$4.50$ and $y = \$5.20$.

Exercises 1.2   Gaussian Elimination

1(b) No, No; no leading 1.

(d) No, Yes; not in reduced form because of the 3 and the top two 1's in the last column.

(f) No, No; the (reduced) row-echelon form would have two rows of zeros.

2(b) $\begin{bmatrix} 0 & -1 & 3 & 1 & 3 & 2 & 1 \\ 0 & -2 & 6 & 1 & -5 & 0 & -1 \\ 0 & 3 & -9 & 2 & 4 & 1 & -1 \\ 0 & 1 & -3 & -1 & 3 & 0 & 1 \end{bmatrix} \rightarrow \begin{bmatrix} 0 & 1 & -3 & -1 & -3 & -2 & -1 \\ 0 & 0 & 0 & -1 & -11 & 4 & -3 \\ 0 & 0 & 0 & 4 & 13 & 7 & 2 \\ 0 & 0 & 0 & 0 & 6 & 2 & 2 \end{bmatrix}$

$\rightarrow \begin{bmatrix} 0 & 1 & -3 & 0 & 8 & 2 & 2 \\ 0 & 0 & 0 & 1 & 11 & 4 & 3 \\ 0 & 0 & 0 & 0 & -42 & -13 & -13 \\ 0 & 0 & 0 & 0 & 6 & 2 & 2 \end{bmatrix} \rightarrow \begin{bmatrix} 0 & 1 & -3 & 0 & 8 & 2 & 2 \\ 0 & 0 & 0 & 1 & 11 & 4 & 3 \\ 0 & 0 & 0 & 0 & 0 & 1 & 1 \\ 0 & 0 & 0 & 0 & 3 & 1 & 1 \end{bmatrix}$

$\rightarrow \begin{bmatrix} 0 & 1 & -3 & 0 & 8 & 0 & 0 \\ 0 & 0 & 0 & 1 & 11 & 0 & -1 \\ 0 & 0 & 0 & 0 & 3 & 0 & 0 \\ 0 & 0 & 0 & 0 & 0 & 1 & 1 \end{bmatrix} \rightarrow \begin{bmatrix} 0 & 1 & -3 & 0 & 0 & 0 & 0 \\ 0 & 0 & 0 & 1 & 0 & 0 & -1 \\ 0 & 0 & 0 & 0 & 1 & 0 & 0 \\ 0 & 0 & 0 & 0 & 0 & 1 & 1 \end{bmatrix}$

3(b) The matrix is already in reduced row-echelon form. The nonleading variables are parameters;
$x_2 = r$, $x_4 = s$ and $x_6 = t$.
The first equation is $x_1 - 2x_2 + 2x_4 + x_6 = 1$, whence $x_1 = 1 + 2r - 2s - t$.
The second equation is $x_3 + 5x_4 - 3x_6 = -1$, whence $x_3 = -1 - 5s + 3t$.
The third equation is $x_5 + 6x_6 = 1$, whence $x_5 = 1 - 6t$.

(d) First carry the matrix to reduced row-echelon form.

$$\begin{bmatrix} 1 & -1 & 2 & 4 & 6 & 2 \\ 0 & 1 & 2 & 1 & -1 & -1 \\ 0 & 0 & 0 & 1 & 0 & 1 \\ 0 & 0 & 0 & 0 & 0 & 0 \end{bmatrix} \rightarrow \begin{bmatrix} 1 & 0 & 4 & 5 & 5 & 1 \\ 0 & 1 & 2 & 1 & -1 & -1 \\ 0 & 0 & 0 & 1 & 0 & 1 \\ 0 & 0 & 0 & 0 & 0 & 0 \end{bmatrix} \rightarrow \begin{bmatrix} 1 & 0 & 4 & 0 & 5 & -4 \\ 0 & 1 & 2 & 0 & -1 & -2 \\ 0 & 0 & 0 & 1 & 0 & 1 \\ 0 & 0 & 0 & 0 & 0 & 0 \end{bmatrix}.$$

The nonleading variables are parameters; $x_3 = s$, $x_5 = t$.
The first equation is $x_1 + 4x_3 + 5x_5 = -4$, whence $x_1 = -4 - 4s - 5t$.
The second equation is $x_2 + 2x_3 - x_5 = -2$, whence $x_2 = -2 - 2s + t$.
The third equation is $x_4 = 1$.

4(b) $\begin{bmatrix} 3 & -1 & 0 \\ 2 & -3 & 1 \end{bmatrix} \rightarrow \begin{bmatrix} 1 & 2 & -1 \\ 2 & -3 & 1 \end{bmatrix} \rightarrow \begin{bmatrix} 1 & 2 & -1 \\ 0 & -7 & 3 \end{bmatrix} \rightarrow \begin{bmatrix} 1 & 2 & -1 \\ 0 & 1 & -\frac{3}{7} \end{bmatrix} \rightarrow \begin{bmatrix} 1 & 0 & -\frac{1}{7} \\ 0 & 1 & -\frac{3}{7} \end{bmatrix}.$
Hence $x = -\frac{1}{7}$, $y = -\frac{3}{7}$.

(d) Note that the variables in the second equation are in the wrong order.
$\begin{bmatrix} 3 & -1 & 2 \\ -6 & 2 & -4 \end{bmatrix} \rightarrow \begin{bmatrix} 3 & -1 & 2 \\ 0 & 0 & 0 \end{bmatrix} \rightarrow \begin{bmatrix} 1 & -\frac{1}{3} & \frac{2}{3} \\ 0 & 0 & 0 \end{bmatrix}$. The nonleading variable $y = t$ is
a parameter; then $x = \frac{2}{3} + \frac{1}{3}t = \frac{1}{3}(t + 2)$.

(f) Again the order of the variables is reversed in the second equation. $\begin{bmatrix} 2 & -3 & 5 \\ -2 & 3 & 2 \end{bmatrix} \rightarrow$
$\begin{bmatrix} 2 & -3 & 5 \\ 0 & 0 & 7 \end{bmatrix}$. There is no solution as the second equation is $0x + 0y = 7$.

5(b) $\begin{bmatrix} -2 & 3 & 3 & -9 \\ 3 & -4 & 1 & 5 \\ -5 & 7 & 2 & -14 \end{bmatrix} \rightarrow \begin{bmatrix} 3 & -4 & 1 & 5 \\ -2 & 3 & 3 & -9 \\ -5 & 7 & 2 & -14 \end{bmatrix} \rightarrow \begin{bmatrix} 1 & -1 & 4 & -4 \\ -2 & 3 & 3 & -9 \\ -5 & 7 & 2 & -14 \end{bmatrix}$

$\rightarrow \begin{bmatrix} 1 & -1 & 4 & -4 \\ 0 & 1 & 11 & -17 \\ 0 & 2 & 22 & -34 \end{bmatrix} \rightarrow \begin{bmatrix} 1 & 0 & 15 & -21 \\ 0 & 1 & 11 & -17 \\ 0 & 0 & 0 & 0 \end{bmatrix}$. Take $z = t$ (the nonleading variable).

The equations give $x = -21 - 15t$, $y = -17 - 11t$.

(d) $\begin{bmatrix} 1 & 2 & -1 & 2 \\ 2 & 5 & -3 & 1 \\ 1 & 4 & -3 & 3 \end{bmatrix} \rightarrow \begin{bmatrix} 1 & 2 & -1 & 2 \\ 0 & 1 & -1 & -3 \\ 0 & 2 & -2 & 1 \end{bmatrix} \rightarrow \begin{bmatrix} 1 & 2 & -1 & 2 \\ 0 & 1 & -1 & -3 \\ 0 & 0 & 0 & 7 \end{bmatrix}$. There is no solution
as the third equation is $0x + 0y + 0z = 7$.

(f) $\begin{bmatrix} 3 & -2 & 1 & | & -2 \\ 1 & -1 & 3 & | & 5 \\ -1 & 1 & 1 & | & -1 \end{bmatrix} \rightarrow \begin{bmatrix} 1 & -1 & 3 & | & 5 \\ 3 & -2 & 1 & | & -2 \\ -1 & 1 & 1 & | & -1 \end{bmatrix} \rightarrow \begin{bmatrix} 1 & -1 & 3 & | & 5 \\ 0 & 1 & -8 & | & -17 \\ 0 & 0 & 4 & | & 4 \end{bmatrix}$

$\rightarrow \begin{bmatrix} 1 & 0 & -5 & | & -12 \\ 0 & 1 & -8 & | & -17 \\ 0 & 0 & 1 & | & 1 \end{bmatrix} \rightarrow \begin{bmatrix} 1 & 0 & 0 & | & -7 \\ 0 & 1 & 0 & | & -9 \\ 0 & 0 & 1 & | & 1 \end{bmatrix}$. Hence $x = -7$, $y = -9$, $z = 1$.

(h) $\begin{bmatrix} 1 & 2 & -4 & 10 \\ 2 & -1 & 2 & 5 \\ 1 & 1 & -2 & 7 \end{bmatrix} \rightarrow \begin{bmatrix} 1 & 2 & -4 & | & 10 \\ 0 & -5 & 10 & | & -15 \\ 0 & -1 & 2 & | & -3 \end{bmatrix} \rightarrow \begin{bmatrix} 1 & 2 & -4 & | & 10 \\ 0 & 1 & -2 & | & 3 \\ 0 & 0 & 0 & | & 0 \end{bmatrix}$

$\rightarrow \begin{bmatrix} 1 & 0 & 0 & | & 4 \\ 0 & 1 & -2 & | & 3 \\ 0 & 0 & 0 & | & 0 \end{bmatrix}$. Hence $z = t$, $x = 4$, $y = 3 + 2t$.

6(b) $\begin{bmatrix} 2 & 1 & -3 & | & -3 \\ 3 & 1 & -5 & | & 5 \\ -2 & 1 & 5 & | & -35 \end{bmatrix} \begin{matrix} R_1 \\ R_2 \\ R_3 \end{matrix} \rightarrow \begin{bmatrix} 3 & 1 & -5 & | & 5 \\ 2 & 1 & -3 & | & -3 \\ -2 & 1 & 5 & | & -35 \end{bmatrix} \begin{matrix} R_2 \\ R_1 \\ R_3 \end{matrix} \rightarrow \begin{bmatrix} 1 & 0 & -2 & | & 8 \\ 2 & 1 & -3 & | & -3 \\ -2 & 1 & 5 & | & -35 \end{bmatrix} \begin{matrix} R_2 - R_1 \\ R_1 \\ R_3 \end{matrix}$

$\rightarrow \begin{bmatrix} 1 & 0 & -2 & | & 8 \\ 0 & 1 & 1 & | & -19 \\ 0 & 1 & 1 & | & -19 \end{bmatrix} \begin{matrix} R_2 - R_1 \\ R_1 - 2(R_2 - R_1) = 3R_1 - 2R_2 \\ R_3 + 2(R_2 - R_1) = R_3 + 2R_2 - 2R_1 \end{matrix}$

$\rightarrow \begin{bmatrix} 1 & 0 & -2 & | & 8 \\ 0 & 1 & 1 & | & -19 \\ 0 & 0 & 0 & | & 0 \end{bmatrix} \begin{matrix} R_2 - R_1 \\ 3R_1 - 2R_2 \\ (R_3 + 2R_2 - 2R_1) - (3R_1 - 2R_2) = R_3 - 5R_1 + 4R_2 \end{matrix}$

Thus $R_3 = 5R_1 - 4R_2$. Hence equation 3 is 5 times equation 1 minus 4 times equation 2. The solution is $x_3 = t$, $x_1 = 8 + 2t$, $x_2 = -19 - t$.

7(b) $\begin{bmatrix} 1 & -1 & 1 & -1 & | & 0 \\ -1 & 1 & 1 & 1 & | & 0 \\ 1 & 1 & -1 & 1 & | & 0 \\ 1 & 1 & 1 & 1 & | & 0 \end{bmatrix} \rightarrow \begin{bmatrix} 1 & -1 & 1 & -1 & | & 0 \\ 0 & 0 & 2 & 0 & | & 0 \\ 0 & 2 & -2 & 2 & | & 0 \\ 0 & 2 & 0 & 2 & | & 0 \end{bmatrix} \rightarrow \begin{bmatrix} 1 & -1 & 1 & -1 & | & 0 \\ 0 & 1 & -1 & 1 & | & 0 \\ 0 & 0 & 1 & 0 & | & 0 \\ 0 & 1 & 0 & 1 & | & 0 \end{bmatrix}$

$\rightarrow \begin{bmatrix} 1 & 0 & 0 & 0 & | & 0 \\ 0 & 1 & -1 & 1 & | & 0 \\ 0 & 0 & 1 & 0 & | & 0 \\ 0 & 0 & 1 & 0 & | & 0 \end{bmatrix} \rightarrow \begin{bmatrix} 1 & 0 & 0 & 0 & | & 0 \\ 0 & 1 & 0 & 1 & | & 0 \\ 0 & 0 & 1 & 0 & | & 0 \\ 0 & 0 & 0 & 0 & | & 0 \end{bmatrix}$. Hence $x_4 = t$; $x_1 = 0$, $x_2 = -t$, $x_3 = 0$.

(d)
$$\begin{bmatrix} 1 & 1 & 2 & -1 & | & 4 \\ 0 & 3 & -1 & 4 & | & 2 \\ 1 & 2 & -3 & 5 & | & 0 \\ 1 & 1 & -5 & 6 & | & -3 \end{bmatrix} \rightarrow \begin{bmatrix} 1 & 1 & 2 & -1 & | & 4 \\ 0 & 3 & -1 & 4 & | & 2 \\ 0 & 1 & -5 & 6 & | & -4 \\ 0 & 0 & -7 & 7 & | & -7 \end{bmatrix} \rightarrow \begin{bmatrix} 1 & 0 & 7 & -7 & | & 8 \\ 0 & 0 & 14 & -14 & | & 14 \\ 0 & 1 & -5 & 6 & | & -4 \\ 0 & 0 & -7 & 7 & | & -7 \end{bmatrix}$$

$$\rightarrow \begin{bmatrix} 1 & 0 & 7 & -7 & | & 8 \\ 0 & 1 & -5 & 6 & | & -4 \\ 0 & 0 & 14 & -14 & | & 14 \\ 0 & 0 & -7 & 7 & | & -7 \end{bmatrix} \rightarrow \begin{bmatrix} 1 & 0 & 0 & 0 & | & 1 \\ 0 & 1 & -5 & 6 & | & -4 \\ 0 & 0 & 1 & -1 & | & 1 \\ 0 & 0 & 0 & 0 & | & 0 \end{bmatrix} \rightarrow \begin{bmatrix} 1 & 0 & 0 & 0 & | & 1 \\ 0 & 1 & 0 & 1 & | & 1 \\ 0 & 0 & 1 & -1 & | & 1 \\ 0 & 0 & 0 & 0 & | & 0 \end{bmatrix}.$$

Hence $x_4 = t$; $x_1 = 1$, $x_2 = 1 - t$, $x_3 = 1 + t$.

8(b) $\begin{bmatrix} 1 & b & | & -1 \\ a & 2 & | & 5 \end{bmatrix} \rightarrow \begin{bmatrix} 1 & b & | & -1 \\ 0 & 2-ab & | & 5+a \end{bmatrix}.$

Case 1 If $ab \neq 2$, it continues $\rightarrow \begin{bmatrix} 1 & b & | & -1 \\ 0 & 1 & | & \frac{5+a}{2-ab} \end{bmatrix} \rightarrow \begin{bmatrix} 1 & 0 & | & \frac{-2-5b}{2-ab} \\ 0 & 1 & | & \frac{5+a}{2-ab} \end{bmatrix}.$

The unique solution is $x = \frac{-2-5b}{2-ab}$, $y = \frac{5+a}{2-ab}$.

Case 2 If $ab = 2$, it is $\begin{bmatrix} 1 & b & | & -1 \\ 0 & 0 & | & 5+a \end{bmatrix}$. Hence there is no solution if $a \neq -5$. If $a = -5$,

then $b = \frac{-2}{5}$ and the matrix is $\begin{bmatrix} 1 & \frac{-2}{5} & | & -1 \\ 0 & 0 & | & 0 \end{bmatrix}$. Then $y = t$, $x = -1 + \frac{2}{5}t$.

8(d) $\begin{bmatrix} a & 1 & | & 1 \\ 2 & 1 & | & b \end{bmatrix} \rightarrow \begin{bmatrix} 1 & \frac{1}{2} & | & \frac{b}{2} \\ a & 1 & | & 1 \end{bmatrix} \rightarrow \begin{bmatrix} 1 & \frac{1}{2} & | & \frac{b}{2} \\ a & 1-\frac{a}{2} & | & 1-\frac{ab}{2} \end{bmatrix} \rightarrow \begin{bmatrix} 1 & \frac{1}{2} & | & \frac{b}{2} \\ 0 & 2-a & | & 2-ab \end{bmatrix}.$

Case 1 If $a \neq 2$ it continues:$\rightarrow \begin{bmatrix} 1 & \frac{1}{2} & | & \frac{b}{2} \\ 0 & 1 & | & \frac{2-ab}{2-a} \end{bmatrix} \rightarrow \begin{bmatrix} 1 & 0 & | & \frac{b-1}{2-a} \\ 0 & 1 & | & \frac{2-ab}{2-a} \end{bmatrix}.$

The unique solution: $x = \frac{b-1}{2-a}$, $y = \frac{2-ab}{2-a}$.

Case 2 If $a = 2$ the matrix is $\begin{bmatrix} 1 & \frac{1}{2} & | & \frac{b}{2} \\ 0 & 0 & | & 2(1-b) \end{bmatrix}.$

Hence there is no solution if $b \neq 1$. If $b = 1$ the matrix is $\begin{bmatrix} 1 & \frac{1}{2} & | & \frac{1}{2} \\ 0 & 0 & | & 0 \end{bmatrix}$, so $y = t$,

$x = \frac{1}{2} - \frac{1}{2}t = \frac{1}{2}(1-t)$.

9(b) $\begin{bmatrix} 2 & 1 & -1 & | & a \\ 0 & 2 & 3 & | & b \\ 1 & 0 & -1 & | & c \end{bmatrix} \rightarrow \begin{bmatrix} 1 & 0 & -1 & | & c \\ 0 & 2 & 3 & | & b \\ 2 & 1 & -1 & | & a \end{bmatrix} \rightarrow \begin{bmatrix} 1 & 0 & -1 & | & c \\ 0 & 2 & 3 & | & b \\ 0 & 1 & 1 & | & a-2c \end{bmatrix}$

$$\rightarrow \begin{bmatrix} 1 & 0 & -1 & c \\ 0 & 1 & 1 & a-2c \\ 0 & 2 & 3 & b \end{bmatrix} \rightarrow \begin{bmatrix} 1 & 0 & -1 & c \\ 0 & 1 & 1 & a-2c \\ 0 & 0 & 1 & b-2a+4c \end{bmatrix} \rightarrow \begin{bmatrix} 1 & 0 & 0 & b-2a+5c \\ 0 & 1 & 0 & 3a-b-6c \\ 0 & 0 & 1 & b-2a+4c \end{bmatrix}$$

Hence, for any values of a, b and c there is a unique solution $x = -2a+b+5c$, $y = 3a-b-6c$, and $z = -2a+b+4c$.

(d) $\begin{bmatrix} 1 & a & 0 & 0 \\ 0 & 1 & b & 0 \\ c & 0 & 1 & 0 \end{bmatrix} \rightarrow \begin{bmatrix} 1 & a & 0 & 0 \\ 0 & 1 & b & 0 \\ 0 & -ac & 1 & 0 \end{bmatrix} \rightarrow \begin{bmatrix} 1 & 0 & -ab & 0 \\ 0 & 1 & b & 0 \\ 0 & 0 & 1+abc & 0 \end{bmatrix}$

Case 1 If $abc \neq -1$, it continues: $\rightarrow \begin{bmatrix} 1 & 0 & -ab & 0 \\ 0 & 1 & b & 0 \\ 0 & 0 & 1 & 0 \end{bmatrix} \rightarrow \begin{bmatrix} 1 & 0 & 0 & 0 \\ 0 & 1 & 0 & 0 \\ 0 & 0 & 1 & 0 \end{bmatrix}$.

Hence we have the unique solution $x = 0$, $y = 0$, $z = 0$.

Case 2 If $abc = -1$, the matrix is $\begin{bmatrix} 1 & 0 & -ab & 0 \\ 0 & 1 & 0 & 0 \\ 0 & 0 & 0 & 0 \end{bmatrix}$, so $z = t$, $x = abt$, $y = -bt$.

Note: It is impossible that there is no solution here: $x = y = z = 0$ always works.

9(f) $\begin{bmatrix} 1 & a & -1 & 1 \\ -1 & a-2 & 1 & -1 \\ 2 & 2 & a-2 & 1 \end{bmatrix} \rightarrow \begin{bmatrix} 1 & a & -1 & 1 \\ 0 & 2(a-1) & 0 & 0 \\ 0 & 2(a-1) & a & -1 \end{bmatrix} \rightarrow \begin{bmatrix} 1 & a & -1 & 1 \\ 0 & a-1 & 0 & 0 \\ 0 & 0 & a & -1 \end{bmatrix}$.

Case 1 If $a = 1$ the matrix is $\begin{bmatrix} 1 & 1 & -1 & 1 \\ 0 & 0 & 0 & 0 \\ 0 & 0 & 1 & -1 \end{bmatrix} \rightarrow \begin{bmatrix} 1 & 1 & 0 & 0 \\ 0 & 0 & 1 & -1 \\ 0 & 0 & 0 & 0 \end{bmatrix}$,

so $y = t$, $x = -t$, $z = -1$.

Case 2 If $a = 0$ the last equation is $0x + 0y + 0z = -1$, so there is no solution.

Case 3 If $a \neq 1$ and $a \neq 0$, there is a unique solution:

$\begin{bmatrix} 1 & a & -1 & 1 \\ 0 & a-1 & 0 & 0 \\ 0 & 0 & a & -1 \end{bmatrix} \rightarrow \begin{bmatrix} 1 & a & -1 & 1 \\ 0 & 1 & 0 & 0 \\ 0 & 0 & 1 & -\frac{1}{a} \end{bmatrix} \rightarrow \begin{bmatrix} 1 & 0 & 0 & 1-\frac{1}{a} \\ 0 & 1 & 0 & 0 \\ 0 & 0 & 1 & -\frac{1}{a} \end{bmatrix}$.

Hence $x = 1 - \frac{1}{a}$, $y = 0$, $z = -\frac{1}{a}$.

10(b) $\begin{bmatrix} 2 & 1 & -1 & 3 \\ 0 & 0 & 0 & 0 \end{bmatrix} \rightarrow \begin{bmatrix} 1 & \frac{1}{2} & -\frac{1}{2} & \frac{3}{2} \\ 0 & 0 & 0 & 0 \end{bmatrix}$; rank is 1.

(d) It is in row-echelon form; rank is 3.

(f) $\begin{bmatrix} 0 & 0 & 1 \\ 0 & 0 & 1 \\ 0 & 0 & 1 \end{bmatrix} \rightarrow \begin{bmatrix} 0 & 0 & 1 \\ 0 & 0 & 0 \\ 0 & 0 & 0 \end{bmatrix}$; rank is 1.

11(b) $\begin{bmatrix} -2 & 3 & 3 \\ 3 & -4 & 1 \\ -5 & 7 & 2 \end{bmatrix} \rightarrow \begin{bmatrix} 1 & -1 & 4 \\ 3 & -4 & 1 \\ -5 & 7 & 2 \end{bmatrix} \rightarrow \begin{bmatrix} 1 & -1 & 4 \\ 0 & -1 & -11 \\ 0 & 2 & 22 \end{bmatrix} \rightarrow \begin{bmatrix} 1 & -1 & 4 \\ 0 & 1 & 11 \\ 0 & 0 & 0 \end{bmatrix}$; rank is 2.

(d) $\begin{bmatrix} 3 & -2 & 1 & -2 \\ 1 & -1 & 3 & 5 \\ -1 & 1 & 1 & -1 \end{bmatrix} \rightarrow \begin{bmatrix} 1 & -1 & 3 & 5 \\ 3 & -2 & 1 & -2 \\ -1 & 1 & 1 & -1 \end{bmatrix} \rightarrow \begin{bmatrix} 1 & -1 & 3 & 5 \\ 0 & 1 & -8 & -17 \\ 0 & 0 & 4 & 4 \end{bmatrix}$

$\rightarrow \begin{bmatrix} 1 & -1 & 3 & 5 \\ 0 & 1 & -8 & -17 \\ 0 & 0 & 1 & 1 \end{bmatrix}$; rank = 3.

(f) $\begin{bmatrix} 1 & 1 & 2 & a^2 \\ 1 & 1-a & 2 & 0 \\ 2 & 2-a & 6-a & 4 \end{bmatrix} \rightarrow \begin{bmatrix} 1 & 1 & 2 & a^2 \\ 0 & -a & 0 & -a^2 \\ 0 & -a & 2-a & 4-2a^2 \end{bmatrix} \rightarrow \begin{bmatrix} 1 & 1 & 2 & a^2 \\ 0 & a & 0 & a^2 \\ 0 & 0 & 2-a & 4-a^2 \end{bmatrix}$.

If $a = 0$ we get $\begin{bmatrix} 1 & 1 & 2 & 0 \\ 0 & 0 & 0 & 0 \\ 0 & 0 & 2 & 4 \end{bmatrix} \rightarrow \begin{bmatrix} 1 & 1 & 2 & 0 \\ 0 & 0 & 1 & 2 \\ 0 & 0 & 0 & 0 \end{bmatrix}$; rank = 2.

If $a = 2$ we get $\begin{bmatrix} 1 & 1 & 2 & 4 \\ 0 & 2 & 0 & 4 \\ 0 & 0 & 0 & 0 \end{bmatrix} \rightarrow \begin{bmatrix} 1 & 1 & 2 & 4 \\ 0 & 1 & 0 & 2 \\ 0 & 0 & 0 & 0 \end{bmatrix}$; rank = 2.

If $a \neq 0$, $a \neq 2$, we get $\begin{bmatrix} 1 & 1 & 2 & a^2 \\ 0 & a & 0 & a^2 \\ 0 & 0 & 2-a & 4-a^2 \end{bmatrix} \rightarrow \begin{bmatrix} 1 & 1 & 2 & a^2 \\ 0 & 1 & 0 & a \\ 0 & 0 & 1 & 2+a \end{bmatrix}$; rank = 3.

12(b) False. $A = \left[\begin{array}{cc|c} 1 & 0 & 1 \\ 0 & 1 & 1 \\ 0 & 0 & 0 \end{array}\right]$

(d) Flase. $A = \left[\begin{array}{cc|c} 1 & 0 & 1 \\ 0 & 1 & 1 \\ 0 & 0 & 0 \end{array}\right]$

(f) True. A has 3 rows so there can be at most 3 leading 1's. Hence the rank of A is at most 3.

14(b) We begin the row reduction $\begin{bmatrix} 1 & a & b+c \\ 1 & b & c+a \\ 1 & c & a+b \end{bmatrix} \rightarrow \begin{bmatrix} 1 & a & b+c \\ 0 & b-a & a-b \\ 0 & c-a & a-c \end{bmatrix}$. Now one of $b-a$ and $c-a$ is nonzero (by hypothesis) so that row provides the second leading 1 (its row becomes $\begin{bmatrix} 0 & 1 & -1 \end{bmatrix}$). Hence further row operations give

$$\rightarrow \begin{bmatrix} 1 & a & b+c \\ 0 & 1 & -1 \\ 0 & 0 & 0 \end{bmatrix} \rightarrow \begin{bmatrix} 1 & 0 & b+c+a \\ 0 & 1 & -1 \\ 0 & 0 & 0 \end{bmatrix}$$

which has the given form.

16(b) Substituting the coordinates of the three points in the equation gives

$$1 + 1 + a + b + c = 0 \qquad a + b + c = -2$$
$$25 + 9 + 5a - 3b + c = 0 \quad 5a - 3b + c = -34$$
$$9 + 9 - 3a - 3b + c = 0 \qquad 3a + 3b - c = 18$$

$$\begin{bmatrix} 1 & 1 & 1 & -2 \\ 5 & -3 & 1 & -34 \\ 3 & 3 & -1 & 18 \end{bmatrix} \rightarrow \begin{bmatrix} 1 & 1 & 1 & -2 \\ 0 & -8 & -4 & -24 \\ 0 & 0 & -4 & 24 \end{bmatrix} \rightarrow \begin{bmatrix} 1 & 1 & 1 & -2 \\ 0 & 1 & \frac{1}{2} & 3 \\ 0 & 0 & 1 & -6 \end{bmatrix}$$

$$\rightarrow \begin{bmatrix} 1 & 0 & \frac{1}{2} & -5 \\ 0 & 1 & \frac{1}{2} & 3 \\ 0 & 0 & 1 & -6 \end{bmatrix} \rightarrow \begin{bmatrix} 1 & 0 & 0 & -2 \\ 0 & 1 & 0 & 6 \\ 0 & 0 & 1 & -6 \end{bmatrix}.$$

Hence $a = -2$, $b = 6$, $c = -6$, so the equation is $x^2 - y^2 - 2x + 6y - 6 = 0$.

18. Let a, b and c denote the fractions of the student population in Clubs A, B and C respectively. The new students in Club A arrived as follows: $\frac{4}{10}$ of those in Club A stayed; $\frac{2}{10}$ of those in Club B go to A, and $\frac{2}{10}$ of those in C go to A. Hence

$$a = \tfrac{4}{10}a + \tfrac{2}{10}b + \tfrac{2}{10}c.$$

Similarly, looking at students in Club B and C.

$$b = \tfrac{1}{10}a + \tfrac{7}{10}b + \tfrac{2}{10}c$$

$$c = \tfrac{5}{10}a + \tfrac{1}{10}b + \tfrac{6}{10}c.$$

Hence

$$-6a + 2b + 2c = 0$$
$$a - 3b + 2c = 0$$
$$5a + b - 4c = 0$$

$$\begin{bmatrix} -6 & 2 & 2 & | & 0 \\ 1 & -3 & 2 & | & 0 \\ 5 & 1 & -4 & | & 0 \end{bmatrix} \rightarrow \begin{bmatrix} 1 & -3 & 2 & | & 0 \\ 0 & -16 & 14 & | & 0 \\ 0 & 16 & -14 & | & 0 \end{bmatrix} \rightarrow \begin{bmatrix} 1 & -3 & 2 & | & 0 \\ 0 & 1 & -\frac{7}{8} & | & 0 \\ 0 & 0 & 0 & | & 0 \end{bmatrix}$$

$$\rightarrow \begin{bmatrix} 1 & 0 & -\frac{5}{8} & | & 0 \\ 0 & 1 & -\frac{7}{8} & | & 0 \\ 0 & 0 & 0 & | & 0 \end{bmatrix}.$$ Thus the solution is $a = \frac{5}{8}t$, $b = \frac{7}{8}t$, $c = t$. However $a + b + c = 1$

(because every student belongs to exactly one club) which gives $t = \frac{2}{5}$. Hence $a = \frac{5}{20}$, $b = \frac{7}{20}$, $c = \frac{8}{20}$.

Exercises 1.3   Homogeneous Equations

1(b) False. $A = \begin{bmatrix} 1 & 0 & 1 & | & 0 \\ 0 & 1 & 1 & | & 0 \end{bmatrix}$

(d) False. $A = \begin{bmatrix} 1 & 0 & 1 & | & 1 \\ 0 & 1 & 1 & | & 0 \end{bmatrix}$

(f) False. $A = \begin{bmatrix} 1 & 0 & | & 0 \\ 0 & 1 & | & 0 \end{bmatrix}$

(h) False. $A = \begin{bmatrix} 1 & 0 & | & 0 \\ 0 & 1 & | & 0 \\ 0 & 0 & | & 0 \end{bmatrix}$

2(b) $\begin{bmatrix} 1 & 2 & 1 & | & 0 \\ 1 & 3 & 6 & | & 0 \\ 2 & 3 & a & | & 0 \end{bmatrix} \rightarrow \begin{bmatrix} 1 & 2 & 1 & | & 0 \\ 0 & 1 & 5 & | & 0 \\ 0 & -1 & a-2 & | & 0 \end{bmatrix} \rightarrow \begin{bmatrix} 1 & 0 & -9 & | & 0 \\ 0 & 1 & 5 & | & 0 \\ 0 & 0 & a+3 & | & 0 \end{bmatrix}.$
Hence there is a nontrivial solution when $a = -3 : x = 9t$, $y = -5t$, $z = t$.

(d) $\begin{bmatrix} a & 1 & 1 & | & 0 \\ 1 & 1 & -1 & | & 0 \\ 1 & 1 & a & | & 0 \end{bmatrix} \rightarrow \begin{bmatrix} 1 & 1 & -1 & | & 0 \\ a & 1 & 1 & | & 0 \\ 1 & 1 & a & | & 0 \end{bmatrix} \rightarrow \begin{bmatrix} 1 & 1 & -1 & | & 0 \\ 0 & 1-a & 1+a & | & 0 \\ 0 & 0 & a+1 & | & 0 \end{bmatrix}.$
Hence if $a \neq 1$ and $a \neq -1$, there is a unique, trivial solution. The other cases are as follows:

$a = 1 : \begin{bmatrix} 1 & 1 & -1 & | & 0 \\ 0 & 0 & 2 & | & 0 \\ 0 & 0 & 2 & | & 0 \end{bmatrix} \rightarrow \begin{bmatrix} 1 & 1 & 0 & | & 0 \\ 0 & 0 & 1 & | & 0 \\ 0 & 0 & 0 & | & 0 \end{bmatrix} ; x = -t, y = t, z = 0.$

$$a = -1: \begin{bmatrix} 1 & 1 & -1 & | & 0 \\ 0 & 2 & 0 & | & 0 \\ 0 & 0 & 0 & | & 0 \end{bmatrix} \rightarrow \begin{bmatrix} 1 & 0 & -1 & | & 0 \\ 0 & 1 & 0 & | & 0 \\ 0 & 0 & 0 & | & 0 \end{bmatrix}; \ x = t, \ y = 0, \ z = t.$$

3(b) The system $\begin{array}{r} x + y = 1 \\ 2x + 2y = 2 \\ -x - y = -1 \end{array}$ has nontrivial solutions with <u>fewer</u> variables than equations.

4(b) There are $n - r = 6 - 1 = 5$ parameters by Theorem 2 §1.2.

(d) The row-echelon form has four rows and, as it has a row of zeros, has at most 3 leading 1's. Hence rank $A = r = 1, 2$ or 3 ($r \neq 0$ because A has nonzero entries). Thus there are $n - r = 6 - r = 5, 4$ or 3 parameters.

6(b) Insisting that the graph of $ax + by + cz + d = 0$ (the plane) contains the three points leads to three linear equations in the four variables a, b, c and d. There is a nontrivial solution by Theorem 1.

Exercises 1.4   An Application to Network Flows

1(b) There are five flow equations, one for each junction:

$$\begin{array}{rcrcrcrcrcrcl}
f_1 & - & f_2 & & & & & & & & & = & 25 \\
f_1 & & & + & f_3 & & & + & f_5 & & & = & 50 \\
& & f_2 & & & + & f_4 & & & + & f_7 & = & 60 \\
& & & - & f_3 & + & f_4 & & & + & f_6 & = & 75 \\
& & & & & & & f_5 & + & f_6 & - & f_7 & = & 40
\end{array}$$

$$\begin{bmatrix} 1 & -1 & 0 & 0 & 0 & 0 & 0 & | & 25 \\ 1 & 0 & 1 & 0 & 1 & 0 & 0 & | & 50 \\ 0 & 1 & 0 & 1 & 0 & 0 & 1 & | & 60 \\ 0 & 0 & -1 & 1 & 0 & 1 & 0 & | & 75 \\ 0 & 0 & 0 & 0 & 1 & 1 & -1 & | & 40 \end{bmatrix} \rightarrow \begin{bmatrix} 1 & -1 & 0 & 0 & 0 & 0 & 0 & | & 25 \\ 0 & 1 & 1 & 0 & 1 & 0 & 0 & | & 25 \\ 0 & 1 & 0 & 1 & 0 & 0 & 1 & | & 60 \\ 0 & 0 & -1 & 1 & 0 & 1 & 0 & | & 75 \\ 0 & 0 & 0 & 0 & 1 & 1 & -1 & | & 40 \end{bmatrix}$$

$$\rightarrow \begin{bmatrix} 1 & 0 & 1 & 0 & 1 & 0 & 0 & | & 50 \\ 0 & 1 & 1 & 0 & 1 & 0 & 0 & | & 25 \\ 0 & 0 & -1 & 1 & -1 & 0 & 1 & | & 35 \\ 0 & 0 & -1 & 1 & 0 & 1 & 0 & | & 75 \\ 0 & 0 & 0 & 0 & 1 & 1 & -1 & | & 40 \end{bmatrix} \rightarrow \begin{bmatrix} 1 & 0 & 0 & 1 & 0 & 0 & 1 & | & 85 \\ 0 & 1 & 0 & 1 & 0 & 0 & 1 & | & 60 \\ 0 & 0 & 1 & -1 & 1 & 0 & -1 & | & -35 \\ 0 & 0 & 0 & 0 & 1 & 1 & -1 & | & 40 \\ 0 & 0 & 0 & 0 & 1 & 1 & -1 & | & 40 \end{bmatrix}$$

$$\rightarrow \begin{bmatrix} 1 & 0 & 0 & 1 & 0 & 0 & 1 & | & 85 \\ 0 & 1 & 0 & 1 & 0 & 0 & 1 & | & 60 \\ 0 & 0 & 1 & -1 & 0 & -1 & 0 & | & -75 \\ 0 & 0 & 0 & 0 & 1 & 1 & -1 & | & 40 \\ 0 & 0 & 0 & 0 & 0 & 0 & 0 & | & 0 \end{bmatrix}$$

If we use f_4, f_6 , and f_7 as parameters, the solution is

$$f_1 = 85 - f_4 - f_7$$
$$f_2 = 60 - f_4 - f_7$$
$$f_3 = -75 + f_4 + f_6$$
$$f_5 = 40 - f_6 + f_7.$$

2(b) The solution to (a) gives $f_1 = 55 - f_4$, $f_2 = 20 - f_4 + f_5$, $f_3 = 15 - f_5$. Closing canal BC means $f_3 = 0$, so $f_5 = 15$. Hence $f_2 = 35 - f_4$, so $f_2 \le 30$ means $f_4 \ge 5$. Similarly $f_1 = 55 - f_4$ so $f_1 \le 30$ implies $f_4 \ge 25$. Hence the range on f_4 is $25 \le f_4 \le 30$.

Exercises 1.5   An Application to Electrical Networks

2. The junction and circuit rules give:

$$
\begin{array}{llrcrcrcr}
\text{Left junction} & & I_1 & - & I_2 & + & I_3 & = & 0 \\
\text{Right junction} & & I_1 & - & I_2 & + & I_3 & = & 0 \\
\text{Top circuit} & & 5I_1 & + & 10I_2 & & & = & 5 \\
\text{Lower circuit} & & & & 10I_2 & + & 5I_3 & = & 10
\end{array}
$$

$$\begin{bmatrix} 1 & -1 & 1 & | & 0 \\ 5 & 10 & 0 & | & 5 \\ 0 & 10 & 5 & | & 10 \end{bmatrix} \rightarrow \begin{bmatrix} 1 & -1 & 1 & | & 0 \\ 0 & 15 & -5 & | & 5 \\ 0 & 10 & 5 & | & 10 \end{bmatrix} \rightarrow \begin{bmatrix} 1 & -1 & 1 & | & 0 \\ 0 & 3 & -1 & | & 1 \\ 0 & 2 & 1 & | & 2 \end{bmatrix}$$

$$\rightarrow \begin{bmatrix} 1 & -1 & 1 & | & 0 \\ 0 & 1 & -2 & | & -1 \\ 0 & 2 & 1 & | & 2 \end{bmatrix} \rightarrow \begin{bmatrix} 1 & 0 & -1 & | & -1 \\ 0 & 1 & -2 & | & -1 \\ 0 & 0 & 5 & | & 4 \end{bmatrix} \rightarrow \begin{bmatrix} 1 & 0 & -1 & | & -1 \\ 0 & 1 & -2 & | & -1 \\ 0 & 0 & 1 & | & \frac{4}{5} \end{bmatrix}$$

$$\rightarrow \begin{bmatrix} 1 & 0 & 0 & | & -\frac{1}{5} \\ 0 & 1 & 0 & | & \frac{3}{5} \\ 0 & 0 & 1 & | & \frac{4}{5} \end{bmatrix}.$$

Hence $I_1 = -\frac{1}{5}$, $I_2 = \frac{3}{5}$ and $I_3 = \frac{4}{5}$.

4. The equations are:

$$\text{Lower left junction} \quad I_1 - I_5 - I_6 = 0$$
$$\text{Top junction} \quad I_2 - I_4 + I_6 = 0$$
$$\text{Middle junction} \quad I_2 + I_3 - I_5 = 0$$
$$\text{Lower right junction} \quad I_1 - I_3 - I_4 = 0$$

Observe that the last of these follows from the others (so may be omitted).

$$\text{Left circuit} \quad 10I_5 - 10I_6 = 10$$
$$\text{Right circuit} \quad -10I_3 + 10I_4 = 10$$
$$\text{Lower circuit} \quad 10I_3 + 10I_5 = 20$$

$$
\begin{bmatrix}
1 & 0 & 0 & 0 & -1 & -1 & 0 \\
0 & 1 & 0 & -1 & 0 & 1 & 0 \\
0 & 1 & 1 & 0 & -1 & 0 & 0 \\
0 & 0 & 0 & 0 & 10 & -1 & 10 \\
0 & 0 & -10 & 10 & 0 & 0 & 10 \\
0 & 0 & 10 & 0 & 10 & 0 & 20
\end{bmatrix}
\rightarrow
\begin{bmatrix}
1 & 0 & 0 & 0 & -1 & -1 & 0 \\
0 & 1 & 0 & -1 & 0 & 1 & 0 \\
0 & 0 & 1 & 1 & -1 & -1 & 0 \\
0 & 0 & 0 & 0 & 1 & -1 & 1 \\
0 & 0 & -1 & 1 & 0 & 0 & 1 \\
0 & 0 & 1 & 0 & 1 & 0 & 2
\end{bmatrix}
$$

$$
\rightarrow
\begin{bmatrix}
1 & 0 & 0 & 0 & -1 & -1 & 0 \\
0 & 1 & 0 & -1 & 0 & 1 & 0 \\
0 & 0 & 1 & 1 & -1 & -1 & 0 \\
0 & 0 & 0 & 0 & 1 & -1 & 1 \\
0 & 0 & 0 & 2 & -1 & -1 & 1 \\
0 & 0 & 0 & -1 & 2 & 1 & 2
\end{bmatrix}
\rightarrow
\begin{bmatrix}
1 & 0 & 0 & 0 & -1 & -1 & 0 \\
0 & 1 & 0 & 0 & -2 & 0 & -2 \\
0 & 0 & 1 & 0 & 1 & 0 & 2 \\
0 & 0 & 0 & 0 & 1 & -1 & 1 \\
0 & 0 & 0 & 0 & 3 & 1 & 5 \\
0 & 0 & 0 & 1 & -2 & -1 & -2
\end{bmatrix}
$$

$$
\rightarrow
\begin{bmatrix}
1 & 0 & 0 & 0 & -1 & -1 & 0 \\
0 & 1 & 0 & 0 & -2 & 0 & -2 \\
0 & 0 & 1 & 0 & 1 & 0 & 2 \\
0 & 0 & 0 & 1 & -2 & -1 & -2 \\
0 & 0 & 0 & 0 & 1 & -1 & 1 \\
0 & 0 & 0 & 0 & 3 & 1 & 5
\end{bmatrix}
\rightarrow
\begin{bmatrix}
1 & 0 & 0 & 0 & 0 & -2 & 1 \\
0 & 1 & 0 & 0 & 0 & -2 & 0 \\
0 & 0 & 1 & 0 & 0 & 1 & 1 \\
0 & 0 & 0 & 1 & 0 & -3 & 0 \\
0 & 0 & 0 & 0 & 1 & -1 & 1 \\
0 & 0 & 0 & 0 & 0 & 4 & 2
\end{bmatrix}
$$

$$
\rightarrow
\begin{bmatrix}
1 & 0 & 0 & 0 & 0 & 0 & 2 \\
0 & 1 & 0 & 0 & 0 & 0 & 1 \\
0 & 0 & 1 & 0 & 0 & 0 & \frac{1}{2} \\
0 & 0 & 0 & 1 & 0 & 0 & \frac{3}{2} \\
0 & 0 & 0 & 0 & 1 & 0 & \frac{3}{2} \\
0 & 0 & 0 & 0 & 0 & 1 & \frac{1}{2}
\end{bmatrix}
$$
. Hence $I_1 = 2$, $I_2 = 1$, $I_3 = \frac{1}{2}$, $I_4 = \frac{3}{2}$, $I_5 = \frac{3}{2}$, $I_6 = \frac{1}{2}$.

Exercises 1.6   An Application to Chemical Reactions

2. Suppose $xNH_3 + yCuO \rightarrow zN_2 + wCu + vH_2O$ where x, y, z, w and v are positive integers. Equating the number of each type of atom on each side gives

$$N : x = 2z \qquad Cu : y = w$$
$$H : 3x = 2v \qquad O : y = v$$

Taking $v = t$ these give $y = t$, $w = t$, $x = \frac{2}{3}t$ and $z = \frac{1}{2}x = \frac{1}{3}t$. The smallest value of t such that there are all integers is $t = 3$, so $x = 2$, $y = 3$, $z = 1$ and $v = 3$. Hence the balanced reaction is

$$2NH_3 + 3CuO \rightarrow N_2 + 3Cu + 3H_2O.$$

Supplementary Exercises   Chapter 1

1(b) No. If the corresonding planes are parallel and distinct, there is no solution. Otherwise they either coincide or have a whole common line of solutions.

2(b)
$$\begin{bmatrix} 1 & 4 & -1 & 1 & | & 2 \\ 3 & 2 & 1 & 2 & | & 5 \\ 1 & -6 & 3 & 0 & | & 1 \\ 1 & 14 & -5 & 2 & | & 3 \end{bmatrix} \rightarrow \begin{bmatrix} 1 & 4 & -1 & 1 & | & 2 \\ 0 & -10 & 4 & -1 & | & -1 \\ 0 & -10 & 4 & -1 & | & -1 \\ 0 & 10 & -4 & 1 & | & 1 \end{bmatrix} \rightarrow \begin{bmatrix} 1 & 0 & \frac{6}{10} & \frac{6}{10} & | & \frac{16}{10} \\ 0 & 1 & -\frac{4}{10} & \frac{1}{10} & | & \frac{1}{10} \\ 0 & 0 & 0 & 0 & | & 0 \\ 0 & 0 & 0 & 0 & | & 0 \end{bmatrix}.$$

Hence $x_3 = s$, $x_4 = t$ are parameters, and the equations give $x_1 = \frac{1}{10}(16 - 6s - 6t)$ and $x_2 = \frac{1}{10}(1 + 4s - t)$.

3(b)
$$\begin{bmatrix} 1 & 1 & 3 & | & a \\ a & 1 & 5 & | & 4 \\ 1 & a & 4 & | & a \end{bmatrix} \rightarrow \begin{bmatrix} 1 & 1 & 3 & | & a \\ 0 & 1-a & 5-3a & | & 4-a^2 \\ 0 & a-1 & 1 & | & 0 \end{bmatrix} \rightarrow \begin{bmatrix} 1 & 1 & 3 & | & a \\ 0 & 1-a & 5-3a & | & 4-a^2 \\ 0 & 0 & 3(2-a) & | & 4-a^2 \end{bmatrix}.$$

If $a = 1$ the matrix is $\begin{bmatrix} 1 & 1 & 3 & | & 1 \\ 0 & 0 & 2 & | & 3 \\ 0 & 0 & 3 & | & 3 \end{bmatrix} \rightarrow \begin{bmatrix} 1 & 1 & 3 & | & 0 \\ 0 & 0 & 1 & | & 1 \\ 0 & 0 & 0 & | & 1 \end{bmatrix}$, so there is no solution.

If $a = 2$ the matrix is $\begin{bmatrix} 1 & 1 & 3 & | & 2 \\ 0 & -1 & -1 & | & 0 \\ 0 & 0 & 0 & | & 0 \end{bmatrix} \rightarrow \begin{bmatrix} 1 & 0 & 2 & | & 2 \\ 0 & 1 & 1 & | & 0 \\ 0 & 0 & 0 & | & 0 \end{bmatrix}$, so $x = 2 - 2t$, $y = -t$, $z = t$.

If $a \neq 1$ and $a \neq 2$ there is a unique solution.

$$\left[\begin{array}{ccc|c} 1 & 1 & 3 & a \\ 0 & 1-a & 5-3a & 4-a^2 \\ 0 & 0 & 3(2-a) & 4-a^2 \end{array}\right] \rightarrow \left[\begin{array}{ccc|c} 1 & 1 & 3 & a \\ 0 & 1 & \frac{3a-5}{a-1} & \frac{a^2-4}{a-1} \\ 0 & 0 & 1 & \frac{a+2}{3} \end{array}\right] \rightarrow \left[\begin{array}{ccc|c} 1 & 0 & \frac{2}{a-1} & \frac{-a+4}{a-1} \\ 0 & 1 & \frac{3a-5}{a-1} & \frac{a^2-4}{a-1} \\ 0 & 0 & 1 & \frac{a+2}{3} \end{array}\right] \rightarrow$$

$$\left[\begin{array}{ccc|c} 1 & 0 & 0 & \frac{-5a+8}{3(a-1)} \\ 0 & 1 & 0 & \frac{-a-2}{3(a-1)} \\ 0 & 0 & 0 & \frac{a+2}{3} \end{array}\right].$$

Hence $x = \frac{8-5a}{3(a-1)}$, $y = \frac{-a-2}{3(a-1)}$, $z = \frac{a+2}{3}$.

4. If R_1 and R_2 denote the two rows, then the following indicate how they can be interchanged using row operations of the other two types:

$$\left[\begin{array}{c} R_1 \\ R_2 \end{array}\right] \rightarrow \left[\begin{array}{c} R_1 + R_2 \\ R_2 \end{array}\right] \rightarrow \left[\begin{array}{c} R_1 + R_2 \\ -R_1 \end{array}\right] \rightarrow \left[\begin{array}{c} R_2 \\ -R_1 \end{array}\right] \rightarrow \left[\begin{array}{c} R_2 \\ R_1 \end{array}\right].$$

Note that only one row operation of Type II was used — a multiplication by -1.

6. Substitute $x = 3$, $y = -1$ and $z = 2$ into the given equations. The result is

$$\begin{array}{llll} 3 - a + 2c = 0 & & a \quad\quad - \ 2c \ = \ 3 \\ 3b - c - 6 = 1 & \text{that is} & \quad\quad 3b \ - \ \ c \ = \ 9 \\ 3a - 2 + 2b = 5 & & 3a \ + \ 2b \quad\quad\quad = \ 7 \end{array}$$

This system of linear equations for a, b and c has unique solution:

$$\left[\begin{array}{ccc|c} 1 & 0 & -2 & 3 \\ 0 & 3 & -1 & 7 \\ 3 & 2 & 0 & 7 \end{array}\right] \rightarrow \left[\begin{array}{ccc|c} 1 & 0 & -2 & 3 \\ 0 & 3 & -1 & 7 \\ 0 & 2 & 6 & -2 \end{array}\right] \rightarrow \left[\begin{array}{ccc|c} 1 & 0 & -2 & 3 \\ 0 & 1 & -7 & 9 \\ 0 & 2 & 6 & -2 \end{array}\right]$$

$$\rightarrow \left[\begin{array}{ccc|c} 1 & 0 & -2 & 3 \\ 0 & 1 & -7 & 9 \\ 0 & 0 & 20 & -20 \end{array}\right] \rightarrow \left[\begin{array}{ccc|c} 1 & 0 & 0 & 1 \\ 0 & 1 & 0 & 2 \\ 0 & 0 & 1 & -1 \end{array}\right]$$

Hence $a = 1$, $b = 2$, $c = -1$.

8. $\left[\begin{array}{ccc|c} 1 & 1 & 1 & 5 \\ 2 & -1 & -1 & 1 \\ -3 & 2 & 2 & 0 \end{array}\right] \rightarrow \left[\begin{array}{ccc|c} 1 & 1 & 1 & 5 \\ 0 & -3 & -3 & -9 \\ 0 & 5 & 5 & 15 \end{array}\right] \rightarrow \left[\begin{array}{ccc|c} 1 & 1 & 1 & 5 \\ 0 & 1 & 1 & 3 \\ 0 & 0 & 0 & 0 \end{array}\right] \rightarrow \left[\begin{array}{ccc|c} 1 & 0 & 0 & 2 \\ 0 & 1 & 1 & 3 \\ 0 & 0 & 0 & 0 \end{array}\right].$

Hence the solution is $x = 2$, $y = 3 - t$, $z = t$. Taking $t = 3 - i$ gives $x = 2$, $y = i$, $z = 3 - i$, as required.

If the real system has a unique solution, the solution is real because all the calculations in the Gaussian algorithm yield real numbers (all entries in the augmented matrix are real).

Chapter 2: Matrix Alegbra

Exercises 2.1   Matrix Addition, Scalar Multiplication, and Transposition

1(b) Equating entries gives four linear equations: $a - b = 2$, $b - c = 2$, $c - d = -6$, $d - a = 2$. The solution is $a = -2 + t$, $b = -4 + t$, $c = -6 + t$, $d = t$.

(d) Equating coefficients gives: $a = b$, $b = c$, $c = d$, $d = a$. The solution is $a = b = c = d = t$, t arbitrary.

2(b) $3 \begin{bmatrix} 3 \\ -1 \end{bmatrix} - 5 \begin{bmatrix} 6 \\ 2 \end{bmatrix} + 7 \begin{bmatrix} 1 \\ -1 \end{bmatrix} = \begin{bmatrix} 9 \\ -3 \end{bmatrix} - \begin{bmatrix} 30 \\ 10 \end{bmatrix} + \begin{bmatrix} 7 \\ -7 \end{bmatrix}$

$= \begin{bmatrix} 9 - 30 + 7 \\ -3 - 10 - 7 \end{bmatrix} = \begin{bmatrix} -14 \\ -20 \end{bmatrix}$

(d) $[3 \quad -1 \quad 2] - 2[9 \quad 3 \quad 4] + [3 \quad 11 \quad -6] = [3 \quad -1 \quad 2] - [18 \quad 6 \quad 8] + [3 \quad 11 \quad -6]$
$= [3 - 18 + 3 \quad -1 - 6 + 11 \quad 2 - 8 - 6] = [-12 \quad 4 \quad -12]$

(f) $\begin{bmatrix} 0 & -1 & 2 \\ 1 & 0 & -4 \\ -2 & 4 & 0 \end{bmatrix}^T = \begin{bmatrix} 0 & 1 & -2 \\ -1 & 0 & 4 \\ 2 & -4 & 0 \end{bmatrix}$

(h) $3 \begin{bmatrix} 2 & 1 \\ -1 & 0 \end{bmatrix}^T - 2 \begin{bmatrix} 1 & -1 \\ 2 & 3 \end{bmatrix} = 3 \begin{bmatrix} 2 & -1 \\ 1 & 0 \end{bmatrix} - 2 \begin{bmatrix} 1 & -1 \\ 2 & 3 \end{bmatrix} = \begin{bmatrix} 6 & -3 \\ 3 & 0 \end{bmatrix} - \begin{bmatrix} 2 & -2 \\ 4 & 6 \end{bmatrix} =$

$\begin{bmatrix} 4 & -1 \\ -1 & -6 \end{bmatrix}$

3(b) $5C - 5 \begin{bmatrix} 3 & -1 \\ 2 & 0 \end{bmatrix} = \begin{bmatrix} 15 & -5 \\ 10 & 0 \end{bmatrix}$

(d) $B + D$ is not defined as B is 2×3 while D is 3×2.

(f) $(A + C)^T = \begin{bmatrix} 2 + 3 & 1 - 1 \\ 0 + 2 & -1 + 0 \end{bmatrix}^T = \begin{bmatrix} 5 & 2 \\ 0 & -1 \end{bmatrix}$

(h) $A - D$ is not defined as A is 2×2 while D is 3×2.

4(b) Given $3A + \begin{bmatrix} 2 \\ 1 \end{bmatrix} = 5A - 2 \begin{bmatrix} 3 \\ 0 \end{bmatrix}$, subtract $3A$ from both sides to get $\begin{bmatrix} 2 \\ 1 \end{bmatrix} = 2A - \begin{bmatrix} 3 \\ 0 \end{bmatrix}$.

Now add $2 \begin{bmatrix} 3 \\ 0 \end{bmatrix}$ to both sides: $2A = \begin{bmatrix} 2 \\ 1 \end{bmatrix} + 2 \begin{bmatrix} 3 \\ 0 \end{bmatrix} = \begin{bmatrix} 8 \\ 1 \end{bmatrix}$. Finally, mulitply both

sides by $\frac{1}{2}$: $A = \frac{1}{2} \begin{bmatrix} 8 \\ 1 \end{bmatrix} = \begin{bmatrix} 4 \\ \frac{1}{2} \end{bmatrix}$.

5(b) Given $2A - B = 5(A + 2B)$, add B to both sides to get

$$2A = 5(A + 2B) + B = 5A + 10B + B = 5A + 11B.$$

Now subtract $5A$ from both sides: $-3A = 11B$. Multiply by $-\frac{1}{3}$ to get $a = -\frac{11}{3}B$.

6(b) Given $4X + 3Y = A$, subtract the first from the second to get $X + Y = B - A$. Now

$$5X + 4Y = B$$

subtract 3 times this equation from the first equation: $X = A - 3(B - A) = 4A - 3B$. Then $X + Y = B - A$ gives $Y = (B - A) - X = (B - A) - (4A - 3B) = 4B - 5A$.

Note that this also follows from the Gaussian Algorithm (with matrix constants):

$$\left[\begin{array}{cc|c} 4 & 3 & A \\ 5 & 4 & B \end{array}\right] \rightarrow \left[\begin{array}{cc|c} 5 & 4 & B \\ 4 & 3 & A \end{array}\right] \rightarrow \left[\begin{array}{cc|c} 1 & 1 & B - A \\ 4 & 3 & A \end{array}\right]$$

$$\rightarrow \left[\begin{array}{cc|c} 1 & 1 & B - A \\ 0 & -1 & 5A - 4B \end{array}\right] \rightarrow \left[\begin{array}{cc|c} 1 & 0 & 4A - 3B \\ 0 & 1 & 4B - 5A \end{array}\right].$$

7(b) Given $2X - 5Y = [1 \quad 2]$ let $Y = T$ where T is an arbitrary 1×2 matrix. Then $2X = 5T + [1 \quad 2]$ so $X = \frac{5}{2}T + \frac{1}{2}[1 \quad 2]$, $Y = T$. If $T = [s \quad t]$, this gives $X = \left[\frac{5}{2}s + \frac{1}{2} \quad \frac{5}{2}t + 1\right]$, $Y = [s \quad t]$, where s and t are arbitrary.

8(b) $5[3(A - B + 2C) - 2(3C - B) - A] + [3(3A - B + C) + 2(B - 2A) - 2C]$
$= 5[3A - 3B + 6C - 6C + 2B - A] + 2[9A - 3B + 3C + 2B - 4A - 2C]$
$= 5[2A - B] + 2[5A - B + C]$
$= 10A - 5B + 10A - 2B + 2C$
$= 20A - 7B + 2C.$

9(b) Write $A = \left[\begin{array}{cc} a & b \\ c & d \end{array}\right]$. We want p, q, r and s such that

$$\left[\begin{array}{cc} a & b \\ c & d \end{array}\right] = p\left[\begin{array}{cc} 1 & 0 \\ 0 & 1 \end{array}\right] + q\left[\begin{array}{cc} 1 & 1 \\ 0 & 0 \end{array}\right] + r\left[\begin{array}{cc} 1 & 0 \\ 1 & 0 \end{array}\right] + s\left[\begin{array}{cc} 0 & 1 \\ 1 & 0 \end{array}\right] = \left[\begin{array}{cc} p+q+r & q+s \\ r+s & p \end{array}\right].$$

Equating components give four linear equations in p, q, r and s:

$$\begin{array}{ccccccc} p & + & q & + & r & & & = & a \\ & & q & & & + & s & = & b \\ & & & & r & + & s & = & c \\ p & & & & & & & = & d \end{array}$$

The solution is $p = d$, $q = \frac{1}{2}(a + b - c - d)$, $r = \frac{1}{2}(a - b + c - d)$, $s = \frac{1}{2}(-a + b + c + d)$.

11(b) $$A + A' = 0$$
$$-A + (A + A') = -A + 0 \quad \text{(add } -A \text{ to both sides)}$$
$$(-A + A) + A' = -A + 0 \quad \text{(associative law}$$
$$0 + A' = -A + 0 \quad \text{(definition of } -A)$$
$$A' = -A \quad \text{(property of 0)}$$

13(b) If $A = \begin{bmatrix} a_1 & 0 & \cdots & 0 \\ 0 & a_2 & \cdots & 0 \\ \vdots & \vdots & & \vdots \\ 0 & 0 & \cdots & a_n \end{bmatrix}$ and $B = \begin{bmatrix} b_1 & 0 & \cdots & 0 \\ 0 & b_2 & \cdots & 0 \\ \vdots & \vdots & & \vdots \\ 0 & 0 & \cdots & b_n \end{bmatrix}$,

then $A - B = \begin{bmatrix} a_1 - b_1 & 0 & \cdots & 0 \\ 0 & a_2 - b_2 & \cdots & 0 \\ \vdots & \vdots & & \vdots \\ 0 & 0 & \cdots & a_n - b_n \end{bmatrix}$ so $A - B$ is also diagonal.

14(b) $\begin{bmatrix} s & t \\ st & 1 \end{bmatrix}$ is symmetric if and only if $t = st$; that is $t(s-1) = 0$; that is $s = 1$ or $t = 0$.

(d) This matrix is symmetric if and only if $2s = s$, $3 = t$, $3 = s + t$; that is $s = 0$ and $t = 3$.

15(b) $\begin{bmatrix} 8 & 0 \\ 3 & 1 \end{bmatrix} = \left(3A^T + 2 \begin{bmatrix} 1 & 0 \\ 0 & 2 \end{bmatrix} \right)^T = (3A^T)^T + \left(2 \begin{bmatrix} 1 & 0 \\ 0 & 2 \end{bmatrix} \right)^T = 3(A^T)^T + 2 \begin{bmatrix} 1 & 0 \\ 0 & 2 \end{bmatrix}^T =$

$3A + 2 \begin{bmatrix} 1 & 0 \\ 0 & 2 \end{bmatrix} = 3A + \begin{bmatrix} 2 & 0 \\ 0 & 4 \end{bmatrix}$. Hence $3A = \begin{bmatrix} 8 & 0 \\ 3 & 1 \end{bmatrix} - \begin{bmatrix} 2 & 0 \\ 0 & 4 \end{bmatrix} = \begin{bmatrix} 6 & 0 \\ 3 & -3 \end{bmatrix}$. Finally

$A = \frac{1}{3} \begin{bmatrix} 6 & 0 \\ 3 & -3 \end{bmatrix} = \begin{bmatrix} 2 & 0 \\ 1 & -1 \end{bmatrix}$.

(d) $4A - 9 \begin{bmatrix} 1 & 1 \\ -1 & 0 \end{bmatrix} = (2A^T)^T - \left(5 \begin{bmatrix} 1 & 0 \\ -1 & 2 \end{bmatrix} \right)^T$ whence $4A = \begin{bmatrix} 9 & 9 \\ -9 & 0 \end{bmatrix} = 2(A^T)^T -$

$5 \begin{bmatrix} 1 & 0 \\ -1 & 2 \end{bmatrix}^T = 2A - 5 \begin{bmatrix} 1 & -1 \\ 0 & 2 \end{bmatrix}$. Then $2A = \begin{bmatrix} 9 & 9 \\ -9 & 0 \end{bmatrix} - \begin{bmatrix} 5 & -5 \\ 0 & 10 \end{bmatrix} = \begin{bmatrix} 4 & 14 \\ -9 & -10 \end{bmatrix}$.

Finally $A = \frac{1}{2} \begin{bmatrix} 4 & 14 \\ -9 & 10 \end{bmatrix} = \begin{bmatrix} 2 & 7 \\ -\frac{9}{2} & -5 \end{bmatrix}$.

16(b) We have $A^T = A$ as A is symmetric. Using Theorem 2: $(kA)^T = kA^T = kA$; so kA is symmetric.

18(c) If $A = S+W$ as in (b), then $A^T = S^T + W^T = S - W$. Hence $A + A^T = 2S$ and $A - A^T = 2W$, so $S = \frac{1}{2}(A + A^T)$ and $W = \frac{1}{2}(A - A^T)$.

20(b) If $A = [a_{ij}]$ then $(kp)A = [(kp)a_{ij}] = [k(pa_{ij})] = k[pa_{ij}] = k(pA)$.

Exercises 2.2   Matrix Multiplication

1(b) $\begin{bmatrix} 1 & -1 & 2 \\ 2 & 0 & 4 \end{bmatrix} \begin{bmatrix} 2 & 3 & 1 \\ 1 & 5 & 7 \\ -1 & 0 & 2 \end{bmatrix} = \begin{bmatrix} 2-1-2 & 3-9+0 & 1-7+4 \\ 4+0-4 & 6+0-0 & 2+0+8 \end{bmatrix} = \begin{bmatrix} -1 & -6 & -2 \\ 0 & 6 & 10 \end{bmatrix}$

(d) $\begin{bmatrix} 1 & 3 & -3 \end{bmatrix} \begin{bmatrix} 3 & 0 \\ -2 & 1 \\ 0 & 6 \end{bmatrix} = \begin{bmatrix} 3-6+0 & 0+3-18 \end{bmatrix} = \begin{bmatrix} -3 & -15 \end{bmatrix}$

(f) $\begin{bmatrix} 1 & -1 & 3 \end{bmatrix} \begin{bmatrix} 2 \\ 1 \\ -8 \end{bmatrix} = \begin{bmatrix} 2-1-24 \end{bmatrix} = \begin{bmatrix} -23 \end{bmatrix}$

(h) $\begin{bmatrix} 3 & 1 \\ 5 & 2 \end{bmatrix} \begin{bmatrix} 2 & -1 \\ -5 & 3 \end{bmatrix} = \begin{bmatrix} 6-5 & -3+3 \\ 10-10 & -5+6 \end{bmatrix} = \begin{bmatrix} 1 & 0 \\ 0 & 1 \end{bmatrix}$

(j) $\begin{bmatrix} a & 0 & 0 \\ 0 & b & 0 \\ 0 & 0 & c \end{bmatrix} \begin{bmatrix} a' & 0 & 0 \\ 0 & b' & 0 \\ 0 & 0 & c' \end{bmatrix} = \begin{bmatrix} aa'+0+0 & 0+0+0 & 0+0+0 \\ 0+0+0 & 0+bb'+0 & 0+0+0 \\ 0+0+0 & 0+0+0 & 0+0+cc' \end{bmatrix} = \begin{bmatrix} aa' & 0 & 0 \\ 0 & bb' & 0 \\ 0 & 0 & cc' \end{bmatrix}$

2(b) A^2, AB, BC and C^2 are all undefined. The other products are

$$BA = \begin{bmatrix} -1 & 4 & -10 \\ 1 & 2 & 4 \end{bmatrix}, \quad B^2 = \begin{bmatrix} 7 & -6 \\ -1 & 6 \end{bmatrix}, \quad CB = \begin{bmatrix} -2 & 12 \\ 2 & -6 \\ 1 & 6 \end{bmatrix}, \quad AC = \begin{bmatrix} 4 & 10 \\ -2 & -1 \end{bmatrix},$$

$$CA = \begin{bmatrix} 2 & 4 & 8 \\ -1 & -1 & -5 \\ 1 & 4 & 2 \end{bmatrix}.$$

3(b) The given matrix equation becomes $\begin{bmatrix} 2a + a_1 & 2b + b_1 \\ -a + 2a_1 & -b + 2b_1 \end{bmatrix} = \begin{bmatrix} 7 & 2 \\ -1 & 4 \end{bmatrix}$.

Equating coefficients gives linear equations

$$2a + a_1 = 7 \qquad 2b + b_1 = 2$$
$$-a + 2a_1 = -1 \qquad -b + 2b_1 = 4$$

The solutions are $a = 3$, $a_1 = 1$; $b = 0$, $b_1 = 2$.

4(b) $A^2 - A - 6I = \begin{bmatrix} 8 & 2 \\ 2 & 5 \end{bmatrix} - \begin{bmatrix} 2 & 2 \\ 2 & -1 \end{bmatrix} - \begin{bmatrix} 6 & 0 \\ 0 & 6 \end{bmatrix} = \begin{bmatrix} 0 & 0 \\ 0 & 0 \end{bmatrix}.$

5(b) $A(BC) = \begin{bmatrix} 1 & -1 \\ 0 & 1 \end{bmatrix} \begin{bmatrix} -9 & -16 \\ 5 & 1 \end{bmatrix} = \begin{bmatrix} -14 & -17 \\ 5 & 1 \end{bmatrix} = \begin{bmatrix} -2 & -1 & -2 \\ 3 & 1 & 0 \end{bmatrix} \begin{bmatrix} 1 & 0 \\ 2 & 1 \\ 5 & 8 \end{bmatrix} =$

$(AB)C.$

6(b) If $A = \begin{bmatrix} a & b \\ c & d \end{bmatrix}$ then $A \begin{bmatrix} 0 & 0 \\ 1 & 0 \end{bmatrix} = \begin{bmatrix} 0 & 0 \\ 1 & 0 \end{bmatrix} A$ becomes $\begin{bmatrix} b & 0 \\ d & 0 \end{bmatrix} = \begin{bmatrix} 0 & 0 \\ a & b \end{bmatrix}$, so $b = 0$

and $d = a$. Hence $A = \begin{bmatrix} a & 0 \\ c & a \end{bmatrix}$ for arbitrary a and c.

7(b) $\begin{bmatrix} -1 & 2 & -1 & 1 \\ 2 & 1 & -1 & 2 \\ 3 & -2 & 0 & 1 \end{bmatrix} \begin{bmatrix} x_1 \\ x_2 \\ x_3 \\ x_4 \end{bmatrix} = \begin{bmatrix} 6 \\ 1 \\ 0 \end{bmatrix}.$

8(b) $\begin{bmatrix} 1 & -1 & -4 & -4 \\ 1 & 2 & 5 & 2 \\ 1 & 1 & 2 & 0 \end{bmatrix} \rightarrow \begin{bmatrix} 1 & -1 & -4 & -4 \\ 0 & 3 & 9 & 6 \\ 0 & 2 & 6 & 4 \end{bmatrix} \rightarrow \begin{bmatrix} 1 & 0 & -1 & -2 \\ 0 & 1 & 3 & 2 \\ 0 & 0 & 0 & 0 \end{bmatrix}$. Hence $x = t - 2$,

$y = 2 - 3t$, $z = t$; that is $\begin{bmatrix} x \\ y \\ z \end{bmatrix} = \begin{bmatrix} -2 + t \\ 2 - 3t \\ t \end{bmatrix} = \begin{bmatrix} -2 \\ 2 \\ 0 \end{bmatrix} + t \begin{bmatrix} 1 \\ -3 \\ 1 \end{bmatrix}$. Observe that $\begin{bmatrix} -2 \\ 2 \\ 0 \end{bmatrix}$

is a solution to the given system of equations, and $\begin{bmatrix} 1 \\ -3 \\ 1 \end{bmatrix}$ is a solution to the associated

homogeneous system.

8(d) $\begin{bmatrix} 2 & 1 & -1 & -1 & -1 \\ 3 & 1 & 1 & -2 & -2 \\ -1 & -1 & 2 & 1 & 2 \\ -2 & -1 & 0 & 2 & 3 \end{bmatrix} \rightarrow \begin{bmatrix} 1 & 1 & -2 & 1 & -2 \\ 0 & -1 & 3 & 1 & 3 \\ 0 & -2 & 7 & 1 & 4 \\ 0 & 1 & -4 & 0 & -1 \end{bmatrix} \rightarrow \begin{bmatrix} 1 & 0 & 1 & 0 & 1 \\ 0 & 1 & -3 & -1 & -3 \\ 0 & 0 & 1 & -1 & -2 \\ 0 & 0 & -1 & 1 & 2 \end{bmatrix}$

$$\rightarrow \left[\begin{array}{cccc|c} 1 & 0 & 0 & 1 & 3 \\ 0 & 1 & 0 & -4 & -9 \\ 0 & 0 & 1 & -1 & -2 \\ 0 & 0 & 0 & 0 & 0 \end{array}\right]. \text{ Hence } x_1 = 3 - t, \ x_2 = 4t - 9, \ x_3 = t - 2, \ x_4 = t,$$

$$\text{so } \left[\begin{array}{c} x_1 \\ x_2 \\ x_3 \\ x_4 \end{array}\right] = \left[\begin{array}{c} 3 - t \\ -9 + 4t \\ -2 + t \\ t \end{array}\right] = \left[\begin{array}{c} 3 \\ -9 \\ -2 \\ 0 \end{array}\right] + t \left[\begin{array}{c} -1 \\ 4 \\ 1 \\ 1 \end{array}\right]. \text{ Here } \left[\begin{array}{c} 3 \\ -9 \\ -2 \\ 0 \end{array}\right] \text{ is a solution to the given}$$

equations, and $\left[\begin{array}{c} -1 \\ 4 \\ 1 \\ 1 \end{array}\right]$ is a solution to the associated homogeneous equations.

10(b) $\left[\begin{array}{ccccc|c} 1 & 1 & -2 & 3 & 2 & 0 \\ 2 & -1 & 3 & 4 & 1 & 0 \\ -1 & -2 & 3 & 1 & 0 & 0 \\ 3 & 0 & 1 & 7 & 3 & 0 \end{array}\right] \rightarrow \left[\begin{array}{ccccc|c} 1 & 1 & -2 & 3 & 2 & 0 \\ 0 & -3 & 7 & -2 & -3 & 0 \\ 0 & -1 & 1 & 4 & 2 & 0 \\ 0 & -3 & 7 & -2 & -3 & 0 \end{array}\right]$

$$\rightarrow \left[\begin{array}{ccccc|c} 1 & 0 & -1 & 7 & 4 & 0 \\ 0 & 0 & 4 & -14 & -9 & 0 \\ 0 & 1 & -1 & -4 & -2 & 0 \\ 0 & 0 & 0 & 0 & 0 & 0 \end{array}\right] \rightarrow \left[\begin{array}{ccccc|c} 1 & 0 & 0 & \frac{7}{2} & \frac{7}{4} & 0 \\ 0 & 1 & 0 & -\frac{15}{2} & -\frac{17}{4} & 0 \\ 0 & 0 & 1 & -\frac{7}{2} & -\frac{9}{4} & 0 \\ 0 & 0 & 0 & 0 & 0 & 0 \end{array}\right]. \text{ Now take } x_4 = s, \ x_5 =$$

$$t. \text{ Then } \left[\begin{array}{c} x_1 \\ x_2 \\ x_3 \\ x_4 \\ x_5 \end{array}\right] = \left[\begin{array}{c} -\frac{7}{2}s - \frac{7}{4}t \\ \frac{15}{2}s + \frac{17}{4}t \\ \frac{7}{2}s + \frac{9}{4}t \\ s \\ t \end{array}\right] = s \left[\begin{array}{c} -\frac{7}{2} \\ \frac{15}{2} \\ \frac{7}{2} \\ 1 \\ 0 \end{array}\right] + t \left[\begin{array}{c} -\frac{7}{4} \\ \frac{17}{2} \\ \frac{9}{4} \\ 0 \\ 1 \end{array}\right].$$

12(b) If A is $m \times n$ and B is $p \times q$ then $n = p$ because AB can be formed and $q = m$ because BA can be formed. So B is $n \times m$, A is $m \times n$.

14(b) BP is the matrix B with columns 2 and 3 interchanged.

16(b) $AB = \left[\begin{array}{ccc|c} 2 & -1 & 3 & 1 \\ 1 & 0 & 1 & 2 \\ 0 & 0 & 1 & 0 \\ \hline 0 & 0 & 0 & 1 \end{array}\right] \left[\begin{array}{cc|c} 1 & 2 & 0 \\ -1 & 0 & 0 \\ 0 & 5 & 1 \\ \hline 1 & -1 & 0 \end{array}\right] = .$

$$\left[\begin{array}{c|c} \begin{pmatrix} 3 & 19 \\ 1 & 7 \\ 0 & 5 \end{pmatrix} + \begin{pmatrix} 1 & -1 \\ 2 & -2 \\ 0 & 0 \end{pmatrix} & \begin{pmatrix} 3 \\ 1 \\ 1 \end{pmatrix} + \begin{pmatrix} 0 \\ 0 \\ 0 \end{pmatrix} \\ \hline (1 \quad -1) & 0 \end{array}\right] = \left[\begin{array}{cc|c} 4 & 18 & 3 \\ 3 & 5 & 1 \\ 0 & 5 & 1 \\ \hline 1 & -1 & 0 \end{array}\right].$$

17(b) Write $A = \begin{bmatrix} P & X \\ 0 & Q \end{bmatrix}$ where $P \begin{bmatrix} 1 & -1 \\ 0 & 1 \end{bmatrix}$, $X = \begin{bmatrix} 2 & -1 \\ 0 & 0 \end{bmatrix}$, and $Q = \begin{bmatrix} -1 & 1 \\ 0 & 1 \end{bmatrix}$. Then

$PX + XQ = \begin{bmatrix} 2 & -1 \\ 0 & 0 \end{bmatrix} + \begin{bmatrix} -2 & 1 \\ 0 & 0 \end{bmatrix} = 0$, so $A^2 = \begin{bmatrix} P^2 & PX + XQ \\ 0 & Q^2 \end{bmatrix} = \begin{bmatrix} P^2 & 0 \\ 0 & Q^2 \end{bmatrix}$.

Then $A^4 = \begin{bmatrix} P^2 & 0 \\ 0 & Q^2 \end{bmatrix}\begin{bmatrix} P^2 & 0 \\ 0 & Q^2 \end{bmatrix} = \begin{bmatrix} P^4 & 0 \\ 0 & Q^4 \end{bmatrix}$, $A^6 = A^4 A^2 = \begin{bmatrix} P^6 & 0 \\ 0 & Q^6 \end{bmatrix}, \ldots$; in general we claim that

$$A^{2k} = \begin{bmatrix} P^{2k} & 0 \\ 0 & Q^{2k} \end{bmatrix} \quad \text{for } k = 1, 2, \ldots \tag{*}$$

This holds for $k = 1$; if it holds for some $k \geq 1$ then

$$A^{2(k+1)} = A^{2k} A^2 = \begin{bmatrix} P^{2k} & 0 \\ 0 & Q^{2k} \end{bmatrix}\begin{bmatrix} P^2 & 0 \\ 0 & Q^2 \end{bmatrix} = \begin{bmatrix} P^{2(k+1)} & 0 \\ 0 & Q^{2(k+1)} \end{bmatrix}$$

Hence (*) follows by induction in k.

Next $P^2 = \begin{bmatrix} 1 & -2 \\ 0 & 1 \end{bmatrix}$, $P^3 = \begin{bmatrix} 1 & -3 \\ 0 & 1 \end{bmatrix}$, and we claim that

$$P^m = \begin{bmatrix} 1 & -m \\ 0 & 1 \end{bmatrix} \quad \text{for } m = 1, 2, \ldots \tag{**}$$

It is true for $m = 1$; if it holds for some $m \geq 1$, then

$$P^{m+1} = P^m P = \begin{bmatrix} 1 & -m \\ 0 & 1 \end{bmatrix}\begin{bmatrix} 1 & -1 \\ 0 & 1 \end{bmatrix} = \begin{bmatrix} 1 & -(m+1) \\ 0 & 1 \end{bmatrix}$$

which proves (**) by induction.

As to Q, $Q^2 = I$ so $Q^{2k} = I$ for all k. Hence (*) and (**) gives

$$A^{2k} = \begin{bmatrix} P^{2k} & 0 \\ 0 & I \end{bmatrix} = \left[\begin{array}{cc|cc} 1 & -2k & 0 & 0 \\ 0 & 1 & 0 & 0 \\ \hline 0 & 0 & 1 & 0 \\ 0 & 0 & 0 & 1 \end{array}\right] \quad \text{for } k \geq 1.$$

Finally

$$A^{2k+1} = A^{2k} \cdot A = \begin{bmatrix} P^{2k} & 0 \\ 0 & I \end{bmatrix} \begin{bmatrix} P & X \\ 0 & Q \end{bmatrix} = \begin{bmatrix} P^{2k+1} & P^{2k}X \\ 0 & Q \end{bmatrix}$$

$$= \left[\begin{array}{cc|cc} 1 & -(2k+1) & 2 & -1 \\ 0 & 1 & 0 & 0 \\ \hline 0 & 0 & -1 & 1 \\ 0 & 0 & 0 & 1 \end{array} \right].$$

18(b) $\begin{bmatrix} I & X \\ 0 & I \end{bmatrix} \begin{bmatrix} I & -X \\ 0 & I \end{bmatrix} = \begin{bmatrix} I^2 + X0 & -IX + XI \\ 0I + I0 & -0X + I^2 \end{bmatrix} = \begin{bmatrix} I & 0 \\ 0 & I \end{bmatrix} = I_{2k}$

(d) $[I \quad X] [-X^T \quad I]^T = [I \quad X] \begin{bmatrix} -X \\ I \end{bmatrix} = -IX + XI = O_k$

(f) $\begin{bmatrix} 0 & X \\ I & 0 \end{bmatrix}^2 = \begin{bmatrix} 0 & X \\ I & 0 \end{bmatrix} \begin{bmatrix} 0 & X \\ I & 0 \end{bmatrix} = \begin{bmatrix} X & 0 \\ 0 & X \end{bmatrix}$

$\begin{bmatrix} 0 & X \\ I & 0 \end{bmatrix}^3 = \begin{bmatrix} 0 & X \\ I & 0 \end{bmatrix} \begin{bmatrix} X & 0 \\ 0 & X \end{bmatrix} = \begin{bmatrix} 0 & X^2 \\ X & 0 \end{bmatrix}$

$\begin{bmatrix} 0 & X \\ I & 0 \end{bmatrix}^4 = \begin{bmatrix} 0 & X \\ I & 0 \end{bmatrix} \begin{bmatrix} 0 & X^2 \\ X & 0 \end{bmatrix} = \begin{bmatrix} X^2 & 0 \\ 0 & X^2 \end{bmatrix}$

Continue. We claim that $\begin{bmatrix} 0 & X \\ I & 0 \end{bmatrix}^{2m} = \begin{bmatrix} X^m & 0 \\ 0 & X^m \end{bmatrix}$ for $m \geq 1$. It is true if $m = 1$ and, if

it holds for some m, $\begin{bmatrix} 0 & X \\ I & 0 \end{bmatrix}^{2(m+1)} = \begin{bmatrix} 0 & X \\ I & 0 \end{bmatrix}^{2m} \begin{bmatrix} 0 & X \\ I & 0 \end{bmatrix}^2 = \begin{bmatrix} X^m & 0 \\ 0 & X^m \end{bmatrix} \begin{bmatrix} X & 0 \\ 0 & X \end{bmatrix} =$

$\begin{bmatrix} X^{m+1} & 0 \\ 0 & X^{m+1} \end{bmatrix}$. Hence the result follows by induction on m. Now $\begin{bmatrix} 0 & X \\ I & 0 \end{bmatrix}^{2m+1} =$

$\begin{bmatrix} 0 & X \\ I & 0 \end{bmatrix}^{2m} \begin{bmatrix} 0 & X \\ I & 0 \end{bmatrix} = \begin{bmatrix} X^m & 0 \\ 0 & X^m \end{bmatrix} \begin{bmatrix} X & 0 \\ 0 & X \end{bmatrix} = \begin{bmatrix} X^{m+1} & 0 \\ 0 & X^{m+1} \end{bmatrix}$ for all $m \geq 1$. It

also holds for $m = 0$ if we take $X^0 = I$.

20(b) If $YA = 0$ for all $1 \times m$ matrices Y, let Y_i denote row i of I_m. Then row i of $I_m A = A$ is $Y_i A = 0$. Thus each row of A is zero, so $A = 0$.

22(b) $A(B + C - D) + B(C - A + D) - (A + B)C + (A - B)D$
$= AB + AC - AD + BC - BA + BD - AC - BC + AD - BD$
$= AB - BA$.

(d) $(A - B)(C - A) + (C - B)(A - C) + (C - A)^2 = [(A - B) - (C - B) + (C - A)](C - A) = 0(C - A) = 0.$

24(b) We are given that $AC = CA$, so $(kA)C = k(AC) = k(CA) = C(kA)$, using Theorem 1. Hence kA commutes with C.

26 Since $A^T = A$ and $B^T = B$, we have $(AB)^T = B^T A^T = BA$. Hence $(AB)^T = AB$ if and only if $BA = AB$.

28(b) Let $A = \begin{bmatrix} a & x & y \\ x & b & z \\ y & z & c \end{bmatrix}$. Then the entries on the main diagonal of A^2 are $a^2 + x^2 + y^2$, $x^2 + b^2 + z^2$, $y^2 + z^2 + c^2$. These are all zero if and only if $a = x = y = b = z = c = 0$; that is if and only if $A = 0$.

30. If $AB = 0$ where $A \neq 0$, suppose $BC = I$. Left multiply by A to get $A = AI = A(BC) = (AB)C = 0C = 0$, a contradiction. So no such matrix C exists.

31(b) If $A = [a_{ij}]$ the sum of the entries in row i is $\sum_{j=1}^{n} a_{ij} = 1$. Similarly for $B = [b_{ij}]$. If $AB = C = [c_{ij}]$ then $c_{ij} = \sum_{k=1}^{n} a_{ik}b_{kj}$. Hence the sum of the entries in row i of C is

$$\sum_{j=1}^{n} c_{ij} = \sum_{j=1}^{n}\sum_{k=1}^{n} a_{ik}b_{kj} = \sum_{k=1}^{n} a_{ik}\left(\sum_{j=1}^{n} b_{kj}\right) = \sum_{k=1}^{n} a_{ik} \cdot 1 = 1.$$

<u>Easier Proof:</u> Let X be the $n \times 1$ matrix with every entry equal to 1. Then the entries of AX are the row sums of A, so these all equal 1 if and only if $AX = X$. But if also $BX = X$ then $(AB)X = A(BX) = AX = X$, as required.

33(b) If $A = [a_{ij}]$ then $\operatorname{tr} A = a_{11} + a_{22} + \cdots + a_{nn}$. Because $ka = [ka_{ij}]$, $\operatorname{tr}(kA) = ka_{11} + ka_{22} + \cdots + ka_{nn} = k(a_{11} + a_{22} + \cdots + a_{nn}) = k \operatorname{tr} A.$

(e) Write $A^T = [a'_{ij}]$ where $a'_{ij} = a_{ji}$. Since the (i, j)-entry of AA^T is $\sum_{k=1}^{n} a_{ik}a'_{kj} = \sum_{k=1}^{n} a_{ik}a_{jk}$, we obtain $\operatorname{tr}(AA^T) = \sum_{i=1}^{n}\left(\sum_{k=1}^{n} a_{ik}a_{ik}\right) = \sum_{i=1}^{n}\sum_{k=1}^{n} a_{ik}^2.$

35(e) We have $Q = P + AP - PAP$ so, since $P^2 = P$,

$$PQ = P^2 + PAP - P^2AP = P + 0 = P.$$

Hence $Q^2 = (P + AP - PAP)Q = PQ + APQ - PAPQ = P + AP - PAP = Q.$

37(b) We always have $(A+B)(A-B) = A^2 + BA - AB - B^2$. If $AB = BA$, this gives $(A+B)(A-B) = A^2 - B^2$. Conversely, if $(A + B)(A - B) = A^2 - B^2$, then

$$A^2 - B^2 = A^2 + BA - AB - B^2.$$

Hence $0 = BA - AB$, whence $AB = BA$.

38(b) Write $A = [a_{ij}]$, $B = [b_{ij}]$, $C = [c_{ij}]$ where A is $n \times k$, B is $k \times \ell$ and C is $\ell \times m$. If $AB = [x_{ij}]$ then $x_{ij} = \sum_{t=1}^{k} a_{it}b_{tj}$. Hence the (i, j)-entry of $(AB)C$ is

$$\sum_{s=1}^{t} x_{is}c_{sj} = \sum_{s=1}^{\ell}\left(\sum_{t=1}^{k} a_{it}b_{ts}\right)c_{sj} = \sum_{s=1}^{\ell}\sum_{t=1}^{k} a_{it}b_{ts}c_{sj}.$$

Similarly if $BC = [y_{ij}]$, the (i,j)-entry of $A(BC)$ is

$$\sum_{t=1}^{k} a_{it} y_{tj} = \sum_{t=1}^{k} a_{it} \sum_{s=1}^{\ell} b_{ts} c_{sj} = \sum_{t=1}^{k} \sum_{s=1}^{\ell} a_{it} b_{ts} c_{sj}.$$

These two double sums are equal for all i and j, so $(AB)C = A(BC)$.

Exercises 2.3   Matrix Inverses

2(b) We begin by subtracting row 2 from row 1.

$$\left[\begin{array}{cc|cc} 4 & 1 & 1 & 0 \\ 3 & 2 & 0 & 1 \end{array}\right] \rightarrow \left[\begin{array}{cc|cc} 1 & -1 & 1 & -1 \\ 3 & 2 & 0 & 1 \end{array}\right] \rightarrow \left[\begin{array}{cc|cc} 1 & -1 & 1 & -1 \\ 0 & 5 & -3 & 4 \end{array}\right] \rightarrow$$

$$\left[\begin{array}{cc|cc} 1 & -1 & 1 & -1 \\ 0 & 1 & -\frac{3}{5} & \frac{4}{5} \end{array}\right] \rightarrow \left[\begin{array}{cc|cc} 1 & 0 & \frac{2}{5} & -\frac{1}{5} \\ 0 & 1 & -\frac{3}{5} & \frac{4}{5} \end{array}\right].$$ Hence the inverse is $\frac{1}{5}\left[\begin{array}{cc} 2 & -1 \\ -3 & 4 \end{array}\right].$

(d) $$\left[\begin{array}{ccc|ccc} 1 & -1 & 2 & 1 & 0 & 0 \\ -5 & 7 & -11 & 0 & 1 & 0 \\ -2 & 3 & -5 & 0 & 0 & 1 \end{array}\right] \rightarrow \left[\begin{array}{ccc|ccc} 1 & -1 & 2 & 1 & 0 & 0 \\ 0 & 2 & -1 & 5 & 1 & 0 \\ 0 & 1 & -1 & 2 & 0 & 1 \end{array}\right] \rightarrow \left[\begin{array}{ccc|ccc} 1 & 0 & 1 & 3 & 0 & 1 \\ 0 & 1 & -1 & 2 & 0 & 1 \\ 0 & 0 & 1 & 1 & 1 & -2 \end{array}\right]$$

$$\rightarrow \left[\begin{array}{ccc|ccc} 1 & 0 & 0 & 2 & -1 & 3 \\ 0 & 1 & 0 & 3 & 1 & -1 \\ 0 & 0 & 1 & 1 & 1 & -2 \end{array}\right].$$ So $A^{-1} = \left[\begin{array}{ccc} 2 & -1 & 3 \\ 3 & 1 & -1 \\ 1 & 1 & -2 \end{array}\right].$

(f) $$\left[\begin{array}{ccc|ccc} 3 & 1 & -1 & 1 & 0 & 0 \\ 2 & 1 & 0 & 0 & 1 & 0 \\ 1 & 5 & -1 & 0 & 0 & 1 \end{array}\right] \rightarrow \left[\begin{array}{ccc|ccc} 1 & 0 & -1 & 1 & -1 & 0 \\ 2 & 1 & 0 & 0 & 1 & 0 \\ 1 & 5 & -1 & 0 & 0 & 1 \end{array}\right] \rightarrow \left[\begin{array}{ccc|ccc} 1 & 0 & -1 & 1 & -1 & 0 \\ 0 & 1 & 2 & -2 & 3 & 0 \\ 0 & 5 & 0 & -1 & 1 & 1 \end{array}\right]$$

$$\rightarrow \left[\begin{array}{ccc|ccc} 1 & 0 & -1 & 1 & -1 & 0 \\ 0 & 1 & 2 & -2 & 3 & 0 \\ 0 & 0 & -10 & 9 & -14 & 1 \end{array}\right] \rightarrow \left[\begin{array}{ccc|ccc} 1 & 0 & 0 & \frac{1}{10} & -\frac{4}{10} & \frac{1}{10} \\ 0 & 1 & 0 & -\frac{2}{10} & \frac{2}{10} & \frac{2}{10} \\ 0 & 0 & 1 & -\frac{9}{10} & \frac{14}{10} & -\frac{1}{10} \end{array}\right].$$

Hence $A^{-1} = \frac{1}{10}\left[\begin{array}{ccc} 1 & -4 & 1 \\ -2 & 2 & 2 \\ -9 & 14 & -1 \end{array}\right].$

(h) We begin by subtracting row 2 from twice row 1:

$$\left[\begin{array}{ccc|ccc} 3 & 1 & -1 & 1 & 0 & 0 \\ 5 & 2 & 0 & 0 & 1 & 0 \\ 1 & 1 & -1 & 0 & 0 & 1 \end{array}\right] \rightarrow \left[\begin{array}{ccc|ccc} 1 & 0 & -2 & 2 & -1 & 0 \\ 5 & 2 & 0 & 0 & 1 & 0 \\ 1 & 1 & -1 & 0 & 0 & 1 \end{array}\right] \rightarrow \left[\begin{array}{ccc|ccc} 1 & 0 & -2 & 2 & -1 & 0 \\ 0 & 2 & 10 & -10 & 6 & 0 \\ 0 & 1 & 1 & -2 & 1 & 1 \end{array}\right]$$

$$\rightarrow \begin{bmatrix} 1 & 0 & -2 & 2 & -1 & 0 \\ 0 & 1 & 5 & -5 & 3 & 0 \\ 0 & 0 & -4 & 3 & -2 & 1 \end{bmatrix} \rightarrow \begin{bmatrix} 1 & 0 & -2 & 2 & -1 & 0 \\ 0 & 1 & 5 & -5 & 3 & 0 \\ 0 & 0 & 1 & -\frac{3}{4} & \frac{2}{4} & -\frac{1}{4} \end{bmatrix}$$

$$\rightarrow \begin{bmatrix} 1 & 0 & 0 & \frac{2}{4} & 0 & -\frac{2}{4} \\ 0 & 1 & 0 & -\frac{5}{4} & \frac{2}{4} & \frac{5}{4} \\ 0 & 0 & 1 & -\frac{3}{4} & \frac{2}{4} & -\frac{1}{4} \end{bmatrix}. \text{ Hence } A^{-1} = \frac{1}{4} \begin{bmatrix} 2 & 0 & -1 \\ -5 & 2 & 5 \\ -3 & 2 & -1 \end{bmatrix}.$$

(j) $\begin{bmatrix} -1 & 4 & 5 & 2 & 1 & 0 & 0 & 0 \\ 0 & 0 & 0 & -1 & 0 & 1 & 0 & 0 \\ 1 & -2 & -2 & 0 & 0 & 0 & 1 & 0 \\ 0 & -1 & -1 & 0 & 0 & 0 & 0 & 1 \end{bmatrix} \rightarrow \begin{bmatrix} 1 & -4 & -5 & -2 & -1 & 0 & 0 & 0 \\ 0 & 0 & 0 & 1 & 0 & -1 & 0 & 0 \\ 0 & 2 & 3 & 2 & 1 & 0 & 1 & 0 \\ 0 & 1 & 1 & 0 & 0 & 0 & 0 & -1 \end{bmatrix}$

$$\rightarrow \begin{bmatrix} 1 & -4 & -5 & -2 & -1 & 0 & 0 & 0 \\ 0 & 1 & 1 & 0 & 0 & 0 & 0 & -1 \\ 0 & 2 & 3 & 2 & 1 & 0 & 1 & 0 \\ 0 & 0 & 0 & 1 & 0 & -1 & 0 & 0 \end{bmatrix} \rightarrow \begin{bmatrix} 1 & 0 & -1 & -2 & -1 & 0 & 0 & -4 \\ 0 & 1 & 1 & 0 & 0 & 0 & 0 & -1 \\ 0 & 0 & 1 & 2 & 1 & 0 & 1 & 2 \\ 0 & 0 & 0 & 1 & 0 & -1 & 0 & 0 \end{bmatrix}$$

$$\rightarrow \begin{bmatrix} 1 & 0 & 0 & 0 & 0 & 0 & 1 & -2 \\ 0 & 1 & 0 & -2 & -1 & 0 & -1 & -3 \\ 0 & 0 & 1 & 2 & 1 & 0 & 1 & 2 \\ 0 & 0 & 0 & 1 & 0 & -1 & 0 & 0 \end{bmatrix} \rightarrow \begin{bmatrix} 1 & 0 & 0 & 0 & 0 & 0 & 1 & -2 \\ 0 & 1 & 0 & 0 & -1 & -2 & -1 & -3 \\ 0 & 0 & 1 & 0 & 1 & 2 & 1 & 2 \\ 0 & 0 & 0 & 1 & 0 & -1 & 0 & 0 \end{bmatrix}$$

Hence $A^{-1} = \begin{bmatrix} 0 & 0 & 1 & -2 \\ -1 & -2 & -1 & -3 \\ 1 & 2 & 1 & 2 \\ 0 & -1 & 0 & 0 \end{bmatrix}.$

3(b) The equations are $AX = B$ where $A = \begin{bmatrix} 2 & -3 \\ 1 & -4 \end{bmatrix}$, $X = \begin{bmatrix} x \\ y \end{bmatrix}$, $B = \begin{bmatrix} 0 \\ 1 \end{bmatrix}$. We have (by the algorithm or Example 4) $A^{-1} = \frac{1}{5} \begin{bmatrix} 4 & -3 \\ 1 & -2 \end{bmatrix}$. Left multiply $AX = B$ by A^{-1} to get

$$X = A^{-1}AX = A^{-1}B = \frac{1}{5} \begin{bmatrix} 4 & -3 \\ 1 & -2 \end{bmatrix} \begin{bmatrix} 0 \\ 1 \end{bmatrix} = \frac{1}{5} \begin{bmatrix} -3 \\ -2 \end{bmatrix}.$$

Hence $x = -\frac{3}{5}$ and $y = -\frac{2}{5}$.

(d) Here $A = \begin{bmatrix} 1 & 4 & 2 \\ 2 & 3 & 3 \\ 4 & 1 & 4 \end{bmatrix}$, $X = \begin{bmatrix} x \\ y \\ z \end{bmatrix}$, $B = \begin{bmatrix} 1 \\ -1 \\ 0 \end{bmatrix}.$

By the algorithm, $A^{-1} = \frac{1}{5} \begin{bmatrix} 9 & -14 & 6 \\ 4 & -4 & 1 \\ -10 & 15 & -5 \end{bmatrix}$.

The equations are $AX = B$ so, left multiplying by A^{-1} gives

$$X = A^{-1}(AX) = A^{-1}B = \frac{1}{5} \begin{bmatrix} 9 & -14 & 6 \\ 9 & -4 & 1 \\ -10 & 15 & -5 \end{bmatrix} \begin{bmatrix} 1 \\ -1 \\ 0 \end{bmatrix} = \frac{1}{5} \begin{bmatrix} 23 \\ 8 \\ -25 \end{bmatrix}.$$

Hence $x = \frac{23}{5}$, $y = \frac{8}{5}$, and $z = -\frac{25}{5} = -5$.

(f) Now $A = \begin{bmatrix} 1 & 1 & 1 & 1 \\ 1 & 1 & 0 & 0 \\ 0 & 1 & 0 & 1 \\ 1 & 0 & 0 & 1 \end{bmatrix}$ so (by the algorithm) $A^{-1} = \frac{1}{2} \begin{bmatrix} 0 & 1 & -1 & 1 \\ 0 & 1 & 1 & -1 \\ 2 & -1 & -1 & -1 \\ 0 & -1 & 1 & 1 \end{bmatrix}$.

Left multiply the equation $AX = B$ by A^{-1} to get

$$X = A^{-1}(AX) = A^{-1}B = \frac{1}{2} \begin{bmatrix} 0 & 1 & -1 & 1 \\ 0 & 1 & 1 & -1 \\ 2 & -1 & -1 & -1 \\ 0 & -1 & 1 & 1 \end{bmatrix} \begin{bmatrix} 1 \\ 0 \\ -1 \\ 2 \end{bmatrix} = \frac{1}{2} \begin{bmatrix} 3 \\ -3 \\ 1 \\ 1 \end{bmatrix}$$

Hence $x = \frac{3}{2}$, $y = -\frac{3}{2}$, $z = \frac{1}{2}$, $w = \frac{1}{2}$.

4(b) We want B such that $AB = P$ where P is given. Since A^{-1} exists left multiply this equation by A^{-1} to get $B = A^{-1}(AB) = A^{-1}P$. [This B will do it because $AB = A(A^{-1}P) = IP = P$]. Explicitly

$$B = A^{-1}P = \begin{bmatrix} 1 & -1 & 3 \\ 2 & 0 & 5 \\ -1 & 1 & 0 \end{bmatrix} \begin{bmatrix} 1 & -1 & 2 \\ 0 & 1 & 1 \\ 1 & 0 & 0 \end{bmatrix} = \begin{bmatrix} 4 & -2 & 1 \\ 7 & -2 & 4 \\ -1 & 2 & -1 \end{bmatrix}.$$

5(b) We have $(2A)^T = \begin{bmatrix} 1 & -1 \\ 2 & 3 \end{bmatrix}^{-1} = \frac{1}{5} \begin{bmatrix} 3 & 1 \\ -2 & 1 \end{bmatrix}$ by Example 4. Since $(2A)^T = 2A^T$, we get

$$2A^T = \frac{1}{5} \begin{bmatrix} 3 & 1 \\ -2 & 1 \end{bmatrix} \text{ so } A^T = \frac{1}{10} \begin{bmatrix} 3 & 1 \\ -2 & 1 \end{bmatrix}. \text{ Finally}$$

$$A = (A^T)^T = \frac{1}{10} \begin{bmatrix} 3 & 1 \\ -2 & 1 \end{bmatrix}^T = \frac{1}{10} \begin{bmatrix} 3 & -2 \\ 1 & 1 \end{bmatrix}.$$

(d) We have $(I - 2A^T)^{-1} = \begin{bmatrix} 2 & 1 \\ 1 & 1 \end{bmatrix}$ so (because $(U^{-1})^{-1} = U$ for any invertible matrix U)

$$(I - 2A^T) = \begin{bmatrix} 2 & 1 \\ 1 & 1 \end{bmatrix}^{-1} = \begin{bmatrix} 1 & -1 \\ -1 & 2 \end{bmatrix}.$$

Thus $2A^T = I - \begin{bmatrix} 1 & -1 \\ -1 & 2 \end{bmatrix} = \begin{bmatrix} 1 & 0 \\ 0 & 1 \end{bmatrix} - \begin{bmatrix} 1 & -1 \\ -1 & 2 \end{bmatrix} = \begin{bmatrix} 0 & 1 \\ 1 & -1 \end{bmatrix}.$

This gives $A^T = \frac{1}{2} \begin{bmatrix} 0 & 1 \\ 1 & -1 \end{bmatrix}$, so

$$A = (A^T)^T = \frac{1}{2} \begin{bmatrix} 0 & 1 \\ 1 & -1 \end{bmatrix}^T = \frac{1}{2} \begin{bmatrix} 0 & 1 \\ 1 & -1 \end{bmatrix}.$$

(f) Given $\left(\begin{bmatrix} 1 & 0 \\ 2 & 1 \end{bmatrix} A \right)^{-1} = \begin{bmatrix} 1 & 0 \\ 2 & 2 \end{bmatrix}$, take inverses to get

$$\begin{bmatrix} 1 & 0 \\ 2 & 1 \end{bmatrix} A = \begin{bmatrix} 1 & 0 \\ 2 & 2 \end{bmatrix}^{-1} = \frac{1}{2} \begin{bmatrix} 2 & 0 \\ -2 & 1 \end{bmatrix}.$$

Now left multiply by $\begin{bmatrix} 1 & 0 \\ 2 & 1 \end{bmatrix}^{-1}$ to obtain

$$A = \begin{bmatrix} 1 & 0 \\ 2 & 1 \end{bmatrix}^{-1} \left(\frac{1}{2} \begin{bmatrix} 2 & 0 \\ -2 & 1 \end{bmatrix} \right) = \begin{bmatrix} 1 & 0 \\ -2 & 1 \end{bmatrix} \left(\frac{1}{2} \begin{bmatrix} 2 & 0 \\ -2 & 1 \end{bmatrix} \right)$$

$$= \frac{1}{2} \begin{bmatrix} 1 & 0 \\ -2 & 1 \end{bmatrix} \begin{bmatrix} 2 & 0 \\ -2 & 1 \end{bmatrix} = \frac{1}{2} \begin{bmatrix} 2 & 0 \\ -6 & 1 \end{bmatrix}.$$

(h) Given $(A^{-1} - 2I)^T = -2 \begin{bmatrix} 1 & 1 \\ 1 & 0 \end{bmatrix}$, take transposes to get

$$A^{-1} - 2I = \left(-2 \begin{bmatrix} 1 & 1 \\ 1 & 0 \end{bmatrix} \right)^T = -2 \begin{bmatrix} 1 & 1 \\ 1 & 0 \end{bmatrix}^T = -2 \begin{bmatrix} 1 & 1 \\ 1 & 0 \end{bmatrix}.$$

Hence $A^{-1} = 2I - 2 \begin{bmatrix} 1 & 1 \\ 1 & 0 \end{bmatrix} = \begin{bmatrix} 2 & 0 \\ 0 & 2 \end{bmatrix} - 2 \begin{bmatrix} 1 & 1 \\ 1 & 0 \end{bmatrix} = \begin{bmatrix} 0 & -2 \\ -2 & 2 \end{bmatrix} = 2 \begin{bmatrix} 0 & -1 \\ -1 & 1 \end{bmatrix}.$

Finally

$$A = (A^{-1})^{-1} = \left(2\begin{bmatrix} 0 & -1 \\ -1 & 1 \end{bmatrix}\right)^{-1} = \tfrac{1}{2}\begin{bmatrix} 0 & -1 \\ -1 & 1 \end{bmatrix}^{-1} = \tfrac{1}{2}\left(\frac{1}{-1}\begin{bmatrix} 1 & 1 \\ 1 & 0 \end{bmatrix}\right) = \frac{-1}{2}\begin{bmatrix} 1 & 1 \\ 1 & 0 \end{bmatrix}.$$

6(b) Have $A = (A^{-1})^{-1} = \begin{bmatrix} 0 & 1 & -1 \\ 1 & 2 & 1 \\ 1 & 0 & 1 \end{bmatrix}^{-1} = \tfrac{1}{2}\begin{bmatrix} 2 & -1 & 3 \\ 0 & 1 & -1 \\ -2 & 1 & -1 \end{bmatrix}$ by the algorithm.

8(b) The equations are $A\begin{bmatrix} x \\ y \end{bmatrix} = \begin{bmatrix} 7 \\ 1 \end{bmatrix}$ and $\begin{bmatrix} x \\ y \end{bmatrix} = B\begin{bmatrix} x' \\ y' \end{bmatrix}$ where $A = \begin{bmatrix} 3 & 4 \\ 4 & 5 \end{bmatrix}$ and

$B = \begin{bmatrix} -5 & 4 \\ 4 & -3 \end{bmatrix}$. Thus $B = A^{-1}$ (by Example 4) so the substitution gives $\begin{bmatrix} 7 \\ 1 \end{bmatrix} =$

$A\begin{bmatrix} x \\ y \end{bmatrix} = AB\begin{bmatrix} x' \\ y' \end{bmatrix} = I\begin{bmatrix} x' \\ y' \end{bmatrix} = \begin{bmatrix} x' \\ y' \end{bmatrix}$. Thus $x' = 7$, $y' = 1$ so $\begin{bmatrix} x \\ y \end{bmatrix} = B\begin{bmatrix} 7 \\ 1 \end{bmatrix} =$

$\begin{bmatrix} -5 & 4 \\ 4 & -3 \end{bmatrix}\begin{bmatrix} 7 \\ 1 \end{bmatrix} = \begin{bmatrix} -31 \\ 25 \end{bmatrix}$.

9(b) False. $A = \begin{bmatrix} 1 & 0 \\ 0 & 1 \end{bmatrix}$ and $B = \begin{bmatrix} 1 & 0 \\ 0 & -1 \end{bmatrix}$.

(d) True. $A^4 = 3I$ gives $A(\tfrac{1}{3}A^3) = I = (\tfrac{1}{3}A^3)A$, so $A^{-1} = \tfrac{1}{3}A^3$.

(f) False. Take $A = \begin{bmatrix} 1 & 1 \\ 0 & 1 \end{bmatrix}$ and $B = \begin{bmatrix} 1 & 0 \\ 0 & 0 \end{bmatrix}$. Then $AB = B$ but A is not invertible

($AX = 0$ where $X = \begin{bmatrix} 1 \\ -1 \end{bmatrix}$ — see Example 11).

10. $B = IB = (CA)B = C(AB) = CI = C$.

11(b) If a solution X to $AX = B$ exists, it can be found by left multiplication by $C : CAX = CB$, $IX = CB$, $X = CB$.

(i) $X = CB = \begin{bmatrix} 3 \\ 0 \end{bmatrix}$ here but $X = \begin{bmatrix} 3 \\ 0 \end{bmatrix}$ is not a solution. So no solution exists.

(ii) $X = CB = \begin{bmatrix} 2 \\ -1 \end{bmatrix}$ in this case and this is indeed a solution.

15(b) $B^2 = \begin{bmatrix} 0 & -1 \\ 1 & 0 \end{bmatrix} \begin{bmatrix} 0 & -1 \\ 1 & 0 \end{bmatrix} = \begin{bmatrix} -1 & 0 \\ 0 & -1 \end{bmatrix}$ so $B^4 = (B^2)^2 = \begin{bmatrix} -1 & 0 \\ 0 & -1 \end{bmatrix} \begin{bmatrix} -1 & 0 \\ 0 & -1 \end{bmatrix} =$

$\begin{bmatrix} 1 & 0 \\ 0 & 1 \end{bmatrix} = I$. Thus $B \cdot B^3 = I = B^3 B$, so $B^{-1} = B^3 = B^2 B = \begin{bmatrix} -1 & 0 \\ 0 & -1 \end{bmatrix} \begin{bmatrix} 0 & -1 \\ 1 & 0 \end{bmatrix} =$

$\begin{bmatrix} 0 & 1 \\ -1 & 0 \end{bmatrix}$.

16(b) By example 4, $\begin{bmatrix} 2 & -c \\ c & 3 \end{bmatrix}^{-1} = \frac{1}{6+c^2} \begin{bmatrix} 3 & c \\ -c & 2 \end{bmatrix}$ for all c (as $6 + c^2 \neq 0$).

(d) $\begin{bmatrix} 1 & 0 & 1 & | & 1 & 0 & 0 \\ c & 1 & c & | & 0 & 1 & 0 \\ 3 & c & 2 & | & 0 & 0 & 1 \end{bmatrix} \rightarrow \begin{bmatrix} 1 & 0 & 1 & | & 1 & 0 & 0 \\ 0 & 1 & 0 & | & -c & 1 & 0 \\ 0 & c & -1 & | & -3 & 0 & 1 \end{bmatrix} \rightarrow \begin{bmatrix} 1 & 0 & 1 & | & 1 & 0 & 0 \\ 0 & 1 & 0 & | & -c & 1 & 0 \\ 0 & 0 & -1 & | & c^2 - 3 & -c & 1 \end{bmatrix}$

$\rightarrow \begin{bmatrix} 1 & 0 & 0 & | & c^2 - 2 & -c & 1 \\ 0 & 1 & 0 & | & -c & 1 & 0 \\ 0 & 0 & 1 & | & 3 - c^2 & c & -1 \end{bmatrix}$. Hence $\begin{bmatrix} 1 & 0 & 1 \\ c & 1 & c \\ 3 & c & 2 \end{bmatrix}^{-1} = \begin{bmatrix} c^2 - 2 & -c & 1 \\ -c & 1 & 0 \\ 3 - c^2 & c & -1 \end{bmatrix}$ for all

values of c.

18. We use the fact that $\cos^2 \theta + \sin^2 \theta = 1$ for any angle θ. Then Example 4 gives

$\begin{bmatrix} \sin \theta & \cos \theta \\ -\cos \theta & \sin \theta \end{bmatrix}^{-1} = \frac{1}{\sin^2 \theta + \cos^2 \theta} \begin{bmatrix} \sin \theta & -\cos \theta \\ \cos \theta & \sin \theta \end{bmatrix} = \begin{bmatrix} \sin \theta & -\cos \theta \\ \cos \theta & \sin \theta \end{bmatrix}$. Note that, in

this case, the inverse is the transpose (such matices are called orthogonal).

19(b) Suppose column j of A consists of zeros. Then $AY = 0$ where Y is the column with 1 in the position j and zeros elsewhere. If A^{-1} exists, left multiply by A^{-1} to get $A^{-1}AY = A^{-1}0$, that is $IY = 0$; a contradiction. So A^{-1} does not exist.

(d) If each column of A sums to 0, then $XA = 0$ where X is the row of 1's. If A^{-1} exists, right multiply by A^{-1} to get $XAA^{-1} = 0A^{-1}$, that is $XI = 0$, $X = 0$, a contradiction. So A^{-1} does not exist.

20(b) (a) Write $A = \begin{bmatrix} 2 & 1 & -1 \\ 1 & 1 & 0 \\ 1 & 0 & -1 \end{bmatrix}$. Then row 1 minus row 2 minus row 3 is zero. If $X =$

$[1 \quad -1 \quad -1]$, this means $XA = 0$. If A^{-1} exists, right multiply by A^{-1} to get $XAA^{-1} = 0A^{-1}$, $XI = 0$, $X = 0$, a contradiction. So A^{-1} does not exist.

21(d) Suppose $PA = QA$. Right multiply by A^{-1} to get $PAA^{-1} = QAA^{-1}$, that is $PI = QI$, so $P = Q$.

23(d) (ii) $\begin{bmatrix} 3 & 1 & 0 \\ 5 & 2 & 0 \\ 0 & 0 & -1 \end{bmatrix}^{-1} = \begin{bmatrix} \begin{bmatrix} 3 & 1 \\ 5 & 2 \end{bmatrix}^{-1} & 0 \\ 0 & 0 & (-1)^{-1} \end{bmatrix} = \begin{bmatrix} 2 & -1 & 0 \\ -5 & 3 & 0 \\ 0 & 0 & 1 \end{bmatrix}.$

(iv) $\begin{bmatrix} 3 & 4 & 0 & 0 \\ 2 & 3 & 0 & 0 \\ 0 & 0 & 1 & 3 \\ 0 & 0 & 0 & -1 \end{bmatrix}^{-1} = \begin{bmatrix} \begin{bmatrix} 3 & 4 \\ 2 & 3 \end{bmatrix}^{-1} & \begin{matrix} 0 & 0 \\ 0 & 0 \end{matrix} \\ \begin{matrix} 0 & 0 \\ 0 & 0 \end{matrix} & \begin{bmatrix} 1 & 3 \\ 0 & -1 \end{bmatrix}^{-1} \end{bmatrix} = \begin{bmatrix} 3 & -4 & 0 & 0 \\ -2 & 3 & 0 & 0 \\ 0 & 0 & 1 & 3 \\ 0 & 0 & 0 & -1 \end{bmatrix}.$

24(b) If A^{-1} and B^{-1} exist, use block multiplication to compute

$$\begin{bmatrix} A & X \\ 0 & B \end{bmatrix}\begin{bmatrix} A^{-1} & -A^{-1}XB^{-1} \\ 0 & B^{-1} \end{bmatrix} = \begin{bmatrix} AA^{-1} & -AA^{-1}XB^{-1} + XB^{-1} \\ 0 & BB^{-1} \end{bmatrix} = \begin{bmatrix} I & 0 \\ 0 & I \end{bmatrix} = I_{2n}$$

where A and B are $n \times n$. The product in the reverse order is also I_{2n} so $\begin{bmatrix} A & X \\ 0 & B \end{bmatrix}^{-1} =$

$\begin{bmatrix} A^{-1} & -A^{-1}XB^{-1} \\ 0 & B^{-1} \end{bmatrix}.$

24(c) (ii) The matrix is $\begin{bmatrix} A & X \\ 0 & B \end{bmatrix}$ where $A = \begin{bmatrix} 3 & 1 \\ 2 & 1 \end{bmatrix}$, $X = \begin{bmatrix} 3 & 0 \\ -1 & 1 \end{bmatrix}$, and $B = \begin{bmatrix} 5 & 2 \\ 3 & 1 \end{bmatrix}$.

Hence $\begin{bmatrix} A & X \\ 0 & B \end{bmatrix}^{-1} = \begin{bmatrix} A^{-1} & -A^{-1}XB^{-1} \\ 0 & B^{-1} \end{bmatrix} = \begin{bmatrix} 1 & -1 & 7 & -13 \\ -2 & 3 & -18 & 33 \\ 0 & 0 & -1 & 2 \\ 0 & 0 & 3 & -5 \end{bmatrix}.$

26(b) If $P = \begin{bmatrix} 1 & 0 \\ 0 & 0 \end{bmatrix}$, $Q = \begin{bmatrix} 1 & 0 \\ 0 & 0 \end{bmatrix}$ and $R = \begin{bmatrix} 1 & 0 \\ 1 & 0 \end{bmatrix}$ then $PQ = PR$ but $Q \neq R$. Of course P is not invertible (by Example 9).

27(d) If $A^n = 0$ write $B = I + A + A^2 + \cdots + A^{n-1}$. Then

$$\begin{aligned} (I - A)B &= (I - A)(I + A + A^2 + \cdots + A^{n-1}) \\ &= I + A + A^2 + \cdots + A^{n-1} - A - A^2 - A^3 - \cdots - A^n \\ &= I - A^n \\ &= I \end{aligned}$$

Similarly $B(I - A) = I$, so $(I - A)^{-1} = B$.

29(b) Assume that AB and BA are both invertible. Then

$$AB(AB)^{-1} = I \text{ so } AX = I \text{ where } X = B(AB)^{-1}$$
$$(BA)^{-1}BA = I \text{ so } YA = I \text{ where } Y = (BA)^{-1}B.$$

But then $X = IX = (YA)X = Y(AX) = YI = Y$, so $X = Y$ is the inverse of A.

30(b) If $A = B$ then $A^{-1}B = A^{-1}A = I$. Conversely, if $A^{-1}B = I$ left multiply by A to get $AA^{-1}B = AI$, $IB = A$, $B = A$.

32(b) $(AB)^2 = A^2B^2$ means $ABAB = AABB$, left multiplication by A^{-1} and right multiplication by B^{-1} yields $BA = AB$.

33. (ii) implies (i). Assume that AB is invertible. If $BX = 0$ then $ABX = 0$. Left multiplication by $(AB)^{-1}$ yields $Y = (AB)^{-1}0 = 0$. Hence B is invertible by Theorem 5. But then we have $A = (AB)B^{-1}$ so A is invertible (B^{-1} and AB are invertible).

34(b) By the hint, $BX = 0$ where $X = \begin{bmatrix} -1 \\ 3 \\ -1 \end{bmatrix}$ so B is not invertible by Theorem 5.

35. If A can be left cancelled then $AX = 0$ implies $AX = A0$ so $X = 0$. Thus A is invertible by Theorem 5. Conversely, if A is invertible, suppose that $AB = AC$. Then left multiplication by A^{-1} yields $A^{-1}AB = A^{-1}AC$, $IB = IC$, $B = C$.

37(b) Write $U = I_n - 2XX^T$. Then U is symmetric because

$$U^T = I_n^T - 2(XX^T)^T = I_n - 2X^{TT}X^T = I_n - 2XX^T = U.$$

Moreover $U^{-1} = U$ because (since $X^TX = I_n$)

$$\begin{aligned} U^2 &= (I_n - 2XX^T)(I - 2XX^T) \\ &= I_n - 2XX^T - 2XX^T + 4XX^TXX^T \\ &= I_n - 4XX^T + 4XI_mX^T \\ &= I_n. \end{aligned}$$

38(b) If $P^2 = P$ then $I - 2P$ is self-inverse because

$$(I - 2P)(I - 2P) = I - 2P - 2P + 4P^2 = I.$$

Conversely, if $I - 2P$ is self-inverse because

$$I = (I - 2P)^2 = I - 4P + 4P^2$$

so $4P = 4P^2$; $P = P^2$.

40(b) We compute

$$A^{-1}(A + B)B^{-1} = A^{-1}AB^{-1} + A^{-1}BB^{-1} = B^{-1} + A^{-1} = A^{-1} + B^{-1}.$$

Hence $A^{-1} + B^{-1}$ is invertible by Theorem 4 because each of A^{-1}, $A + B$, and B^{-1} is invertible. Furthermore

$$(A^{-1} + B^{-1})^{-1} = [A^{-1}(A + B)B^{-1}]^{-1} = (B^{-1})^{-1}(A + B)^{-1}(A^{-1})^{-1} = B(A + B)^{-1}A$$

gives the desired formula.

Exercises 2.4   Elementary Matrices

1(b) Interchange rows 1 and 3 of I, $E^{-1} = E$.

(d) Add (-2) times row 1 of I to row 2. $E^{-1} = \begin{bmatrix} 1 & 0 & 0 \\ 2 & 1 & 0 \\ 0 & 0 & 1 \end{bmatrix}$.

(f) Multiply row 3 of I by 5. $E^{-1} = \begin{bmatrix} 1 & 0 & 0 \\ 0 & 1 & 0 \\ 0 & 0 & \frac{1}{5} \end{bmatrix}$.

2(b) $A \to B$ is accomplished by negating row 1, so $E = \begin{bmatrix} -1 & 0 \\ 0 & 0 \end{bmatrix}$.

(d) $A \to B$ is accomplished by subtracting row 2 from row 1, so $E = \begin{bmatrix} 1 & -1 \\ 0 & 1 \end{bmatrix}$.

(f) $A \to B$ is acomplished by interchanging rows 1 and 2, so $E = \begin{bmatrix} 0 & 1 \\ 1 & 0 \end{bmatrix}$.

3(b) The possibilities for E are $\begin{bmatrix} 0 & 1 \\ 1 & 0 \end{bmatrix}$, $\begin{bmatrix} k & 0 \\ 0 & 1 \end{bmatrix}$, $\begin{bmatrix} 1 & 0 \\ 0 & k \end{bmatrix}$. $\begin{bmatrix} 1 & k \\ 0 & 1 \end{bmatrix}$ and $\begin{bmatrix} 1 & 0 \\ k & 1 \end{bmatrix}$. In each case EA has a row different from C.

4. If E is Type I, EA and A differ only in the interchanged rows.
   If E is of Type II, EA and A differ only in the row multiplied by a nonzero constant.
   If E is of Type II, EA and A differ only in the row to which a multiple of a row is added.

6(b) $\begin{bmatrix} 1 & 2 & 1 & | & 1 & 0 \\ 5 & 12 & -1 & | & 0 & 1 \end{bmatrix} \to \begin{bmatrix} 1 & 2 & 1 & | & 1 & 0 \\ 0 & 2 & -6 & | & -5 & 1 \end{bmatrix} \to \begin{bmatrix} 1 & 2 & 1 & | & 1 & 0 \\ 0 & 1 & -3 & | & -\frac{5}{2} & \frac{1}{2} \end{bmatrix}$

$\to \begin{bmatrix} 1 & 0 & 7 & | & \frac{12}{2} & -1 \\ 0 & 1 & -3 & | & -\frac{5}{2} & \frac{1}{2} \end{bmatrix}$ so $UA = R = \begin{bmatrix} 1 & 0 & 7 \\ 0 & 1 & -3 \end{bmatrix}$ where $U = \frac{1}{2}\begin{bmatrix} 12 & -2 \\ -5 & 1 \end{bmatrix}$. This matrix U is the product of the elementary matrices used at each stage:

$$\begin{bmatrix} 1 & 2 & 1 \\ 5 & 12 & -1 \end{bmatrix} = A$$

$$\downarrow$$

$$\begin{bmatrix} 1 & 2 & 1 \\ 0 & 2 & -6 \end{bmatrix} = E_1 A \qquad \text{where } E_1 = \begin{bmatrix} 1 & 0 \\ -5 & 1 \end{bmatrix}$$

$$\downarrow$$

$$\begin{bmatrix} 1 & 2 & 1 \\ 0 & 1 & -3 \end{bmatrix} = E_2 E_1 A \qquad \text{where } E_2 = \begin{bmatrix} 1 & 0 \\ 0 & \frac{1}{2} \end{bmatrix}$$

$$\downarrow$$

$$\begin{bmatrix} 1 & 0 & 7 \\ 0 & 1 & -3 \end{bmatrix} = E_3 E_2 E_1 A \qquad \text{where } E_3 = \begin{bmatrix} 1 & -2 \\ 0 & 1 \end{bmatrix}$$

(d) Analogous to (b).

7(b) $\begin{bmatrix} 2 & -1 & 0 & | & 1 & 0 \\ 1 & 1 & 1 & | & 0 & 1 \end{bmatrix} \rightarrow \begin{bmatrix} 1 & 1 & 1 & | & 0 & 1 \\ 2 & -1 & 0 & | & 1 & 0 \end{bmatrix} \rightarrow \begin{bmatrix} 3 & 0 & 1 & | & 1 & 1 \\ 2 & -1 & 0 & | & 1 & 0 \end{bmatrix}$. So $U = \begin{bmatrix} 1 & 1 \\ 1 & 0 \end{bmatrix}$.

8(b) $\begin{bmatrix} 2 & 3 \\ 1 & 2 \end{bmatrix} = A$

$$\downarrow$$

$$\begin{bmatrix} 1 & 2 \\ 2 & 3 \end{bmatrix} = E_1 A \qquad \text{where } E_1 = \begin{bmatrix} 0 & 1 \\ 1 & 0 \end{bmatrix}$$

$$\downarrow$$

$$\begin{bmatrix} 1 & 2 \\ 0 & -1 \end{bmatrix} = E_2 E_1 A \qquad \text{where } E_2 = \begin{bmatrix} 1 & 0 \\ -2 & 1 \end{bmatrix}$$

$$\downarrow$$

$$\begin{bmatrix} 1 & 2 \\ 0 & 1 \end{bmatrix} = E_3 E_2 E_1 A \qquad \text{where } E_3 = \begin{bmatrix} 1 & 0 \\ 0 & -1 \end{bmatrix}$$

$$\downarrow$$

$$\begin{bmatrix} 1 & 0 \\ 0 & 1 \end{bmatrix} = E_4 E_3 E_2 E_1 A \qquad \text{where } E_4 = \begin{bmatrix} 1 & -2 \\ 0 & 1 \end{bmatrix}$$

Thus $E_4 E_3 E_2 E_1 A = I$ so

$$A = (E_4 E_3 E_2 E_1)^{-1}$$
$$= E_1^{-1} E_2^{-2} E_3^{-1} E_4^{-1}$$
$$= \begin{bmatrix} 0 & 1 \\ 1 & 0 \end{bmatrix} \begin{bmatrix} 1 & 0 \\ 2 & 1 \end{bmatrix} \begin{bmatrix} 1 & 0 \\ 0 & -1 \end{bmatrix} \begin{bmatrix} 1 & 2 \\ 0 & 1 \end{bmatrix}.$$

Of course a different sequence of row operations yields a different factorization of A.

(d) Analogous to (b).

10. By Theorem 3, $UA = R$ for some invertible matrix U. Hence $A = U^{-1}R$ where U^{-1} is invertible.

17. Write $U^{-1} = E_k E_{k-1} \ldots E_2 E_1$. E_i elementary. Then

$$[I \quad U^{-1}A] = U^{-1}[U \quad A] = E_1 E_2 \ldots E_k [U \quad A].$$

Hence a sequence of k row operations carries $[U \quad A]$ to $[I \quad U^{-1}A]$. Clearly $[I \quad U^{-1}A]$ is in reduced row-echelon form.

18(b) $A \overset{r}{\sim} A$ because $A = IA$. If $A \overset{r}{\sim} B$, let $A = UB$, U invertible. Then $B = U^{-1}A$ so $B \overset{r}{\sim} A$. Finally if $A \overset{r}{\sim} B$ and $B \overset{r}{\sim} C$, let $A = UB$ and $B = VC$ where U and V are invertible. Hence $A = U(VC) = (UV)C$ so $A \overset{r}{\sim} C$.

20(b) The matrices row-equivalent to $A = \begin{bmatrix} 0 & 0 & 0 \\ 0 & 0 & 1 \end{bmatrix}$ are the matrices UA whre U is invertible. If

$U = \begin{bmatrix} a & b \\ c & d \end{bmatrix}$ then $UA = \begin{bmatrix} 0 & 0 & b \\ 0 & 0 & d \end{bmatrix}$ where b and d are not both zero (as U is invertible).

Every such matrix arises — use $U = \begin{bmatrix} a & b \\ -b & a \end{bmatrix}$ — it is invertible as $a^2 + b^2 \neq 0$ (Example 4 §2.3).

Exercises 2.5   LU-factorization

1(b) $\begin{bmatrix} 2 & 4 & 2 \\ 1 & -1 & 3 \\ -1 & 7 & -7 \end{bmatrix} \rightarrow \begin{bmatrix} 1 & 2 & 1 \\ 1 & -1 & 3 \\ -1 & 7 & -7 \end{bmatrix} \rightarrow \begin{bmatrix} 1 & 2 & 1 \\ 0 & -3 & 2 \\ 0 & 9 & -6 \end{bmatrix} \rightarrow \begin{bmatrix} 1 & 2 & 1 \\ 0 & 1 & -\frac{2}{3} \\ 0 & 0 & 0 \end{bmatrix} = U.$ Hence

$A = LU$ where U is above and $L = \begin{bmatrix} 2 & 0 & 0 \\ 1 & -3 & 0 \\ -1 & 9 & 1 \end{bmatrix}.$

(d) $\begin{bmatrix} \boxed{-1} & -3 & 1 & 0 & -1 \\ 1 & 4 & 1 & 1 & 1 \\ 1 & 2 & -3 & -1 & 1 \\ 0 & -2 & -4 & -2 & 0 \end{bmatrix} \rightarrow \begin{bmatrix} 1 & 3 & -1 & 0 & 1 \\ 0 & \boxed{1} & 2 & 1 & 0 \\ 0 & \boxed{-1} & -2 & -1 & 0 \\ 0 & \boxed{-2} & -4 & -2 & 0 \end{bmatrix} \rightarrow \begin{bmatrix} 1 & 3 & -1 & 0 & 1 \\ 0 & 1 & 2 & 1 & 0 \\ 0 & 0 & 0 & 0 & 0 \\ 0 & 0 & 0 & 0 & 0 \end{bmatrix} = U.$

Hence $A = LU$ where U is as above and $L = \begin{bmatrix} -1 & 0 & 0 & 0 \\ 1 & 1 & 0 & 0 \\ 1 & -1 & 1 & 0 \\ 0 & -2 & 0 & 1 \end{bmatrix}.$

(f) $\begin{bmatrix} \boxed{2} & 2 & -2 & 4 & 2 \\ 1 & -1 & 0 & 2 & 1 \\ 3 & 1 & -2 & 6 & 3 \\ 1 & 3 & -2 & 2 & 1 \end{bmatrix} \rightarrow \begin{bmatrix} 1 & 1 & -1 & 2 & 1 \\ 0 & \boxed{-2} & 1 & 0 & 0 \\ 0 & \boxed{-2} & 1 & 0 & 0 \\ 0 & \boxed{2} & -1 & 0 & 0 \end{bmatrix} \rightarrow \begin{bmatrix} 1 & 1 & -1 & 2 & 1 \\ 0 & 1 & -\frac{1}{2} & 0 & 0 \\ 0 & 0 & 0 & 0 & 0 \\ 0 & 0 & 0 & 0 & 0 \end{bmatrix} = U.$

Hence $A = LU$ where U is above and $L = \begin{bmatrix} 2 & 0 & 0 & 0 \\ 1 & -2 & 0 & 0 \\ 3 & -2 & 1 & 0 \\ 1 & 2 & 0 & 1 \end{bmatrix}.$

2(b) The reduction to row-echelon form requires two row interchanges:

$$\begin{bmatrix} 0 & -1 & 2 \\ 0 & 0 & 4 \\ -1 & 2 & 1 \end{bmatrix} \rightarrow \begin{bmatrix} 0 & -1 & 2 \\ 0 & 0 & 4 \\ -1 & 2 & 1 \end{bmatrix} \rightarrow \begin{bmatrix} -1 & 2 & 1 \\ 0 & -1 & 2 \\ 0 & 0 & 4 \end{bmatrix} \rightarrow \cdots$$

The elementary matrices corresponding (in order) to the interchanges are

$P_1 = \begin{bmatrix} 1 & 0 & 0 \\ 0 & 0 & 1 \\ 0 & 1 & 0 \end{bmatrix}$ and $P_2 = \begin{bmatrix} 0 & 1 & 0 \\ 1 & 0 & 0 \\ 0 & 0 & 1 \end{bmatrix}$, so take $P = P_2 P_1 = \begin{bmatrix} 0 & 0 & 1 \\ 1 & 0 & 0 \\ 0 & 1 & 0 \end{bmatrix}.$

We apply the LU-algorithm to PA:

$$PA = \begin{bmatrix} \boxed{-1} & 2 & 1 \\ 0 & -1 & 2 \\ 0 & 0 & 4 \end{bmatrix} \rightarrow \begin{bmatrix} 1 & -2 & -1 \\ 0 & \boxed{-1} & 2 \\ 0 & \boxed{0} & 4 \end{bmatrix} \rightarrow \begin{bmatrix} 1 & -2 & -1 \\ 0 & 1 & -2 \\ 0 & 0 & \boxed{4} \end{bmatrix} \rightarrow \begin{bmatrix} 1 & -2 & -1 \\ 0 & 1 & -2 \\ 0 & 0 & 1 \end{bmatrix} = U.$$

Hence $PA = LU$ where U is as above and $L = \begin{bmatrix} -1 & 0 & 0 \\ 0 & -1 & 0 \\ 0 & 0 & 4 \end{bmatrix}.$

(d) The reduction to row-echelon form requires two row interchanges:

$$\begin{bmatrix} -1 & -2 & 3 & 0 \\ 2 & 4 & -6 & 5 \\ 1 & 1 & -1 & 3 \\ 2 & 5 & -10 & 1 \end{bmatrix} \rightarrow \begin{bmatrix} 1 & 2 & -3 & 0 \\ 0 & 0 & 0 & 5 \\ 0 & -1 & 2 & 3 \\ 0 & 1 & -4 & 1 \end{bmatrix} \rightarrow \begin{bmatrix} 1 & 2 & -3 & 0 \\ 0 & 0 & 0 & 5 \\ 0 & 1 & -2 & -3 \\ 0 & 0 & -2 & 4 \end{bmatrix}$$

$$\rightarrow \begin{bmatrix} 1 & 2 & -3 & 0 \\ 0 & 1 & -2 & -3 \\ 0 & 0 & 0 & 5 \\ 0 & 0 & -2 & 4 \end{bmatrix} \rightarrow \begin{bmatrix} 1 & 2 & -3 & 0 \\ 0 & 1 & -2 & -3 \\ 0 & 0 & -2 & 4 \\ 0 & 0 & 0 & 5 \end{bmatrix}.$$

The elementary matrices corresponding (in order) to the interchanges are

$$P_1 = \begin{bmatrix} 1 & 0 & 0 & 0 \\ 0 & 0 & 1 & 0 \\ 0 & 1 & 0 & 0 \\ 0 & 0 & 0 & 1 \end{bmatrix} \text{ and } P_2 = \begin{bmatrix} 1 & 0 & 0 & 0 \\ 0 & 1 & 0 & 0 \\ 0 & 0 & 0 & 1 \\ 0 & 0 & 1 & 0 \end{bmatrix} \text{ so } P = P_2 P_1 = \begin{bmatrix} 1 & 0 & 0 & 0 \\ 0 & 0 & 1 & 0 \\ 0 & 0 & 0 & 1 \\ 0 & 1 & 0 & 0 \end{bmatrix}.$$

We apply the LU-algorithm to PA:

$$PA = \begin{bmatrix} -1 & -2 & 3 & 0 \\ 1 & 1 & -1 & 3 \\ 2 & 5 & -10 & 1 \\ 2 & 4 & -6 & 5 \end{bmatrix} \rightarrow \begin{bmatrix} 1 & 2 & -3 & 0 \\ 0 & -1 & 2 & 3 \\ 0 & 1 & -4 & 1 \\ 0 & 0 & 0 & 5 \end{bmatrix} \rightarrow \begin{bmatrix} 1 & 2 & -3 & 0 \\ 0 & 1 & -2 & -3 \\ 0 & 0 & -2 & 4 \\ 0 & 0 & 0 & 5 \end{bmatrix}$$

$$\rightarrow \begin{bmatrix} 1 & 2 & -3 & 0 \\ 0 & 1 & -2 & -3 \\ 0 & 0 & 1 & -2 \\ 0 & 0 & 0 & 5 \end{bmatrix} \rightarrow \begin{bmatrix} 1 & 2 & -3 & 0 \\ 0 & 1 & -2 & -3 \\ 0 & 0 & 1 & -2 \\ 0 & 0 & 0 & 1 \end{bmatrix} = U.$$

Hence $PA = LU$ where U is as above and $\dot{L} = \begin{bmatrix} -1 & 0 & 0 & 0 \\ 1 & -1 & 0 & 0 \\ 2 & 1 & -2 & 0 \\ 2 & 0 & 0 & 5 \end{bmatrix}.$

3(b) Write $L = \begin{bmatrix} 2 & 0 & 0 \\ 1 & 3 & 0 \\ -1 & 2 & 1 \end{bmatrix}$, $U = \begin{bmatrix} 1 & 1 & 0 & -1 \\ 0 & 1 & 0 & 1 \\ 0 & 0 & 0 & 0 \end{bmatrix}$, $X = \begin{bmatrix} x_1 \\ x_2 \\ x_3 \\ x_4 \end{bmatrix}$, $Y = \begin{bmatrix} y_1 \\ y_2 \\ y_3 \end{bmatrix}$. The sys-

tem $LY = B$ is
$$2y_1 \qquad\qquad = -2$$
$$y_1 + 3y_2 \qquad = -1$$
$$-y_1 + 2y_2 + y_3 = 1$$
and we solve this by forward substitu-

tion: $y_1 = -1$, $y_2 = \frac{1}{3}(-1 - y_1) = 0$, $y_3 = 1 + y_1 - 2y_2 = 0$. The system $UX = Y$ is $x_1 + x_2 - x_4 = -1$ and we solve this by back substitution: $x_4 = t$, $x_3 = 5$,

$$x_2 + x_4 = 0$$
$$0 = 0$$

$x_2 = -x_4 = -t$, $x_1 = -1 + x_4 - x_2 = -1 + 2t$.

(d) Analogous to (a).

6. If the rows in question are R_1 and R_2, they can be interchanged thus:

$$\begin{bmatrix} R_1 \\ R_2 \end{bmatrix} \rightarrow \begin{bmatrix} R_1 + R_2 \\ R_2 \end{bmatrix} \rightarrow \begin{bmatrix} R_1 + R_2 \\ -R_1 \end{bmatrix} \rightarrow \begin{bmatrix} R_1 \\ -R_1 \end{bmatrix} \rightarrow \begin{bmatrix} R_2 \\ R_1 \end{bmatrix}.$$

7(b) Let $A = LU = L_1U_1$ be LU-factorizations of the invertible matrix A. Then U and U_1 have no row of zeros so (being row-echelon) are upper triangular with 1's on the main diagonal. Thus $L_1^{-1}L = U_1U^{-1}$ is both lower triangular ($L_1^{-1}L$) and upper triangular (U_1U^{-1}) and so is diagonal. But it has 1's on the diagonal (U_1 and U do) so it is I. Hence $L_1 = L$ and $U_1 = U$.

8. We proceed by induction on n where A and B are $n \times n$. It is clear if $n = 1$. In general, write $A = \begin{bmatrix} a & 0 \\ X & A_1 \end{bmatrix}$ and $B = \begin{bmatrix} b & 0 \\ Y & B_1 \end{bmatrix}$ where A_1 and B_1 are lower triangular. Then

$$AB = \begin{bmatrix} ab & 0 \\ Xb + A_1Y & A_1B_1 \end{bmatrix}$$ by Theorem 4 §2.2, and A_1B_1 is upper triangular by induction.

Hence AB is upper triangular.

Exercises 2.6   Input-Output Economic Models

1(b) $I - E = \begin{bmatrix} .5 & 0 & -.5 \\ -.1 & .1 & -.2 \\ -.4 & -.1 & .7 \end{bmatrix} \rightarrow \begin{bmatrix} 1 & 0 & -1 \\ 0 & 1 & -3 \\ 0 & -1 & 3 \end{bmatrix} \rightarrow \begin{bmatrix} 1 & 0 & -1 \\ 0 & 1 & -3 \\ 0 & 0 & 0 \end{bmatrix}$. The equilibrium price

structure P is the solution to $(I - E)P = 0$: $p_1 = t$, $p_2 = 3t$, $p_3 = t$.

(d) $I - E = \begin{bmatrix} .5 & 0 & -.1 & -.1 \\ -.2 & .3 & 0 & -.1 \\ -.1 & -.2 & .2 & -.2 \\ -.2 & -.1 & -.1 & .4 \end{bmatrix} \rightarrow \begin{bmatrix} 1 & 2 & -2 & 2 \\ 5 & 0 & -1 & -1 \\ -2 & 3 & 0 & -1 \\ -2 & -1 & -1 & 4 \end{bmatrix} \rightarrow \begin{bmatrix} 1 & 2 & -2 & 2 \\ 0 & -10 & 9 & -11 \\ 0 & 7 & -4 & 3 \\ 0 & 3 & -5 & 8 \end{bmatrix}.$

Now add 3 times row 4 to row 2:

$$\rightarrow \begin{bmatrix} 1 & 2 & -2 & 2 \\ 0 & -1 & -6 & 13 \\ 0 & 7 & -4 & 3 \\ 0 & 3 & -5 & 8 \end{bmatrix} \rightarrow \begin{bmatrix} 1 & 0 & -14 & 28 \\ 0 & 1 & 6 & -13 \\ 0 & 0 & -46 & 94 \\ 0 & 0 & -23 & 47 \end{bmatrix} \rightarrow \begin{bmatrix} 1 & 0 & -14 & 28 \\ 0 & 1 & 6 & -13 \\ 0 & 0 & 1 & -\frac{47}{23} \\ 0 & 0 & 0 & 0 \end{bmatrix} \rightarrow \begin{bmatrix} 1 & 0 & 0 & -\frac{14}{23} \\ 0 & 1 & 0 & -\frac{17}{23} \\ 0 & 0 & 1 & -\frac{47}{23} \\ 0 & 0 & 0 & 0 \end{bmatrix}.$$

The equilibrium price structure P is the solution to $(I - E)P = 0$. The solution is $p_1 = 14t$, $p_2 = 17t$, $p_3 = 47t$, $p_4 = 23t$.

2. Here the input-output matrix is $E = \begin{bmatrix} 0 & 0 & 1 \\ 1 & 0 & 0 \\ 0 & 1 & 0 \end{bmatrix}$ so we get

$$I - E = \begin{bmatrix} 1 & 0 & -1 \\ -1 & 1 & 0 \\ 0 & -1 & 1 \end{bmatrix} \rightarrow \begin{bmatrix} 1 & 0 & -1 \\ 0 & 1 & -1 \\ 0 & -1 & 1 \end{bmatrix} \rightarrow \begin{bmatrix} 1 & 0 & -1 \\ 0 & 1 & -1 \\ 0 & 0 & 0 \end{bmatrix}.$$ Thus the solution to

$(I - E)P$ is $p_1 = p_2 = p_3 = t$. Thus all three industries produce the same output.

4. $I - E = \begin{bmatrix} 1-a & -b \\ -1+a & b \end{bmatrix} \rightarrow \begin{bmatrix} 1-a & -b \\ 0 & 0 \end{bmatrix}$ so the possible equilibrium price structures are

$P = \begin{bmatrix} bt \\ (1-a)t \end{bmatrix}$, t arbitrary. This is nonzero for some t unless $b = 0$ and $a = 1$, and in

that case $P = \begin{bmatrix} 1 \\ 1 \end{bmatrix}$ is a solution. If the entries of A are positive then $P = \begin{bmatrix} b \\ 1-a \end{bmatrix}$ has

positive entries.

7(b) $E = \begin{bmatrix} .4 & .8 \\ .7 & .2 \end{bmatrix}$ and $(I - E)^{-1} = -\frac{10}{8} \begin{bmatrix} 8 & 8 \\ 7 & 6 \end{bmatrix}$.

8. If $E = \begin{bmatrix} a & b \\ c & d \end{bmatrix}$ then $I - E = \begin{bmatrix} 1-a & -b \\ -c & 1-d \end{bmatrix}$. We have $\det(I - E) = (1-a)(1-d) - bc =$
$1 - (a + d) + (ad - bc) = 1 - \operatorname{tr} E + \det E$. If $\det(I - E) \neq 0$ then Example 4 §2.3 gives
$(I - E)^{-1} = \frac{1}{\det(I-E)} \begin{bmatrix} 1-d & b \\ c & 1-a \end{bmatrix}$. The entries $1 - d$, b, c, and $1 - a$ are all between 0
and 1 so $(I - E)^{-1} \geq 0$ if $\det(I - E) > 0$, that is if $\operatorname{tr} E < 1 + \det E$.

9(b) If $P = \begin{bmatrix} 3 \\ 2 \\ 1 \end{bmatrix}$ then $P > EP$ so Theorem 2 applies.

(d) If $P = \begin{bmatrix} 3 \\ 2 \\ 2 \end{bmatrix}$ then $P > EP$ so Theorem 2 applies.

Exercises 2.7   Markov Chains

1(b) Not regular. Every power of P has the $(1,2)$- and $(3,2)$-entries zero.

2(b) $I - P = \begin{bmatrix} \frac{1}{2} & -1 \\ -\frac{1}{2} & 1 \end{bmatrix} \to \begin{bmatrix} 1 & -2 \\ 0 & 0 \end{bmatrix}$ so $(I-P)S = 0$ has solutions $S = \begin{bmatrix} 2t \\ t \end{bmatrix}$. The entries

of S sum to 1 if $t = \frac{1}{3}$, so $S = \begin{bmatrix} \frac{1}{2} \\ \frac{1}{3} \end{bmatrix}$ is the steady state vector. Given $S_0 = \begin{bmatrix} 1 \\ 0 \end{bmatrix}$, we get

$S_1 = PS_0 = \begin{bmatrix} \frac{1}{2} \\ \frac{1}{2} \end{bmatrix}$, $S_2 = PS_1 \begin{bmatrix} \frac{3}{4} \\ \frac{1}{4} \end{bmatrix}$, $S_3 = PS_2 = \begin{bmatrix} \frac{5}{8} \\ \frac{3}{8} \end{bmatrix}$. So it is in state 2 after three transitions with probability $\frac{3}{8}$.

(d) $I - P = \begin{bmatrix} .6 & -.1 & -.5 \\ -.2 & .4 & -.2 \\ -.4 & -.3 & .7 \end{bmatrix} \to \begin{bmatrix} 1 & -2 & 1 \\ 0 & 11 & -11 \\ 0 & -11 & 11 \end{bmatrix} \to \begin{bmatrix} 1 & 0 & -1 \\ 0 & 1 & -1 \\ 0 & 0 & 0 \end{bmatrix}$ so $(I-P)S = 0$ has

solution $S = \begin{bmatrix} t \\ t \\ t \end{bmatrix}$. The entries sum to 1 if $t = \frac{1}{3}$ so the steady state vector is $S = \begin{bmatrix} \frac{1}{3} \\ \frac{1}{3} \\ \frac{1}{3} \end{bmatrix}$.

Given $S_0 = \begin{bmatrix} 0 \\ 0 \\ 0 \end{bmatrix}$, $S_1 = PS_0 = \begin{bmatrix} .4 \\ .2 \\ .4 \end{bmatrix}$, $S_2 = PS_1 = \begin{bmatrix} .38 \\ .28 \\ .34 \end{bmatrix}$, $S_3 = PS_2 = \begin{bmatrix} .350 \\ .312 \\ .338 \end{bmatrix}$.

Hence it is in state 2 after three transitions with prability .312.

(f) $I - P = \begin{bmatrix} .9 & -.3 & -.3 \\ -.3 & .9 & -.6 \\ -.6 & -.6 & .9 \end{bmatrix} \to \begin{bmatrix} 1 & -3 & 2 \\ 0 & 24 & -21 \\ 0 & -24 & 21 \end{bmatrix} \to \begin{bmatrix} 1 & 0 & -\frac{5}{8} \\ 0 & 1 & -\frac{7}{8} \\ 0 & 0 & 0 \end{bmatrix}$, so $(I-P)S = 0$ has

solution $S = \begin{bmatrix} 5t \\ 7t \\ 8t \end{bmatrix}$. The entries sum to 1 if $t = 20$ so the steady state vector is $S = \begin{bmatrix} \frac{5}{20} \\ \frac{7}{20} \\ \frac{8}{20} \end{bmatrix}$.

Given $S_0 = \begin{bmatrix} 1 \\ 0 \\ 0 \end{bmatrix}$, $S_1 = PS_0 = \begin{bmatrix} .1 \\ .3 \\ .6 \end{bmatrix}$, $S_2 = PS_1 = \begin{bmatrix} .28 \\ .42 \\ .30 \end{bmatrix}$, $S_3 = PS_1 = \begin{bmatrix} .244 \\ .306 \\ .450 \end{bmatrix}$.

Hence it is in state 2 after three transitions with probability .306.

4(b) The transition matrix is $P = \begin{bmatrix} .7 & .1 & .1 \\ .1 & .8 & .3 \\ .2 & .1 & .6 \end{bmatrix}$ where the columns (and rows) represent the up-

per, middle and lower classes respectively and, for example, the last column asserts that, for children of lower class people, 10% become upper class, 30% become middle class and 60% remain lower class. Hence $I - P = \begin{bmatrix} .3 & -.1 & -.1 \\ -.1 & .2 & -.3 \\ -.2 & -.1 & -.4 \end{bmatrix} \rightarrow \begin{bmatrix} 1 & -2 & 3 \\ 0 & 5 & -10 \\ 0 & -5 & 10 \end{bmatrix} \rightarrow \begin{bmatrix} 1 & 0 & -1 \\ 0 & 1 & -2 \\ 0 & 0 & 0 \end{bmatrix}$.

Thus the general solution to $(I - P)S = 0$ is $S = \begin{bmatrix} t \\ 2t \\ t \end{bmatrix}$, so $S = \begin{bmatrix} \frac{1}{4} \\ \frac{1}{2} \\ \frac{1}{4} \end{bmatrix}$ is the steady state

solution. Eventually, upper, middle and lower classes will comprixe 25%, 50% and 25% of this society respectively.

6. Let States 1 and 2 be "late" and "on time" respectively. Then the transition matrix in
$P = \begin{bmatrix} \frac{1}{3} & \frac{1}{2} \\ \frac{2}{3} & \frac{1}{2} \end{bmatrix}$. Here column 1 describes what happens if he was late one day: the two entries sum to 1 and the top entry is twice the bottom entry by the information we are given. Column 2 is determined similarly. Now if Monday is the initial state, we are given that
$S_0 = \begin{bmatrix} \frac{3}{4} \\ \frac{1}{4} \end{bmatrix}$. Hence $S_1 = PS_0 = \begin{bmatrix} \frac{3}{8} \\ \frac{5}{8} \end{bmatrix}$ and $S_2 = PS_1 = \begin{bmatrix} \frac{7}{16} \\ \frac{9}{16} \end{bmatrix}$. Hence the probabilities
that he is late and on time Wednesdays are $\frac{7}{16}$ and $\frac{9}{16}$ respectively.

8. Let the states be the five compartments. Since each tunnel entry is equally likely,

$$P = \begin{bmatrix} 0 & \frac{1}{2} & \frac{1}{5} & 0 & \frac{1}{3} \\ \frac{1}{3} & 0 & 0 & \frac{1}{5} & 0 \\ \frac{1}{3} & 0 & \frac{2}{5} & \frac{1}{5} & \frac{1}{3} \\ 0 & \frac{1}{2} & \frac{1}{5} & \frac{2}{5} & \frac{1}{3} \\ \frac{1}{3} & 0 & \frac{1}{5} & \frac{1}{5} & 0 \end{bmatrix} = \frac{1}{30} \begin{bmatrix} 0 & 15 & 6 & 0 & 10 \\ 10 & 0 & 0 & 6 & 0 \\ 10 & 0 & 12 & 6 & 10 \\ 0 & 15 & 6 & 12 & 10 \\ 10 & 0 & 6 & 6 & 0 \end{bmatrix}.$$

(a) Since he starts in Compartment 1,

$$S_0 = \begin{bmatrix} 1 \\ 0 \\ 0 \\ 0 \\ 0 \end{bmatrix}, \ S_1 = PS_0 = \frac{1}{30} \begin{bmatrix} 0 \\ 10 \\ 10 \\ 0 \\ 10 \end{bmatrix}, \ S_2 = PS_1 = \frac{1}{90} \begin{bmatrix} 31 \\ 0 \\ 22 \\ 31 \\ 6 \end{bmatrix}, \ S_3 = \frac{1}{2700} \begin{bmatrix} 192 \\ 496 \\ 820 \\ 564 \\ 628 \end{bmatrix}.$$

Hence the probability that he is in compartment 4 after three moves is $\frac{564}{2700} = \frac{94}{450}$.

(b) Equal time in 3 and 4. Steady state is $S = \frac{1}{18} \begin{bmatrix} 3 \\ 2 \\ 5 \\ 5 \\ 3 \end{bmatrix}$ (solution to $(I - P)S = 0$).

12(a) $\begin{bmatrix} 1-p & q \\ p & 1-q \end{bmatrix} \cdot \frac{1}{p+q} \begin{bmatrix} q \\ p \end{bmatrix} = \frac{1}{p+q} \begin{bmatrix} (1-p)q + qp \\ pq + (1-q)p \end{bmatrix} = \frac{1}{p+q} \begin{bmatrix} q \\ p \end{bmatrix}$. Since the entries of

$\frac{1}{p+q} \begin{bmatrix} q \\ p \end{bmatrix}$ add to 1, it is the steady state vector.

(b) If $m = 1$

$$\cdot \frac{1}{p+q} \begin{bmatrix} q & q \\ p & p \end{bmatrix} + \frac{1-p-q}{p+q} \begin{bmatrix} p & -q \\ -p & q \end{bmatrix} = \frac{1}{p+q} \begin{bmatrix} q+p-p^2-pq & q-q+pq+q^2 \\ p-p\_p^2+pq & p+q-pq-q^2 \end{bmatrix}$$

$$= \frac{1}{p+q} \begin{bmatrix} (p+q)(1-p) & (p+q)q \\ (p+q)p & (p+q)(1-q) \end{bmatrix}$$

$$= P$$

In general, write $X = \begin{bmatrix} q & q \\ p & p \end{bmatrix}$ and $Y = \begin{bmatrix} p & -q \\ -p & q \end{bmatrix}$. Then $PX = X$ and

$PY = (1-p-q)Y$. Hence if $P^m = \frac{1}{p+q}X + \frac{(1-0-q)^m}{p+1}Y$ for some $m \geq 1$, then

$$P^{m+1} = PP^m = \frac{1}{p+q}PX + \frac{(1-p-q)^m}{p+1}PY$$

$$= \frac{1}{p+q}X + \frac{(1-p-q)^m}{p+q}(1-p-q)Y$$

$$= \frac{1}{p+q}X + \frac{(1-p-q)^{m+1}}{p+q}Y.$$

Hence the formula holds for all $m \geq 1$ by induction.

Now $0 < p < 1$ and $0 < q < 1$ imply $0 < p+q < 2$, so that $-1 < (p+q-1) < 1$. Multiplying through by -1 gives $1 > (1-p-q) > -1$, so $(1-p-q)^m$ converges to zero as m increases.

Supplementary Exercises   Chapter 2

2(b) We have $0 = p(U) = U^3 - 5U^2 + 11U - 4I$ so that $U(U^2 - 5U + 11I) = 4I = (U^2 - 5U + 11I)U$. Hence $U^{-1} = \frac{1}{4}(U^2 - 5U + 11I)$.

4(b) If $X_h = X_m$, then $Y = k(Y - Z) = Y + m(Y - Z)$, whence $(k-m)(Y - Z) = 0$. But the matrix $Y - Z \neq 0$ (because $Y \neq Z$) so $k - m = 0$ by Example 7 §2.1.

6(d) Using (c), $I_{pq}AI_{rs} = \sum_{i=1}^{n}\sum_{j=1}^{n} a_{ij}I_{pq}I_{ij}I_{rs}$. Now (b) shows that $I_{pq}I_{ij}I_{rs} = 0$ unless $i = q$ and $j = r$, when it equals I_{ps}. Hence the double sum for $I_{pq}AI_{rs}$ has only one nonzero term — the one for which $i = q$, $j = r$. Hence $I_{pq}AI_{rs} = a_{qr}I_{ps}$.

7(b) If $n = 1$ it is clear. If $n > 1$, Exercise 6(d) gives

$$a_{qr}I_{ps} = I_{pq}AI_{rs} = I_{pq}I_{rs}A$$

because $AI_{rs} = I_{rs}A$. Hence $a_{qr} = 0$ if $q \neq r$ by Exercise 6(b). If $r = q$ then $a_{qq}I_{ps} = I_{ps}A$ is the same for each value of q. Hence $a_{11} = a_{22} = \cdots = a_{nn}$, so A is a scalar matrix.

Chapter 3: Determinants and Diagonalization

Exercises 3.1   The Laplace Expansion

1(b) $\begin{vmatrix} 6 & 9 \\ 8 & 12 \end{vmatrix} = 3\begin{vmatrix} 2 & 3 \\ 8 & 12 \end{vmatrix} = 3\begin{vmatrix} 2 & 3 \\ 0 & 0 \end{vmatrix} = 0$

(d) $\begin{vmatrix} a+1 & a \\ a & a-1 \end{vmatrix} = \begin{vmatrix} 1 & 1 \\ a & a-1 \end{vmatrix} = (a-1) - a = -1$

(f) $\begin{vmatrix} 2 & 0 & -3 \\ 1 & 2 & 5 \\ 0 & 3 & 0 \end{vmatrix} = \begin{vmatrix} 0 & -4 & -13 \\ 1 & 2 & 5 \\ 0 & 3 & 0 \end{vmatrix} = -\begin{vmatrix} -4 & -13 \\ 3 & 0 \end{vmatrix} = -39$

(h) $\begin{vmatrix} 0 & a & 0 \\ b & c & d \\ 0 & e & 0 \end{vmatrix} = -a\begin{vmatrix} b & d \\ 0 & 0 \end{vmatrix} = -a(0) = 0$

(j) Expand along row 1:

$\begin{vmatrix} 0 & a & b \\ a & 0 & c \\ b & c & 0 \end{vmatrix} = -a\begin{vmatrix} a & c \\ b & 0 \end{vmatrix} + b\begin{vmatrix} a & 0 \\ b & c \end{vmatrix} = -a(-bc) + b(ac) = 2abc$

(l) $\begin{vmatrix} 1 & 0 & 3 & 1 \\ 2 & 2 & 6 & 0 \\ -1 & 0 & -3 & 1 \\ 4 & 1 & 12 & 0 \end{vmatrix} = \begin{vmatrix} 1 & 0 & 3 & 1 \\ 0 & 2 & 0 & -2 \\ 0 & 0 & 0 & 2 \\ 0 & 1 & 0 & -4 \end{vmatrix} = \begin{vmatrix} 2 & 0 & -2 \\ 0 & 0 & 2 \\ 1 & 0 & -4 \end{vmatrix} = 0$

(n) Subtract multiplies of row 4 from rows 1 and 2.

$\begin{vmatrix} 4 & -1 & 3 & -1 \\ 3 & 1 & 0 & 2 \\ 0 & 1 & 2 & 2 \\ 1 & 2 & -1 & 1 \end{vmatrix} = \begin{vmatrix} 0 & -9 & 7 & -5 \\ 0 & -5 & 3 & -1 \\ 0 & 1 & 2 & 2 \\ 1 & 2 & -1 & 1 \end{vmatrix} = -\begin{vmatrix} -9 & 7 & -5 \\ -5 & 3 & -1 \\ 1 & 2 & 2 \end{vmatrix} = -\begin{vmatrix} 0 & 25 & 13 \\ 0 & 13 & 9 \\ 1 & 2 & 2 \end{vmatrix} =$

$\begin{vmatrix} 25 & 13 \\ 13 & 9 \end{vmatrix} = -\begin{vmatrix} -1 & -5 \\ 13 & 9 \end{vmatrix} = -(-9 + 65) = -56$

(p) Keep expanding along row 1:

$$
\begin{vmatrix} 0 & 0 & 0 & a \\ 0 & 0 & b & p \\ 0 & c & q & k \\ d & s & t & u \end{vmatrix} = -a \begin{vmatrix} 0 & 0 & b \\ 0 & v & q \\ d & s & t \end{vmatrix} = -a \left(b \begin{vmatrix} 0 & c \\ d & s \end{vmatrix} \right) = -ab(-cd) = abcd.
$$

4. I is triangular with diagonal entries 1, so Theorem 4 applies.

5(b) $\begin{vmatrix} -1 & 3 & 1 \\ 2 & 5 & 3 \\ 1 & -2 & 1 \end{vmatrix} = \begin{vmatrix} -1 & 3 & 1 \\ 0 & 11 & 5 \\ 0 & 1 & 2 \end{vmatrix} = - \begin{vmatrix} -1 & 3 & 1 \\ 0 & 1 & 2 \\ 0 & 11 & 5 \end{vmatrix} = - \begin{vmatrix} -1 & 3 & 1 \\ 0 & 1 & 2 \\ 0 & 0 & -17 \end{vmatrix} = -(-1)(1)(-17) = -17$

(d) $\begin{vmatrix} 2 & 3 & 1 & 1 \\ 0 & 2 & -1 & 3 \\ 0 & 5 & 1 & 1 \\ 1 & 1 & 2 & 5 \end{vmatrix} = - \begin{vmatrix} 1 & 1 & 2 & 5 \\ 0 & 2 & -1 & 3 \\ 0 & 5 & 1 & 1 \\ 2 & 3 & 1 & 1 \end{vmatrix} = - \begin{vmatrix} 1 & 1 & 2 & 5 \\ 0 & 2 & -1 & 3 \\ 0 & 5 & 1 & 1 \\ 0 & 1 & -3 & -9 \end{vmatrix} = \begin{vmatrix} 1 & 1 & 2 & 5 \\ 0 & 1 & -3 & -9 \\ 0 & 5 & 1 & 1 \\ 0 & 2 & -1 & 3 \end{vmatrix} =$

$\begin{vmatrix} 1 & 1 & 2 & 5 \\ 0 & 1 & -3 & -9 \\ 0 & 0 & 15 & 46 \\ 0 & 0 & 5 & 21 \end{vmatrix} = \begin{vmatrix} 1 & 1 & 2 & 5 \\ 0 & 1 & -3 & -9 \\ 0 & 0 & 1 & -17 \\ 0 & 0 & 0 & 106 \end{vmatrix} = 106$

6(b) $\begin{vmatrix} a & b & c \\ a+b & 2b & c+b \\ 2 & 2 & 2 \end{vmatrix} = \begin{vmatrix} a & b & c \\ a & b & c \\ 2 & 2 & 2 \end{vmatrix} = 0$

7(b) $\begin{vmatrix} -2a & -2b & -2c \\ 2p+x & 2q+y & 2r+z \\ 3x & 3y & 3z \end{vmatrix} = -6 \begin{vmatrix} a & b & c \\ 2p+x & 2q+y & 2r+z \\ x & y & z \end{vmatrix}$

$= -6 \begin{vmatrix} a & b & c \\ 2p & 2q & 2r \\ x & y & z \end{vmatrix} = -12 \begin{vmatrix} a & b & c \\ p & q & r \\ x & y & z \end{vmatrix} = 12.$

8(b) First add rows 2 and 3 to row 1:

$\begin{vmatrix} 2a+p & 2b+q & 2c+r \\ 2p+x & 2q+y & 2r+z \\ 2x+a & 2y+b & 2z+c \end{vmatrix} = \begin{vmatrix} 3a+3p+3x & 3b+3q+3y & 3c+3r+3z \\ 2p+x & 2q+y & 2r+z \\ 2x+a & 2y+b & 2z+c \end{vmatrix}$

$= 3 \begin{vmatrix} a+p+x & b+q+y & c+r+z \\ 2p+x & 2q+y & 2r+z \\ 2x+a & 2y+b & 2z+c \end{vmatrix}$

$$= 3 \begin{vmatrix} a+p+x & b+q+y & c+r+z \\ p-a & q-b & r-c \\ x-p & y-q & z-r \end{vmatrix} \quad \text{(subtract row 1 from rows 2 and 3)}$$

$$= 3 \begin{vmatrix} 3x & 3y & 3z \\ p-a & q-b & r-c \\ x-p & y-q & z-r \end{vmatrix} \quad \text{(add row 2 plus twice row 3 to row 1)}$$

$$= 9 \begin{vmatrix} x & y & z \\ p-a & q-b & r-c \\ -p & -q & -r \end{vmatrix} = 9 \begin{vmatrix} x & y & z \\ -a & -b & -c \\ -p & -q & -r \end{vmatrix} = -9 \begin{vmatrix} -p & -q & -r \\ -a & -b & -c \\ x & y & z \end{vmatrix}$$

$$= 9 \begin{vmatrix} -a & -b & -c \\ -p & -q & -r \\ x & y & z \end{vmatrix} = 9 \begin{vmatrix} a & b & c \\ p & q & r \\ x & y & z \end{vmatrix}$$

9(b) Partition the matrix as follows and use Theorem 5:

$$\begin{vmatrix} 1 & 2 & 0 & 3 & 0 \\ -1 & 3 & 1 & 4 & 0 \\ 0 & 0 & 2 & 1 & 1 \\ 0 & 0 & -1 & 0 & 2 \\ 0 & 0 & 3 & 0 & 1 \end{vmatrix} = \begin{vmatrix} 1 & 2 \\ -1 & 3 \end{vmatrix} \begin{vmatrix} 2 & 1 & 1 \\ -1 & 0 & 2 \\ 3 & 0 & 1 \end{vmatrix} = 5 \left(-1 \begin{vmatrix} -1 & 2 \\ 3 & 1 \end{vmatrix} \right) = -5(-7) = 35.$$

10(b) Use Theorem 5 twice:

$$\begin{vmatrix} A & 0 & 0 \\ X & B & 0 \\ Y & Z & 0 \end{vmatrix} = \det \begin{bmatrix} A & 0 \\ X & B \end{bmatrix} \det C = (\det A \det B) \det C = 2(-1)3 = -6$$

(d)
$$\begin{vmatrix} A & X & 0 \\ 0 & B & 0 \\ Y & Z & C \end{vmatrix} = \det \begin{bmatrix} A & X \\ 0 & B \end{bmatrix} \det C = (\det A \det B) \det C = 2(-1)3 = -6$$

12(b)
$$\det \begin{vmatrix} x-1 & -3 & 1 \\ 2 & -1 & x-1 \\ -3 & x+2 & -2 \end{vmatrix} = \begin{vmatrix} x-2 & x-2 & x-2 \\ 2 & -1 & x-1 \\ -3 & x+2 & -2 \end{vmatrix} = (x-2) \begin{vmatrix} 1 & 1 & 1 \\ 2 & -1 & x-1 \\ -3 & x+2 & -2 \end{vmatrix}$$

$$= (x-2) \begin{vmatrix} 1 & 0 & 0 \\ 2 & -3 & x-3 \\ -3 & x+5 & 1 \end{vmatrix} = (x-2) \begin{vmatrix} -3 & x-3 \\ x+5 & 1 \end{vmatrix}$$

$$= (x-2)[-x^2 - 2x + 12] = -(x-2)(x^2 + 2x - 12).$$

13(b) If we expand along column 2, the coefficient of z is $-\begin{vmatrix} 2 & -1 \\ 1 & 3 \end{vmatrix} = -(6+1) = -7.$ So $c = -7$.

14(b) $\det A = \begin{vmatrix} 1 & x & x \\ -x & -2 & x \\ -x & -x & -3 \end{vmatrix} = \begin{vmatrix} 1 & x & x \\ 0 & x^2-2 & x^2+x \\ 0 & x^2-x & x^2-3 \end{vmatrix} = \begin{vmatrix} x^2-2 & x^2+x \\ x^2-x & x^2-3 \end{vmatrix}$

$= (x^2-2)(x^2-3) - (x^2+x)(x^2-x) = (x^4 - 5x^2 + 6) - x^2(x^2-1) = 6 - 4x^2$

Hence $\det A = 0$ means $x^2 = \frac{3}{2} = \frac{6}{4}$, $x = \pm\frac{\sqrt{6}}{2}$.

(d) $\det A = \begin{vmatrix} x & y & 0 & 0 \\ 0 & x & y & 0 \\ 0 & 0 & x & y \\ y & 0 & 0 & x \end{vmatrix} = x\begin{vmatrix} x & y & 0 \\ 0 & x & y \\ 0 & 0 & x \end{vmatrix} - y\begin{vmatrix} y & 0 & 0 \\ x & y & 0 \\ 0 & x & y \end{vmatrix} = x \cdot x^3 - y \cdot y^3$

$= x^4 - y^4 = (x^2 - y^2)(x^2 + y^2) = (x - y)(x + y)(x^2 + y^2).$

Hence $\det A = 0$ means $x = y$ or $x = -y$ $(x^2 + y^2 = 0$ only if $x = y = 0).$

18. Begin by adding rows 2 and 3 to row 1:

$\begin{vmatrix} a+x & b+x & c+x \\ b+x & c+x & a+x \\ c+x & a+x & b+c \end{vmatrix} = \begin{vmatrix} a+b+c+3x & a+b+c+3x & a+b+c+3x \\ b+x & c+x & a+x \\ c+x & a+x & b+x \end{vmatrix}$

$= (a+b+c+3x)\begin{vmatrix} 1 & 1 & 1 \\ b+x & c+x & a+x \\ c+x & a+x & b+x \end{vmatrix}$

$= (a+b+c+3x)\begin{vmatrix} 1 & 0 & 0 \\ b+x & c-b & a-b \\ c+x & a-c & b-c \end{vmatrix}$

$= (a+b+c+3x)[(c-b)(b-c) - (a-b)(a-c)]$

$= (a+b+c+3x)[ab + ac + bc - (a^2 + b^2 + c^2)]$

22. Suppose A is $n \times n$. B can be found from A by interchanging the following pairs of columns: 1 and n, 2 and $n-1,\ldots$. There are two cases:

Case 1 $n = 2k$

Then we interchange columns 1 and n, 2 and $n-1, \ldots, k$ and $k+1$, k interchanges in all. Thus $\det B = (-1)^k \det A$.

Case 2 $n = 2k+1$

Now we interchange columns 1 and n, 2 and $n-1, \ldots k$ and $k+2$, leaving column k fixed. Again k interchanges are used so $\det B = (-1)^k \det A$.

Thus $\det B = (-1)^k \det A$ where A is $n \times n$ and $n = 2k$ or $n = 2k+1$.

Exercises 3.2   Determinants and Matrix Inverses

1(b) The cofactor matrix is
$$
\begin{bmatrix}
\begin{vmatrix} 1 & 0 \\ -1 & 1 \end{vmatrix} & -\begin{vmatrix} 3 & 0 \\ 0 & 1 \end{vmatrix} & \begin{vmatrix} 3 & 1 \\ 0 & -1 \end{vmatrix} \\[12pt]
-\begin{vmatrix} -1 & 2 \\ -1 & 1 \end{vmatrix} & \begin{vmatrix} 1 & 2 \\ 0 & 1 \end{vmatrix} & -\begin{vmatrix} 1 & -1 \\ 0 & -1 \end{vmatrix} \\[12pt]
\begin{vmatrix} -1 & 2 \\ 1 & 0 \end{vmatrix} & -\begin{vmatrix} 1 & 2 \\ 3 & 0 \end{vmatrix} & \begin{vmatrix} 1 & -1 \\ 3 & 1 \end{vmatrix}
\end{bmatrix}
=
\begin{bmatrix}
1 & -3 & -3 \\
-1 & 1 & 1 \\
-2 & 6 & 4
\end{bmatrix}
$$

The adjoint is the transpose of the cofactor matrix: $\begin{bmatrix} 1 & -1 & -2 \\ -3 & 1 & 6 \\ -3 & 1 & 4 \end{bmatrix}$.

(d) In computing the cofactor matrix, we use the fact that $\det\left[\frac{1}{3}M\right] = \frac{1}{9}\det M$ for any 2×2 matrix M. Thus the cofactor matrix is

$$
\begin{bmatrix}
\frac{1}{9}\begin{vmatrix} -1 & 2 \\ 2 & -1 \end{vmatrix} & -\frac{1}{9}\begin{vmatrix} 2 & 2 \\ 2 & -1 \end{vmatrix} & \frac{1}{9}\begin{vmatrix} 2 & -1 \\ 2 & 2 \end{vmatrix} \\[12pt]
-\frac{1}{9}\begin{vmatrix} 2 & 2 \\ 2 & -1 \end{vmatrix} & \frac{1}{9}\begin{vmatrix} -1 & 2 \\ 2 & -1 \end{vmatrix} & -\frac{1}{9}\begin{vmatrix} -1 & 2 \\ 2 & 2 \end{vmatrix} \\[12pt]
\frac{1}{9}\begin{vmatrix} 2 & 2 \\ -1 & 2 \end{vmatrix} & -\frac{1}{9}\begin{vmatrix} -1 & 2 \\ 2 & 2 \end{vmatrix} & \frac{1}{9}\begin{vmatrix} -1 & 2 \\ 2 & -1 \end{vmatrix}
\end{bmatrix}
= \frac{1}{9}\begin{bmatrix} -3 & 6 & 6 \\ 6 & -3 & 6 \\ 6 & 6 & -3 \end{bmatrix} = \frac{1}{3}\begin{bmatrix} -1 & 2 & 2 \\ 2 & -1 & 2 \\ 2 & 2 & -1 \end{bmatrix}.
$$

The adjoint is the transpose of the cofactor matrix: $\frac{1}{3}\begin{bmatrix} -1 & 2 & 2 \\ 2 & -1 & 2 \\ 2 & 2 & -1 \end{bmatrix}$. Note that this is the same here as the cofactor matrix is symmetric. Note also that the adjoint actually equals the original matrix in this case — the matrix is said to be selfadjoint.

2(b) We compute the determinant by first adding column 3 to column 2:
$$
\begin{vmatrix} 0 & c & -c \\ -1 & 2 & -1 \\ c & -c & c \end{vmatrix} = \begin{vmatrix} 0 & 0 & -c \\ -1 & 1 & -1 \\ c & 0 & c \end{vmatrix} = (-c)\begin{vmatrix} -1 & 1 \\ c & 0 \end{vmatrix} = (-c)(-c) = c^2.
$$
This is zero if and only if $c = 0$, so the matrix is invertible if and only if $c \neq 0$.

(d) $\begin{vmatrix} 4 & c & 3 \\ c & 2 & c \\ 5 & c & 4 \end{vmatrix} = \begin{vmatrix} 4 & c & 3 \\ c & 2 & c \\ 1 & 0 & 1 \end{vmatrix} = \begin{vmatrix} 4 & c & -1 \\ c & 2 & 0 \\ 1 & 0 & 0 \end{vmatrix} = 1 \begin{vmatrix} c & -1 \\ 2 & 0 \end{vmatrix} = 2.$

This is nonzero for *all* values of c, so the matrix is invertible for all c.

(f) $\begin{vmatrix} 1 & c & -1 \\ c & 1 & 1 \\ 0 & 1 & c \end{vmatrix} = \begin{vmatrix} 1 & c & -1 \\ 0 & 1-c^2 & 1+c \\ 0 & 1 & c \end{vmatrix} = \begin{vmatrix} 1-c^2 & 1+c \\ 1 & c \end{vmatrix} = \begin{vmatrix} (1+c)(1-c) & 1+c \\ 1 & c \end{vmatrix}$

$= (1+c) \begin{vmatrix} 1-c & 1 \\ 1 & c \end{vmatrix} = (1+c)[c(1-c)-1] = -(1+c)(c^2-c+1) = -(c^3+1).$

This is zero if and only if $c = -1$ (the roots of $c^2 - c + 1$ are not real). Hence the matrix is invertible if and only if $c \neq -1$.

3(b) $\begin{aligned} \det(B^2 C^{-1} A B^{-1} C^T) &= \det B^2 \det C^{-1} \det A \det B^{-1} \det C^T \\ &= (\det B)^2 \frac{1}{\det C} \det A \frac{1}{\det B} \det C \\ &= \det B \det A \\ &= -2 \end{aligned}$

4(b) $\det(A^{-1}B^{-1}AB) = \det A^{-1} \det B^{-1} \det A \det B = \frac{1}{\det A} \frac{1}{\det B} \det A \det B = 1.$
Note the following proof is *wrong*:
$\det(A^{-1}B^{-1}AB) = \det(A^{-1}AB^{-1}B) = \det(I \cdot I) = \det I = 1.$
The reason is that $A^{-1}B^{-1}AB$ may not equal $A^{-1}AB^{-1}B$ because $B^{-1}A$ need not equal AB^{-1}.

6(b) Since C is 3×3, the same is true for C^{-1}, so $\det(2C^{-1}) = 2^3 \cdot \det C^{-1} = \frac{8}{\det C}$. Now

$\det C = \begin{vmatrix} 2p & -a+u & 3u \\ 2q & -b+v & 3v \\ 2r & -c+w & 3w \end{vmatrix} = 6 \begin{vmatrix} p & -a+u & u \\ q & -b+v & v \\ r & -c+w & w \end{vmatrix} = 6 \begin{vmatrix} p & -a & u \\ q & -b & v \\ r & -c & w \end{vmatrix}$

$= -6 \begin{vmatrix} p & a & u \\ q & b & v \\ r & c & w \end{vmatrix} = 6 \begin{vmatrix} a & p & u \\ b & q & v \\ c & r & w \end{vmatrix} = 6 \begin{vmatrix} a & b & c \\ p & q & r \\ u & v & w \end{vmatrix} = 6 \cdot 3 = 18.$

Finally $\det 2C^{-1} = \frac{8}{\det C} = \frac{8}{18} = \frac{4}{9}.$

7(b) $\begin{vmatrix} 2b & 0 & 4d \\ 1 & 2 & -2 \\ a+1 & 2 & 2(c-1) \end{vmatrix} = \begin{vmatrix} 2b & 0 & 4d \\ 1 & 2 & -2 \\ a & 0 & 2c \end{vmatrix} = 2 \begin{vmatrix} 2b & 4d \\ a & 2c \end{vmatrix} = 4 \begin{vmatrix} b & 2d \\ a & 2c \end{vmatrix} = 8 \begin{vmatrix} b & d \\ a & c \end{vmatrix}$

$= -8 \begin{vmatrix} a & c \\ b & d \end{vmatrix} = -8 \begin{vmatrix} a & b \\ c & d \end{vmatrix} = -8(-2) = 16.$

8(b) $x = \dfrac{\begin{vmatrix} 9 & 4 \\ -1 & -1 \end{vmatrix}}{\begin{vmatrix} 3 & 4 \\ 2 & -1 \end{vmatrix}} = \dfrac{-5}{-11} = \dfrac{5}{11}, \ y = \dfrac{\begin{vmatrix} 3 & 9 \\ 2 & -1 \end{vmatrix}}{\begin{vmatrix} 3 & 4 \\ 2 & -1 \end{vmatrix}} = \dfrac{-21}{-11} = \dfrac{21}{11}$

(d) The coefficient matrix has determinant:

$$\begin{vmatrix} 4 & -1 & 3 \\ 6 & 2 & -1 \\ 3 & 3 & 2 \end{vmatrix} = \begin{vmatrix} 0 & -1 & 0 \\ 14 & 2 & 5 \\ 15 & 3 & 11 \end{vmatrix} = -(-1) \begin{vmatrix} 14 & 5 \\ 15 & 11 \end{vmatrix} = 79$$

$$x = \frac{1}{79} \begin{vmatrix} 1 & -1 & 3 \\ 0 & 2 & -1 \\ -1 & 3 & 2 \end{vmatrix} = \frac{1}{79} \begin{vmatrix} 1 & -1 & 3 \\ 0 & 2 & -1 \\ 0 & 2 & 5 \end{vmatrix} = \frac{1}{79} \begin{vmatrix} 2 & -1 \\ 2 & 5 \end{vmatrix} = \frac{12}{79}$$

$$y = \frac{1}{79} \begin{vmatrix} 4 & 1 & 3 \\ 6 & 0 & -1 \\ 3 & -1 & 2 \end{vmatrix} = \frac{1}{79} \begin{vmatrix} 4 & 1 & 3 \\ 6 & 0 & -1 \\ 7 & 0 & 5 \end{vmatrix} = -\frac{1}{79} \begin{vmatrix} 6 & -1 \\ 7 & 5 \end{vmatrix} = -\frac{37}{79}$$

$$x = \frac{1}{79} \begin{vmatrix} 4 & -1 & 1 \\ 6 & 2 & 0 \\ 3 & 3 & -1 \end{vmatrix} = \frac{1}{79} \begin{vmatrix} 4 & -1 & 1 \\ 6 & 2 & 0 \\ 7 & 2 & 0 \end{vmatrix} = \frac{1}{79} \begin{vmatrix} 6 & 2 \\ 7 & 2 \end{vmatrix} = -\frac{2}{79}.$$

9(b) $A^{-1} = \frac{1}{\det A} \operatorname{adj} A = \frac{1}{\det A} [C_{ij}]^T$ where $[C_{ij}]$ is the cofactor matrix. Hence the $(2,3)$-entry of A^{-1} is $\frac{1}{\det A} C_{32}$. Now

$$C_{32} = -\begin{vmatrix} 1 & -1 \\ 3 & 1 \end{vmatrix} = -4.$$

$$\det A = \begin{vmatrix} 1 & 2 & -1 \\ 3 & 1 & 1 \\ 0 & 4 & 7 \end{vmatrix} = \begin{vmatrix} 1 & 2 & -1 \\ 0 & -5 & 4 \\ 0 & 4 & 7 \end{vmatrix} = \begin{vmatrix} -5 & 4 \\ 4 & 7 \end{vmatrix} = -51.$$

Finally the $(2,3)$ entry of A^{-1} is $\frac{-4}{-51} = \frac{4}{51}$.

10(b) If $A^2 = I$ then $\det A^2 = 1$, that is $(\det A)^2 = 1$. Hence $\det A = 1$ or $\det A = -1$.

(d) If $PA = P$, P invertible, then $\det P \det A = \det P$. Since $\det P \neq 0$ (as P is invertible), this gives $\det A = 1$.

(f) If $A = -A^T$, A is $n \times n$, then A^T is also $n \times n$ so $\det A = \det(-A^T) = (-1)^n \det A^T = (-1)^n \det A$. If n is even this is $\det A = \det A$ and so gives no information about $\det A$. But if n is odd, it reads $\det A = -\det A$, so $\det A = 0$ in this case.

16. Here $A^T = A^{-1} = \frac{1}{d} \operatorname{adj} A = \frac{1}{d} [C_{ij}]^T$ where $d = \det A$ and $[C_{ij}]$ is the cofactor matrix of A. Take transposes to get $A = \frac{1}{d} [C_{ij}]$, whence $[C_{ij}] = dA$.

20(b) Write $A = \begin{bmatrix} 0 & c & -c \\ -1 & 2 & -1 \\ c & -c & c \end{bmatrix}$. Then $\det A = c^2$ (Exercise 2) and the cofactor matrix is

$$[C_{ij}] = \begin{bmatrix} \begin{vmatrix} 2 & -1 \\ -c & c \end{vmatrix} & -\begin{vmatrix} -1 & -1 \\ c & c \end{vmatrix} & \begin{vmatrix} -1 & 2 \\ c & -c \end{vmatrix} \\[12pt] -\begin{vmatrix} c & -c \\ -c & c \end{vmatrix} & \begin{vmatrix} 0 & -c \\ c & c \end{vmatrix} & -\begin{vmatrix} 0 & c \\ c & -c \end{vmatrix} \\[12pt] \begin{vmatrix} c & -c \\ 2 & -1 \end{vmatrix} & -\begin{vmatrix} 0 & -c \\ -1 & -1 \end{vmatrix} & \begin{vmatrix} 0 & c \\ -1 & 2 \end{vmatrix} \end{bmatrix} = \begin{bmatrix} c & 0 & -c \\ 0 & c^2 & c^2 \\ c & c & c \end{bmatrix}$$

Hence $A^{-1} = \frac{1}{\det A}\operatorname{adj} A = \frac{1}{c^2}[C_{ij}]^T = \frac{1}{c^2}\begin{bmatrix} c & 0 & c \\ 0 & c^2 & c \\ -c & c^2 & c \end{bmatrix} = \frac{1}{c}\begin{bmatrix} 1 & 0 & 1 \\ 0 & c & 1 \\ -1 & c & 1 \end{bmatrix}$ for any $c \neq 0$.

(d) Write $A = \begin{bmatrix} 4 & c & 3 \\ c & 2 & c \\ 5 & c & 4 \end{bmatrix}$. Then $A = 2$ (Exercise 2) and the cofactor matrix is

$$[C_{ij}] = \begin{bmatrix} \begin{vmatrix} 2 & c \\ c & 4 \end{vmatrix} & -\begin{vmatrix} c & c \\ 5 & 4 \end{vmatrix} & \begin{vmatrix} c & 2 \\ 5 & c \end{vmatrix} \\[12pt] -\begin{vmatrix} c & 3 \\ c & 4 \end{vmatrix} & \begin{vmatrix} 4 & 3 \\ 5 & 4 \end{vmatrix} & -\begin{vmatrix} 4 & c \\ 5 & c \end{vmatrix} \\[12pt] \begin{vmatrix} c & 3 \\ 2 & c \end{vmatrix} & -\begin{vmatrix} 4 & 3 \\ c & c \end{vmatrix} & \begin{vmatrix} 4 & c \\ c & 2 \end{vmatrix} \end{bmatrix} = \begin{bmatrix} 8-c^2 & c & c^2-10 \\ -c & 1 & c \\ c^2-6 & -c & 8-c^2 \end{bmatrix}.$$

Hence $A^{-1} = \frac{1}{\det A}\operatorname{adj} A = \frac{1}{2}[C_{ij}]^T = \frac{1}{2}\begin{bmatrix} 8-c^2 & -c & c^2-6 \\ c & 1 & -c \\ c^2-10 & c & 8-c^2 \end{bmatrix}.$

(f) Write $A = \begin{bmatrix} 1 & c & -1 \\ c & 1 & 1 \\ 0 & 1 & c \end{bmatrix}$. Then $\det A = -(c^2+1)$ (Exercise 2) and the cofactor matrix is

$$[C_{ij}] = \begin{bmatrix} \begin{vmatrix} 1 & 1 \\ 1 & c \end{vmatrix} & -\begin{vmatrix} c & 1 \\ 0 & c \end{vmatrix} & \begin{vmatrix} c & 1 \\ 0 & 1 \end{vmatrix} \\ -\begin{vmatrix} c & -1 \\ 1 & c \end{vmatrix} & \begin{vmatrix} 1 & -1 \\ 0 & c \end{vmatrix} & -\begin{vmatrix} 1 & c \\ 0 & 1 \end{vmatrix} \\ \begin{vmatrix} c & -1 \\ 1 & 1 \end{vmatrix} & -\begin{vmatrix} 1 & -1 \\ c & 1 \end{vmatrix} & \begin{vmatrix} 1 & c \\ c & 1 \end{vmatrix} \end{bmatrix} = \begin{bmatrix} c-1 & -c^2 & c \\ -(c^2+1) & c & -1 \\ c+1 & -(1+c) & 1-c^2 \end{bmatrix}.$$

Hence $A^{-1} = \frac{1}{\det A}\,\text{adj}\,A = \frac{-1}{c^3+1}[C_{ij}]^T = \frac{-1}{c^3+1}\begin{bmatrix} c-1 & -(c^2+1) & c+1 \\ -c^2 & c & -(c+1) \\ c & -1 & 1-c^2 \end{bmatrix}$

$$= \frac{1}{c^3+1}\begin{bmatrix} 1-c & c^2+1 & -c-1 \\ c^2 & -c & c+1 \\ -c & 1 & c^2-1 \end{bmatrix}.$$

22(b) Let A be an upper triangular, invertible, $n \times n$ matrix. We use induction on n. If $n = 1$ it is clear (every 1×1 matrix is upper triangular). If $n > 1$ write $A = \begin{bmatrix} a & X \\ 0 & B \end{bmatrix}$ and $A^{-1} = \begin{bmatrix} b & Y \\ Z & C \end{bmatrix}$ in block form. Then

$$\begin{bmatrix} 1 & 0 \\ 0 & I \end{bmatrix} = AA^{-1} = \begin{bmatrix} ab+XZ & aY+XC \\ BZ & BC \end{bmatrix}.$$

So $BC = I$, $BZ = 0$. Thus $C = B^{-1}$ is upper triangular by induction (B is upper triangular because A is) and $BZ = 0$ gives $Z = 0$ because B is invertible. Hence $A^{-1} = \begin{bmatrix} b & Y \\ 0 & C \end{bmatrix}$ is upper triangular.

24. We have $A \cdot \text{adj}\,A = dI$ where $d = \det A = -21$. Hence $\left(\frac{1}{d}A\right)\text{adj}\,A = I$, $(\text{adj}\,A)^{-1} = \frac{1}{d}A = \frac{-1}{21}\begin{bmatrix} 3 & 0 & 1 \\ 0 & 2 & 3 \\ 3 & 1 & -1 \end{bmatrix}$.

30(b) Write $d = \det A$ so $\det A^{-1} = \frac{1}{d}$. Now the adjoint formula gives

$$(\text{adj}\,A)A = dI \quad \text{and} \quad A^{-1}(\text{adj}\,A^{-1}) = \frac{1}{d}I.$$

Take inverses in the second of these to get

$$(\text{adj}\,A^{-1})^{-1}A = dI = (\text{adj}\,A)A.$$

Since A is invertible, $\left[\operatorname{adj} A^{-1}\right]^{-1} = \operatorname{adj} A$, and the result follows by taking inverses again.

(d) The adjoint formula gives

$$AB \operatorname{adj}(AB) = \det AB \cdot I = \det A \cdot \det B \cdot I.$$

On the other hand

$$
\begin{aligned}
AB \operatorname{adj} B \cdot \operatorname{adj} A &= A[(\det B)I]\operatorname{adj} A \\
&= A \cdot \operatorname{adj} A \cdot (\det B)I \\
&= (\det A)I \cdot (\det B)I \\
&= \det A \det B \cdot I.
\end{aligned}
$$

Thus $AB \operatorname{adj}(AB) = AB \cdot \operatorname{adj} B \cdot \operatorname{adj} A$, and the result follows because AB is invertible.

Exercises 3.3   Diagonalization and Eigenvalues

1(b) $c_A(x) = \begin{vmatrix} x-2 & 4 \\ 1 & x+1 \end{vmatrix} = x^2 - x - 6 = (x-3)(x+2)$; hence the eigenvalues are $\lambda_1 = 3$, and

$\lambda_2 = -2$. Take these values for x in the matrix $XI = A$ for $c_A(x)$:

$$\lambda_1 = 3: \quad \begin{bmatrix} 1 & 4 \\ 1 & 4 \end{bmatrix} \rightarrow \begin{bmatrix} 1 & 4 \\ 0 & 0 \end{bmatrix}; \ X_1 = \begin{bmatrix} 4 \\ -1 \end{bmatrix}.$$

$$\lambda_2 = -2: \quad \begin{bmatrix} -4 & 4 \\ 1 & -1 \end{bmatrix} \rightarrow \begin{bmatrix} 1 & -1 \\ 0 & 0 \end{bmatrix}; \ X_2 = \begin{bmatrix} 1 \\ 1 \end{bmatrix}.$$

So $P = [X_1 \ X_2] = \begin{bmatrix} 4 & 1 \\ -1 & 1 \end{bmatrix}$ has $P^{-1}AF = \begin{bmatrix} 3 & 0 \\ 0 & -2 \end{bmatrix}$.

(d) $c_A(x) = \begin{vmatrix} x-1 & -1 & 3 \\ -2 & x & -6 \\ -1 & 1 & x-5 \end{vmatrix} = \begin{vmatrix} x-1 & -1 & 3 \\ -2 & x & -6 \\ x-2 & 0 & x-2 \end{vmatrix} = \begin{vmatrix} x-1 & -1 & -x+4 \\ -2 & x & -4 \\ x-2 & 0 & 0 \end{vmatrix}$

$= (x-2)\begin{vmatrix} -1 & -x+4 \\ x & -4 \end{vmatrix} = (x-2)[x^2 - 4x + 4] = (x-2)^3$; hence the eigenvalue is $\lambda_1 = 2$ of

multiplicity 3. Take this value for x in the matrix $xI - A$ for $c_A(x)$:

$$\begin{bmatrix} 1 & -1 & 3 \\ -2 & 2 & 6 \\ -1 & 1 & -3 \end{bmatrix} \rightarrow \begin{bmatrix} 1 & -1 & 3 \\ 0 & 0 & 0 \\ 0 & 0 & 0 \end{bmatrix}, \ X = \begin{bmatrix} s - 3t \\ s \\ t \end{bmatrix}, \ X_1 = \begin{bmatrix} 1 \\ 1 \\ 0 \end{bmatrix}, \ X_2 = \begin{bmatrix} -3 \\ 0 \\ 1 \end{bmatrix}.$$

Hence there are not $n = 3$ basic eigenvectors, so A is not diagonalizable.

(f) To get $c_A(x)$, add rows 2 and 3 to row 1:

$$c_A(x) = \begin{vmatrix} x & -1 & -1 \\ -1 & x & -1 \\ -1 & -1 & x \end{vmatrix} = \begin{vmatrix} x-2 & x-2 & x-2 \\ -1 & x & -1 \\ -1 & -1 & x \end{vmatrix} = \begin{vmatrix} x-2 & 0 & 0 \\ -1 & x+1 & 0 \\ -1 & 0 & x+1 \end{vmatrix}$$

$= (x+1)^2(x-2)$.

Hence, the eigenvalues are $\lambda_1 = -1$ and $\lambda_2 = 2$. Substitute in the matrix $xI = A$ for $c_A(x)$:

$$\lambda_1 = -1: \quad \begin{bmatrix} -1 & -1 & -1 \\ -1 & -1 & -1 \\ -1 & -1 & -1 \end{bmatrix} \rightarrow \begin{bmatrix} 1 & 1 & 1 \\ 0 & 0 & 0 \\ 0 & 0 & 0 \end{bmatrix}; \text{ hence } X_1 = \begin{bmatrix} -1 \\ 1 \\ 0 \end{bmatrix} \text{ are } Y_1 = \begin{bmatrix} -1 \\ 0 \\ 1 \end{bmatrix} \text{ are}$$

two basic λ_1-eigenvectors.

$$\lambda_2 = 2: \quad \begin{bmatrix} 2 & -1 & -1 \\ -1 & 2 & -1 \\ -1 & -1 & 2 \end{bmatrix} \rightarrow \begin{bmatrix} 1 & -2 & 1 \\ 0 & 3 & -3 \\ 0 & -3 & 3 \end{bmatrix} \rightarrow \begin{bmatrix} 1 & 0 & -1 \\ 0 & 1 & -1 \\ 0 & 0 & 0 \end{bmatrix}; X_2 = \begin{bmatrix} 1 \\ 1 \\ 1 \end{bmatrix}. \text{ So}$$

there are $n = 3$ basic eigenvectors and $P = [X_1 \ Y_1 \ X_2]$ is invertible. Thus $P^{-1}AF =$

$$\begin{bmatrix} -1 & 0 & 0 \\ 0 & -1 & 0 \\ 0 & 0 & 2 \end{bmatrix}.$$

(h) $c_A(x) = \begin{vmatrix} x-2 & -1 & -1 \\ 0 & x-1 & 0 \\ -1 & 1 & x-2 \end{vmatrix} = (x-1) \begin{vmatrix} x-2 & -1 \\ -1 & x-2 \end{vmatrix} = (x-1)^2(x-3)$. Hence the

eigenvalues are $\lambda_1 = 1$, $\lambda_2 = 3$. Take these values for x in the matrix $xI - A$ for $c_A(x)$.

$$\lambda_1 - 1: \quad \begin{bmatrix} -1 & -1 & -1 \\ 0 & 0 & 0 \\ -1 & 1 & -1 \end{bmatrix} \rightarrow \begin{bmatrix} 1 & 1 & 1 \\ 0 & 2 & 0 \\ 0 & 0 & 0 \end{bmatrix} \rightarrow \begin{bmatrix} 1 & 0 & 1 \\ 0 & 1 & 0 \\ 0 & 0 & 0 \end{bmatrix}; X_1 = \begin{bmatrix} -1 \\ 0 \\ 1 \end{bmatrix}$$

$$\lambda_2 = 3: \quad \begin{bmatrix} 1 & -1 & -1 \\ 0 & 2 & 0 \\ -1 & 1 & 1 \end{bmatrix} \rightarrow \begin{bmatrix} 1 & 0 & -1 \\ 0 & 1 & 0 \\ 0 & 0 & 0 \end{bmatrix}; X_2 = \begin{bmatrix} 1 \\ 0 \\ 1 \end{bmatrix}. \text{ Since } n = 3 \text{ and there are}$$

only two basic eigenvectors, A is not diagonalizable.

2(b) $\lambda_1 = 2$ and $\lambda_2 = -1$; $X_1 = \begin{bmatrix} 2 \\ 1 \end{bmatrix}$ and $X_2 = \begin{bmatrix} 1 \\ 2 \end{bmatrix}$. $F^{-1}V_0 = \frac{1}{3}\begin{bmatrix} 7 \\ -5 \end{bmatrix}$ so $b_1 = \frac{7}{3}$. Hence

$$V_k \cong b_1\lambda_1^k X_1 = \frac{7}{3}2^k \begin{bmatrix} 2 \\ 1 \end{bmatrix}.$$

(d) $\lambda_1 = 3$, $\lambda_2 = -2$ and $\lambda_3 = 1$; $X_1 = \begin{bmatrix} 1 \\ 0 \\ 1 \end{bmatrix}$, $X_2 = \begin{bmatrix} 1 \\ 1 \\ -3 \end{bmatrix}$ and $X_3 = \begin{bmatrix} 1 \\ -2 \\ 3 \end{bmatrix}$. $P^{-1}V_0 =$

$\frac{1}{6}\begin{bmatrix} 9 \\ 2 \\ 1 \end{bmatrix}$ so $b_1 = \frac{3}{2}$. Hence $V_k \cong \frac{3}{2}3^k \begin{bmatrix} 1 \\ 0 \\ 1 \end{bmatrix}$.

4. If λ is an eigenvalue for A, let $AX = \lambda X$, $X \neq 0$. Then

$$A_1 X = (A - \alpha I)X = AX - \alpha X = \lambda \equiv \alpha X = (\lambda - \alpha)X.$$

So $\lambda - \alpha$ is an eigenvalue of $A_1 = A - \alpha I$ (with the same eigenvector). Conversely, if $\lambda - \alpha$ is an eigenvector of A_1, let $A_1 Y = (\lambda - \alpha)Y$, $Y \neq 0$. Thus, $(A - \alpha I)Y = (\lambda - \alpha)Y$, whence $AY = \lambda Y$ and λ is an eigenvalue of A.

6. $c_I(x) = \det(xI - I) = \det[(x - 1)I] = (x - 1)^n$ as $I = I_n$ is $n \times n$. Hence, the only eigenvalue is $\lambda = 1$ so $(\lambda I - I)X = 0$ for *every* column $X \neq 0$.

8(b) $P^{-1}AF = \begin{bmatrix} 1 & 0 \\ 0 & 2 \end{bmatrix}$, so $A^n = P \begin{bmatrix} 1 & 0 \\ 0 & 2^n \end{bmatrix} P^{-1} = \begin{bmatrix} 9 - 8 \cdot 2^n & 12(1 - 2^n) \\ 6(2^n - 1) & 9 \cdot 2^n - 8 \end{bmatrix}$.

9. $c_A(x) = x(x-1)$ so A has eigenvalues $\lambda_1 = 0$ and $\lambda_2 = 1$ with basic eigenvectors $X_1 = \begin{bmatrix} -1 \\ 1 \end{bmatrix}$

and $X_2 = \begin{bmatrix} 1 \\ 0 \end{bmatrix}$. Since $[X_1 \quad X_2] = \begin{bmatrix} -1 & 1 \\ 1 & 0 \end{bmatrix}$ is invertible, it is a diagonalizing matrix

for A. Similarly B has diagonalizing matrix $\begin{bmatrix} 1 & 1 \\ 0 & 1 \end{bmatrix}$. However $AB = \begin{bmatrix} 0 & 2 \\ 0 & 0 \end{bmatrix}$ is not

diagonalizable: It has only one eigenvalue $\lambda = 0$ of multiplicity 2, but has only one basic

eigenvector $X = \begin{bmatrix} 1 \\ 0 \end{bmatrix}$.

11(b) Let $P^{-1}AF = D$ be diagonal. Then $P^{-1}(kA)P = d(P^{-1}AF) = dD$ is also diagonal, so kA is diagonalizable too.

12. $\begin{bmatrix} 1 & 1 \\ 0 & 1 \end{bmatrix} = \begin{bmatrix} 2 & 1 \\ 0 & -1 \end{bmatrix} + \begin{bmatrix} -1 & 0 \\ 0 & 2 \end{bmatrix}$ and both $\begin{bmatrix} 2 & 1 \\ 0 & -1 \end{bmatrix}$ and $\begin{bmatrix} -1 & 0 \\ 0 & 2 \end{bmatrix}$ are diagonaliz-

able. However, $\begin{bmatrix} 1 & 1 \\ 0 & 1 \end{bmatrix}$ is not diagonalizable by Example 8.

14. If A is $n \times n$, let $\lambda_1 = \cdots = \lambda_k = 1$, $\lambda_{k+1} = \cdots = \lambda_n = -1$ be the eigenvalues. Since A is diagonalizable (by hypothesis), we have $P^{-1}AF = D$ where $D = \text{diag}(\lambda_1, \ldots, \lambda_n)$ is the diagonal matrix with $\lambda_1, \ldots, \lambda_n$ down the main diagonal. Thus $D^2 = \text{diag}(\lambda_1^2, \ldots, \lambda_n^2) = \text{diag}(1, \ldots, 1) = I$. But $A = PDF^{-1}$, so $A^2 = (PDF^{-1})(PDF^{-1}) = PD^2 P^{-1} = PIF^{-1} = I$.

18(b) $c_A(x) = \det[xI - rA] = \det[r\left(\frac{x}{r}I - A\right)] = r^n \det\left[\frac{x}{r}I - A\right]$. As $c_A(x) = \det[xI - A]$, this shows that $c_{rA}(x) = r^n c_A\left(\frac{x}{r}\right)$.

20(b) If μ is an eigenvalue of A^{-1} then $A^{-1}X = \mu X$ for some column $X \neq 0$. Left multiplication by A gives $X = \mu AX$, whence $AX = \frac{1}{\mu}X$. Thus, $\frac{1}{\mu}$ is an eigenvalue of A; call it $\frac{1}{\mu} = \lambda$. Hence, $\mu = \frac{1}{\lambda}$ as required. Conversely, if λ is any eigenvalue of A then $\lambda \neq 0$ by (a) and we claim that $\frac{1}{\lambda}$ is an eigenvalue of A^{-1}. We have $AX = \lambda X$ for some column $x \neq 0$. Multiply on the left by A^{-1} to get $X = \lambda A^{-1}X$; whence $A^{-1}X = \frac{1}{\lambda}X$. Thus $\frac{1}{\lambda}$ is indeed an eigenvalue of A^{-1}.

21(b) We have $AX = \lambda X$ for some column $X \neq 0$. Hence, $A^2X = \lambda AX = \lambda^2X$, $A^3 = \lambda^2 AX = \lambda^3 X$, so

$$(A^3 - 2A + 3I)X = A^3X - 2AX + 3X = \lambda^3 X - 2\lambda X + 3X = (\lambda^3 - 2\lambda + 3)X.$$

23(b) If λ is an eigenvalue of A, let $AX = \lambda X$ for some $X \neq 0$. Then $A^2X = \lambda AX = \lambda^2X$, $A^3X = \lambda^2 AX = \lambda^3 X, \ldots$. We claim that $A^kX = \lambda^kX$ holds for every $k \geq 1$. We have already checked this for $k = 1$. If it holds for some $k \geq 1$, then $A^kX = \lambda^kX$, so

$$A^{k+1}X = A(A^kX) = A(\lambda^kX) = \lambda^k AX = \lambda^k(\lambda X) = \lambda^{k+1}X.$$

Hence, it also holds for $k+1$, and so A holds for all $k \geq 1$ by induction. In particular, if $A^m = 0$, $m \geq 1$, then $\lambda^mX = A^mX = 0X = 0$. As $X \neq 0$, this implies that $\lambda^m = 0$, so $\lambda = 0$.

25(a) Suppose A is diagonalizable with equal eigenvalues $\lambda_1 = \cdots = \lambda_n = \lambda$. Then an invertible matrix P exists such that $P^{-1}AF = \text{diag}\{\lambda_1, \ldots, \lambda_n\} = \text{diag}\{\lambda, \ldots, \lambda\} = \lambda I$. Hence, $A = P(\lambda I)P^{-1} = \lambda PIF = \lambda I$.

(b) No. Here the characteristic polynomial is $(x-1)^2$ so $\lambda = 1$ is the only eigenvalue. Use (a).

Exercises 3.4   Polynomial Interpolation

1(b) If $p(x) = r_0 + r_1x + r_2x^2$, the conditions give linear equations for r_0, r_1, r_2.

$$
\begin{array}{ccccccccc}
r_0 & & & & & = & p(0) & = & 5 \\
r_0 & + & r_1 & + & r_3 & = & p(1) & = & 3 \\
r_0 & + & 2r_1 & + & 4r_2 & = & p(2) & = & 5.
\end{array}
$$

The solution is $r_0 = 5$, $r_1 = -4$, $r_2 = 2$, so $p(x) = 5 - 4x + 2x^2$.

2(b) If $p(x) = r_0 + r_1x + r_2x^2 + r_3x^3$, the conditions give linear equations for a, b, c, d:

$$
\begin{array}{ccccccccccc}
r_0 & & & & & & & = & p(0) & = & 1 \\
r_0 & + & r_1 & + & r_2 & + & r_3 & = & p(1) & = & 1 \\
r_0 & - & r_1 & + & r_2 & - & r_3 & = & p(-1) & = & 2 \\
r_0 & - & 2r_1 & + & 4r_2 & - & 8r_3 & = & p(-2) & = & -3.
\end{array}
$$

The solution is $r_0 = 1$, $r_1 = \frac{-5}{3}$, $r_2 = \frac{1}{2}$, $r_3 = \frac{7}{6}$, so $p(x) = 1 - \frac{5}{3}X + \frac{1}{2}x^2 + \frac{7}{6}x^3$.

3(b) If $p(x) = r_0 + r_1x + r_2x^2 + r_3x^3$, the data give

$$
\begin{array}{ccccccccccc}
r_0 & & & & & & & = & p(0) & = & 1 \\
r_0 & + & r_1 & + & r_2 & + & r_3 & = & p(1) & = & 1.49 \\
r_0 & + & 2r_1 & + & 4r_2 & + & 8r_3 & = & p(2) & = & -0.42 \\
r_0 & + & 3r_1 & + & 9r_2 & + & 27r_3 & = & p(3) & = & -11.33.
\end{array}
$$

The solution is $r_0 = 1$, $r_1 = -0.51$, $r_2 = 2.1$ and $r_3 = -1.1$. Hence

$$p(x) = 1 - 0.51x + 2.1x^2 - 1.1x^3.$$

The estimate for the value of y corresponding to $x = 1.5$ is

$$y = p(1.5) = 1 - 0.51(1.5) + 2.1(1.5)^2 - 1.1(1.5)^3 = 1.25$$

to two decimals.

4(b) The approximation to $\sin(.7)$ is

$$p(7) = 1.0203(.7) - 0.0652(.7)^2 - 0.1140(.7)^3 = 0.6432$$

to four figures. The true value of $\sin(.7)$ is 0.6442 to four figures.

Exercises 3.5   Linear Recurrences

1(b) In this case $x_{k+2} = 2x_k - x_{k+1}$, we have $V_{k+1} = \begin{bmatrix} x_{k+1} \\ 2x_k - x_k + 1 \end{bmatrix} = \begin{bmatrix} 0 & 1 \\ 2 & -1 \end{bmatrix} \begin{bmatrix} x_k \\ x_{k+1} \end{bmatrix} =$

AV_k. Diagonalizing A gives $P = \begin{bmatrix} 1 & 1 \\ 1 & -2 \end{bmatrix}$ and $D = \begin{bmatrix} 1 & 0 \\ 0 & -2 \end{bmatrix}$, so $\begin{bmatrix} b_1 \\ b_2 \end{bmatrix} = P^{-1}V_0 =$

$\frac{1}{3}\begin{bmatrix} 2 & 1 \\ 1 & -1 \end{bmatrix}\begin{bmatrix} 1 \\ 2 \end{bmatrix} = \begin{bmatrix} \frac{4}{3} \\ -\frac{1}{3} \end{bmatrix}$. Hence $\begin{bmatrix} x_k \\ x_{k+1} \end{bmatrix} = \frac{4}{3}1^k\begin{bmatrix} 1 \\ 1 \end{bmatrix} - \frac{1}{3}(-2)^k\begin{bmatrix} 1 \\ -2 \end{bmatrix}$ for each k.
Comparing top entries gives

$$x_k = \frac{4}{3} - \frac{1}{3}(-2)^k = -\frac{(-2)^k}{3}\left[1 - 4\left(\frac{1}{-2}\right)^k\right] \approx -\frac{(-2)^k}{3} \quad \text{for large } k.$$

(d) Here $x_{k+2} = 6x_k - x_{k+1}$, so $V_{k+1} = \begin{bmatrix} x_{k+1} \\ 6x_k - x_{k+1} \end{bmatrix} = \begin{bmatrix} 0 & 1 \\ 6 & -1 \end{bmatrix} \begin{bmatrix} x_k \\ x_{k+1} \end{bmatrix} = AV_k.$

Diagonalizing A gives $P = \begin{bmatrix} 1 & 1 \\ 2 & -3 \end{bmatrix}$ and $D = \begin{bmatrix} 2 & 0 \\ 0 & -3 \end{bmatrix}$, so $\begin{bmatrix} b_1 \\ b_2 \end{bmatrix} = P^{-1}V_0 =$

$\frac{1}{5}\begin{bmatrix} 3 & 1 \\ 2 & -1 \end{bmatrix}\begin{bmatrix} 1 \\ 1 \end{bmatrix} = \begin{bmatrix} \frac{4}{5} \\ \frac{1}{5} \end{bmatrix}$. Hence $\begin{bmatrix} x_k \\ x_{k+1} \end{bmatrix} = \frac{4}{5}2^k\begin{bmatrix} 1 \\ 2 \end{bmatrix} + \frac{1}{5}(-3)^k\begin{bmatrix} 1 \\ -3 \end{bmatrix}$, and so

$$x_k = \frac{4}{5}2^k + \frac{1}{5}(-3)^k = \frac{(-3)^k}{5}\left[1 + r\left(\frac{2}{-3}\right)^k\right] \approx \frac{(-3)^k}{5} \quad \text{for large } k.$$

2(b) Let $V_k = \begin{bmatrix} x_k \\ x_{k+1} \\ x_{k+2} \end{bmatrix}$. Then $A = \begin{bmatrix} 0 & 1 & 0 \\ 0 & 0 & 1 \\ 2 & 1 & -2 \end{bmatrix}$, and diagonalization gives $P = \begin{bmatrix} 1 & 1 & 1 \\ -1 & -2 & 1 \\ 1 & 4 & 1 \end{bmatrix}$

and $D = \begin{bmatrix} -1 & 0 & 0 \\ 0 & -2 & 0 \\ 0 & 0 & 1 \end{bmatrix}$. Then $\begin{bmatrix} b_1 \\ b_2 \\ b_3 \end{bmatrix} = P^{-1}V_0 = \begin{bmatrix} 1 & -\frac{1}{2} & -\frac{1}{2} \\ -\frac{1}{3} & 0 & \frac{1}{3} \\ \frac{1}{3} & \frac{1}{2} & \frac{1}{6} \end{bmatrix} \begin{bmatrix} 1 \\ 0 \\ 1 \end{bmatrix} = \begin{bmatrix} \frac{1}{2} \\ 0 \\ \frac{1}{2} \end{bmatrix}$,

giving the general formula $V_k = \frac{1}{2}(-1)1^k \begin{bmatrix} 1 \\ -1 \\ 1 \end{bmatrix} + (0)(-2)^k \begin{bmatrix} 1 \\ -2 \\ 4 \end{bmatrix} + \frac{1}{2}1^k \begin{bmatrix} 1 \\ 1 \\ 1 \end{bmatrix}$. Thus

equating first entries give $x_k = \frac{1}{2}(-1)^k + \frac{1}{2}1^k = \frac{1}{2}[(-1)^k + 1]$. Note that the sequence x_k here is $0, 1, 0, 1, 0, 1, \ldots$ which does not converge to any fixed value for large k.

5. Let x_k denote the number of ways to form words of k letters. A word of $k + 2$ letters must end in either a or b. The number of words that end in b is x_{k+1} — just add a b to a $(k + 1)$-letter word. But the number ending in a is x_k since the second-last letter must be a b (no adjacent a's) so we simply add ba to any k-letter word. This gives the recurrence $x_{k+2} = x_{k+1} + x_k$ which is the same as in Example 2, but with different initial conditions: $x_0 = 1$ (since the "empty" word is the only one formed with no letters) and $x_1 = 2$. The eigenvalues, eigenvectors, and diagonalization remain the same, and so

$$V_k = b_1 \lambda_1^k \begin{bmatrix} 1 \\ \lambda_1 \end{bmatrix} + b_2 \lambda_2^k \begin{bmatrix} 1 \\ \lambda_2 \end{bmatrix}$$

where $\lambda_1 = \frac{1}{2}(1 + \sqrt{5})$ and $\lambda_2 = \frac{1}{2}(1 - \sqrt{5})$. Comparing top entires gives

$$x_k = b_1 \lambda_1^k + b_2 \lambda_2^k.$$

By Theorem 1 §2.4, the constants b_1 and b_2 come from $\begin{bmatrix} b_1 & b_2 \end{bmatrix}^T = P^{-1}V_0$.] However, we vary the method and use the initial conditions to determine the values of b_1 and b_2 directly. More precisely, $x_0 = 1$ means $1 = b_1 + b_2$ while $x_1 = 2$ means $2 = b_1\lambda_1 + b_2\lambda_2$. These equations have unique solution $b_1 = \frac{\sqrt{5}-3}{2\sqrt{5}}$ and $b_2 = \frac{\sqrt{5}-3}{2\sqrt{5}}$. It follows that

$$x_k = \frac{1}{2\sqrt{5}}\left[(3 + \sqrt{5})\left(\frac{1+\sqrt{5}}{2}\right)^k + (-3 + \sqrt{5})\left(\frac{1-\sqrt{5}}{2}\right)^k \right] \qquad \text{for each } k \geq 0.$$

7. In a stack of $k + 2$ chips, if the last chip is gold then (to avoid having two gold chips together) the second last chip must be either red or blue. This can happen in $2x_k$ ways. But there are x_{k+1} ways that the last chip is red (or blue) so there are $2x_{k+1}$ ways these possibilities can occur. Hence $x_{k+2} = 2x_k + 2x_{k+1}$. The matrix is $A = \begin{bmatrix} 0 & 1 \\ 2 & 2 \end{bmatrix}$ with eigenvalues $\lambda_1 = 1 + \sqrt{3}$

and $\lambda_2 = 1 - \sqrt{3}$ and corresponding eigenvectors $X_1 = \begin{bmatrix} 1 \\ \lambda_1 \end{bmatrix}$ and $X_2 = \begin{bmatrix} 1 \\ \lambda_2 \end{bmatrix}$. Given the initial conditions $x_0 = 1$ and $x_1 = 3$, we get

$$\begin{bmatrix} b_1 \\ b_2 \end{bmatrix} = P^{-1}V_0 = \frac{1}{\sqrt{3}} \begin{bmatrix} \lambda_2 & -1 \\ -\lambda_1 & 1 \end{bmatrix} \begin{bmatrix} 1 \\ 3 \end{bmatrix} = \frac{1}{-2\sqrt{3}} \begin{bmatrix} -2 - \sqrt{3} \\ 2 - \sqrt{3} \end{bmatrix} = \frac{1}{2\sqrt{3}} \begin{bmatrix} 2 + \sqrt{3} \\ -2 + \sqrt{3} \end{bmatrix}.$$

Since Theorem 1 §2.4 gives

$$V_k = b_1 \lambda_1^k \begin{bmatrix} 1 \\ \lambda_1 \end{bmatrix} + b_2 \lambda_2^k \begin{bmatrix} 1 \\ \lambda_2 \end{bmatrix},$$

comparing top entries gives

$$x_k = b_1 \lambda_1^k + b_2 \lambda_2^k + \frac{1}{2\sqrt{3}} \left[(2 + \sqrt{3})(1 + \sqrt{3})^k + (-2 + \sqrt{3})(1 - \sqrt{3})^k \right].$$

9. Let y_k be the yield for year k. Then the yield for year $k+2$ is $y_{k+2} = \frac{y_k + y_{k+1}}{2} = \frac{1}{2}y_k + \frac{1}{2}y_{k+1}$.

The eigenvalues are $\lambda_1 = 1$ and $\lambda_2 = -\frac{1}{2}$, with corresponding eigenvectors $X_1 = \begin{bmatrix} 1 \\ 1 \end{bmatrix}$ and

$X_2 = \begin{bmatrix} -2 \\ 1 \end{bmatrix}$. Given that $k = 0$ for the year 1990, we have the initial conditions $y_0 = 10$ and $y_1 = 12$. Thus

$$\begin{bmatrix} b_1 \\ b_2 \end{bmatrix} = P^{-1}V_0 = \frac{1}{3} \begin{bmatrix} 1 & 2 \\ -1 & 1 \end{bmatrix} \begin{bmatrix} 10 \\ 12 \end{bmatrix} = \frac{1}{3} \begin{bmatrix} 34 \\ 2 \end{bmatrix}.$$

Since

$$V_k = \frac{34}{3}(1)^k \begin{bmatrix} 1 \\ 1 \end{bmatrix} + \frac{2}{3} \left(-\frac{1}{2}\right)^k \begin{bmatrix} -2 \\ 1 \end{bmatrix}$$

then

$$y_k = \frac{34}{3}(1)^k + \frac{2}{3}(-2) \left(-\frac{1}{2}\right)^k = \frac{34}{3} - \frac{4}{3} \left(-\frac{1}{2}\right)^k.$$

For large k, $y_k \approx \frac{34}{3}$ so the long term yield is $11\frac{1}{3}$ million tons of wheat.

13(a) If p_k is a solution of (*) and q_k is a solution of (**) then

$$q_{k+2} = aq_{k+1} + bq_k$$
$$p_{k+2} = ap_{k+1} + bp_k + c(k).$$

for all k. Adding these equations we obtain

$$p_{k+2} + q_{k+2} = a(p_{k+1} + q_{k+1}) + b(p_k + q_k) + c(k)$$

that is $p_k + q_k$ is also a solution of (*).

(b) If r_k is any solution of (*) then $r_{k+2} = ar_{k+1} + br_k + c(k)$. Define $q_k = r_k - p_k$ for each k. Then it suffices to show that q_k is a solution of (**). But

$$q_{k+2} = r_{k+2} - p_{k+2} = (ar_{k+1} + br_k + c(k)) - (ap_{k+1} + bp_k + c(k)) = aq_{k+1} + bq_k$$

which is what we wanted.

Exercises 3.6   Population Growth

1(b) $A = \begin{bmatrix} \frac{1}{4} & \frac{1}{4} \\ 3 & 0 \end{bmatrix}$ so $c_A(x) = x^2 - \frac{1}{4}x - \frac{3}{4}$. The roots are 1 and $-\frac{3}{4}$ so the dominant eigenvalue is 1 and the population stabilizes.

(d) $A = \begin{bmatrix} \frac{3}{5} & \frac{1}{5} \\ 3 & 0 \end{bmatrix}$ so $c_A(x) = x^2 - \frac{3}{5}x - \frac{3}{5}$. The roots are $\frac{1}{10}(3 \pm \sqrt{69})$ so the dominant eigenvalue is $\frac{1}{10}\left(3 + \sqrt{69}\right) = 1.13 > 0$. Hence the population diverges.

4. $A = \begin{bmatrix} \alpha & \frac{2}{5} \\ 2 & 0 \end{bmatrix}$ so $c_A(x) = x^2 - \alpha x - \frac{4}{5}$. The roots are $\frac{1}{2}\left[\alpha \pm \sqrt{\alpha^2 + \frac{16}{5}}\right]$. Then $\lambda_1 < 1$ means $\alpha + \sqrt{\alpha^2 + \frac{16}{5}} < 2$, that is $\sqrt{\alpha^2 + \frac{16}{5}} < 2 - \alpha$. Squaring gives $\alpha^2 + \frac{16}{5} < 4 - 4\alpha + \alpha^2$ that is $\frac{4}{5} < 1 - \alpha$. So $\alpha < \frac{1}{5}$. So it becomes extinct if and only if $\alpha < \frac{1}{5}$. Similarly, it stabilizes (diverges) if and only if $\alpha = \frac{1}{5}$ $\left(\alpha > \frac{1}{5}\right)$.

Exercises 3.7   Proof of the Laplace Expansion

2. Consider the rows $R_p, R_{p+1}, \ldots, R_{q-1}, R_q$. Using adjacent interchanges we have

$$\begin{bmatrix} R_p \\ R_{p+1} \\ \vdots \\ R_{q-1} \\ R_q \end{bmatrix} \xrightarrow[\text{interchanges}]{q-p} \begin{bmatrix} R_{p+1} \\ \vdots \\ R_{q-1} \\ R_q \\ R_p \end{bmatrix} \xrightarrow[\text{interchanges}]{q-p-1} \begin{bmatrix} R_q \\ R_{p+1} \\ \vdots \\ R_{q-1} \\ R_p \end{bmatrix}.$$

Hence $2(q - p) - 1$ interchanges are used in all.

Supplementary Exercises   Chapter 3

2(b) Proceed by induction on n where A is $n \times n$. If $n = 1$, $A^T = A$. In general, induction and (a) give

$$\det[A_{ij}] = \det[(A_{ij})^T] = \det[(A^T)_{ij}].$$

Write $A^T = [a'_{ij}]$ where $a'_{ij} = a_{ji}$, and expand $\det(A^T)$ along column 1:

$$\det(A^T) = \sum_{j=1}^{n} a'_{j1}(-1)^{j+1} \det[(A^T)_{j1}] = \sum_{j=1}^{n} a_{1j}(-1)^{1+j} \det[A_{1j}] = \det A$$

where the last equality is the expansion of $\det A$ along row 1.

Chapter 4: Vector Geometry

Exercises 4.1   Vectors and Lines

3(b) In the diagram, let E and F be the midpoints of sides BC and AC respectively. Then $\overrightarrow{FC} = \frac{1}{2}\overrightarrow{AC}$ and $\overrightarrow{CE} = \frac{1}{2}\overrightarrow{CB}$. Hence

$$\overrightarrow{FE} = \overrightarrow{FC} + \overrightarrow{CE} = \tfrac{1}{2}\overrightarrow{AC} + \tfrac{1}{2}\overrightarrow{CB} = \tfrac{1}{2}(\overrightarrow{AC} + \overrightarrow{CB}) = \tfrac{1}{2}\overrightarrow{AB}$$

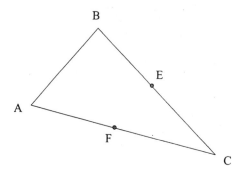

4(b) $8(2\mathbf{u} - \mathbf{v} + 3\mathbf{w}) + 3(5\mathbf{v} - 6\mathbf{w}) - 2(8\mathbf{u} + 3\mathbf{v} + 3\mathbf{w})$
     $= 16\mathbf{u} - 8\mathbf{v} + 24\mathbf{w} + 15\mathbf{v} - 18\mathbf{w} - 16\mathbf{u} - 6\mathbf{v} - 6\mathbf{w}$
     $= \mathbf{v}$

5(b) $\tfrac{1}{3}(3\mathbf{u} - \mathbf{v} + 4\mathbf{w}) \;=\; \tfrac{1}{3}[(-3,3,6) - (2,0,3) + (-4,12,36)]$
                                 $=\; \tfrac{1}{3}(-9,15,39)$
                                   $=\; (-3,5,13).$

(d) $2\mathbf{v} - 3(\mathbf{u} + \mathbf{w}) = 2(2,0,3) - 3(-2,4,11) = (4,0,6) - (-6,12,33) = (10,-12,-27)$

(f) $2(\mathbf{u} + \mathbf{v}) - (\mathbf{v} + \mathbf{w} - \mathbf{u}) = 2(1,1,5) - (2,2,10) = (2,2,10) - (2,2,10) = (0,0,0)$

6(b) Yes, they are parallel: $\mathbf{u} = (-3)\mathbf{v}$

(d) Yes, they are parallel: $\mathbf{v} = (-4)\mathbf{u}$

7(b) $\overrightarrow{QR} = \mathbf{u}$ because $OPQR$ is a parallelogram

(d) $\overrightarrow{RO} = -(\mathbf{u} + \mathbf{v})$ because $\overrightarrow{OR} = \mathbf{u} + \mathbf{v}$

8(b) $\overrightarrow{PQ} = (1 - 2, -1 - 0, 6 - 1) = (-1, -1, 5)$

(d) $\overrightarrow{PQ} = (1 - 1, -1 - (-1), 2 - 2) = (0,0,0) = \mathbf{0}$

(f) $\overrightarrow{PQ} = (1 - 3, 1 - (-1), 4 - 6) = (-2, 2, -2)$

9(b) Given $Q(x,y,z)$ let $\mathbf{q} = (x,y,z)$ and $\mathbf{p} = (3,0,-1)$ be the position vectors of Q and P. Then $\overrightarrow{PQ} = \mathbf{q} - \mathbf{p}$

    (i) If $PQ = \mathbf{v}$ then $\mathbf{q} - \mathbf{p} = \mathbf{v}$, so $\mathbf{q} = \mathbf{p} + \mathbf{v} = (5,-1,2)$. Thus $Q = Q(5,-1,2)$

(ii) If $PQ = -\mathbf{v}$ then $\mathbf{q} - \mathbf{p} = -\mathbf{v}$, so $\mathbf{q} = \mathbf{p} - \mathbf{v} = (1, 1, -4)$. Thus $Q = Q(1, 1, -4)$.

10(b) If $2(3\mathbf{v} - \mathbf{x}) = 5\mathbf{w} + \mathbf{u} - 3\mathbf{x}$ then $6\mathbf{v} - 2\mathbf{x} = 5\mathbf{w} + \mathbf{u} - 3\mathbf{x}$, so $\mathbf{x} = 5\mathbf{w} + \mathbf{u} - 6\mathbf{v} = (5, 5, 15) + (3, -1, 0) - (24, 0, 6) = (-16, 4, 9)$

11(b) $a\mathbf{u} + b\mathbf{v} + c\mathbf{w} = (a, a, 2a) + (0, b, 2b) + (c, 0, -c) = (a+c, a+b, 2a+2b-c)$. Hence $a\mathbf{u} + b\mathbf{v} + c\mathbf{w} = \mathbf{x} = (1, 3, 0)$ gives equations

$$
\begin{array}{rcrcrcr}
a & & & + & c & = & 1 \\
a & + & b & & & = & 3 \\
2a & + & 2b & - & c & = & 0.
\end{array}
$$

The solution is $a = -5$, $b = 8$, $c = 6$.

12(b) Suppose $(5, 6, -1) = a\mathbf{u} + b\mathbf{v} + c\mathbf{w} = (3a + 4b + c, \ -a + c, \ b + c)$. Equating coefficients gives linear equations for a, b, c:

$$
\begin{array}{rcrcrcr}
3a & + & 4b & + & c & = & 5 \\
-a & & & + & c & = & 6 \\
& & b & + & c & = & -1
\end{array}
$$

This system has no solution, so no such a, b, c exist.

13(b) Write $P = P(x, y, z)$ and let $\mathbf{p} = (x, y, z)$, $\mathbf{p}_1 = (2, 1, -2)$ and $\mathbf{p}_2 = (1, -2, 0)$ be the position vectors of P, P_1 and P_2 respectively. Then $\mathbf{p} = \mathbf{p}_2 = \overrightarrow{P_2P} = \mathbf{p}_2 + \frac{1}{4}(\overrightarrow{P_2P_1}) = \mathbf{p}_2 + \frac{1}{4}(\mathbf{p}_1 - \mathbf{p}_2) = \frac{1}{4}\mathbf{p}_1 + \frac{3}{4}\mathbf{p}_2$. Since $\mathbf{p}_1$ and $\mathbf{p}_2$ are known, this gives

$$\mathbf{p} = \tfrac{1}{4}(2, 1, -2) + \tfrac{3}{4}(1, -2, 0) = \tfrac{1}{4}(5, -5, -2).$$

Hence $P = P\left(\frac{5}{4}, -\frac{5}{4}, -\frac{1}{2}\right)$.

16(b) One direction vector is $\mathbf{d} = \overrightarrow{QP} = (2, -1, 5)$. Let $\mathbf{p}_0 = (3, -1, 4)$ be the position vector of P. Then the vector equation of the line is

$$\mathbf{p} = \mathbf{p}_0 + t\mathbf{d} = (3, -1, 4) + t(2, -1, 5).$$

when $\mathbf{p} = (x, y, z)$ is the position vector of an arbitrary point on the line. Equating coefficients gives the parametric equations of the line

$$
\begin{aligned}
x &= 3 + 2t \\
y &= -1 - t \\
z &= 4 + 5t.
\end{aligned}
$$

(d) Now $\mathbf{p}_0 = (1, 1, 1)$ because $P_1(1, 1, 1)$ is on the line, and take $\mathbf{d} = (1, 1, 1)$ because the line is to be parallel to $\mathbf{d}$. Hence the vector equation is

$$\mathbf{p} = \mathbf{p}_0 + t\mathbf{d} = (1, 1, 1) + t(1, 1, 1).$$

The scalar equations are (taking $\mathbf{p} = (x, y, z)$)

$$
\begin{aligned}
x &= 1 + t \\
y &= 1 + t \\
z &= 1 + t.
\end{aligned}
$$

(f) The line with parametric equations

$$x = 2 - t$$
$$y = 1$$
$$z = \quad t$$

has direction vector $\mathbf{d} = (-1, 0, 1)$ — the components are the coefficients of t. Since our line is parallel to this one, $\mathbf{d}$ will do as direction vector. We are given the position vector $\mathbf{p}_0 = (2, -1, 1)$ of a point on the line, so the vector equation is

$$\mathbf{p} = \mathbf{p}_0 + t\mathbf{d} = (2, -1, 1) + t(-1, 0, 1).$$

The scalar equations are

$$x = 2 - t$$
$$y = -1$$
$$z = 1 + t.$$

17(b) $P(2, 3, -3)$ lies on the line $\begin{bmatrix} x \\ y \\ z \end{bmatrix} = \begin{bmatrix} 4 - t \\ 3 \\ 1 - 2t \end{bmatrix}$ since it corresponds to $t = 2$. Similarly

$Q(-1, 3, -9)$ corresponds to $t = 5$, so Q lies on the line too.

18(b) If $P = P(x, y, z)$ is a point on both lines then

$$x = 1 - t$$
$$y = 2 + 2t \qquad \text{for some } t \text{ because } P \text{ lies on the first line.}$$
$$z = -1 + 3t$$

$$x = 2s$$
$$y = 1 + s \qquad \text{for some } s \text{ because } P \text{ lies on the second line.}$$
$$z = 3$$

If we eliminate x, y, and z we get three equations for s and t:

$$1 - t = 2s$$
$$2 + 2t = 1 + s$$
$$-1 + 3t = 3.$$

The last two equations require $t = \frac{4}{3}$ and $s = \frac{11}{3}$, but these values do *not* satisfy the first equation. Hence no such s and t exist, so the lines do *not* intersect.

(d) If (x, y, z) is the position vector of a position on both lines, then

$$(x, y, z) = (4, -1, 5) + t(1, 0, 1) \qquad \text{for some } t \quad \text{(first line)}$$
$$(x, y, z) = (2, -7, 12) + s(0, -2, 3) \quad \text{for some } s \quad \text{(second line)}.$$

Eliminating (x, y, z) gives

$$(4, -1, 5) + t(1, 0, 1) = (2, -7, 12) + s(0, -2, 3).$$

Equating coefficients gives three equations for s and t:

$$4 + t = 2$$
$$-1 = -7 - 2s$$
$$5 + t = 12 + 3s.$$

This has a (unique) solution $t = -2$, $s = -3$ so the lines *do* intersect. The point of intersection has position vector

$$(4, -1, 5) + t(1, 0, 1) = (4, -1, 5) - 2(1, 0, 1) = (2, -1, 3)$$

(equivalently $(2, -7, 12) + s(0, -2, 3) = (2, -7, 12) - 3(0, -2, 3) = (2, -1, 3)$).

22(b) Let $\mathbf{u} = (x, y, z)$. Then $\mathbf{u} = x\mathbf{i} + y\mathbf{j} + z\mathbf{k}$ by equation (*) (preceding Theorem 3) so Theorem 1 gives

$$a\mathbf{u} + a(x\mathbf{i} + y\mathbf{j} + z\mathbf{k}) = a(x\mathbf{i}) + a(y\mathbf{j}) + a(z\mathbf{k}) = (ax)\mathbf{i} + (ay)\mathbf{j} + (az)\mathbf{k}.$$

Hence $a\mathbf{u} = (ax, ay, az)$, again by equation (*).

24. Let $\mathbf{a} = (1, -1, 2)$ and $\mathbf{b} = (2, 0, 1)$ be the position vectors of A and B. Then $\mathbf{d} = \mathbf{b} - \mathbf{a} = (1, 1, -1)$ is a direction vector for the line through A and B, so the position vector $\mathbf{c}$ of C is given by $\mathbf{c} = \mathbf{a} + t\mathbf{d}$ for some t. Then

$$\left\|\overrightarrow{AC}\right\| = \|\mathbf{c} - \mathbf{a}\| = \|t\mathbf{d}\| = |t| \, \|\mathbf{d}\| \quad \text{and} \quad \left\|\overrightarrow{BC}\right\| = \|\mathbf{c} - \mathbf{b}\| = \|(t-1)\mathbf{d}\| = |t - 1| \, \|\mathbf{d}\|.$$

Hence $\left\|\overrightarrow{AC}\right\| = 2\left\|\overrightarrow{BC}\right\|$ means $|t| = 2|t - 1|$, so $t^2 = 4(t - 1)^2$, whence $0 = 3t^2 - 8t + 4 = (t - 2)(3t - 2)$. Thus $t = 2$ or $t = \frac{2}{3}$, that is $\mathbf{c} = \mathbf{a} + t\mathbf{d} = (3, 1, 0)$ or $\left(\frac{5}{3}, \frac{-1}{3}, \frac{4}{3}\right)$.

25(b)

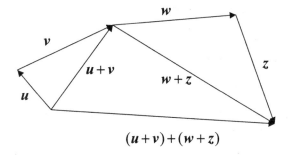

$(\mathbf{u} + \mathbf{v}) + (\mathbf{w} + \mathbf{z})$

28(b) If there are $2n$ points, then P_k and P_{n+k} are opposite ends of a diameter of the circle for each $k = 1, 2, \ldots$. Hence $\overrightarrow{CP}_k = -\overrightarrow{CP}_{n+k}$ so these terms cancel in the sum $\overrightarrow{CP}_1 + \overrightarrow{CP}_2 + \cdots + \overrightarrow{CP}_{2n}$. Thus all terms cancel and the sum in $\mathbf{0}$.

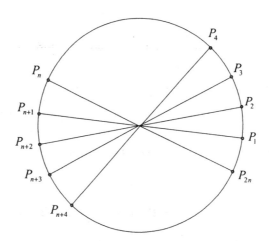

30. We have $2\overrightarrow{EA} = \overrightarrow{DA}$ because E is the midpoint of side AD and $2\overrightarrow{AF} = \overrightarrow{FC}$ because F is $\frac{1}{3}$ the way from A to C. Thus

$$2\overrightarrow{EF} = 2(\overrightarrow{EA} + \overrightarrow{AF}) = 2\overrightarrow{EA} + 2\overrightarrow{AF} = \overrightarrow{DA} + \overrightarrow{FC} = \overrightarrow{CB} + \overrightarrow{FC} = \overrightarrow{FB}.$$

Hence $\overrightarrow{EF} = \frac{1}{2}\overrightarrow{FB}$ so F is in the line segment EB, $\frac{1}{3}$ the way from E to B.

Exercises 4.2   The Dot Product and Projections

1(b) $\|(1,-1,2)\| = \sqrt{1^2 + (-1)^2 + 2^2} = \sqrt{6}$

(d) $\|(-1,0,2)\| = \sqrt{(-1)^2 + 0^2 + 2^2} = \sqrt{5}$

(f) $\|-3(1,1,2)\| = |-3|\ \|(1,1,2)\| = 3\sqrt{1^2 + 1^2 + 2^2} = 3\sqrt{6}$

2(b) If $\mathbf{v} = (-2,-1,2)$ then $\|\mathbf{v}\| = \sqrt{4+1+4} = 3$. Hence the desired unit vector is $\frac{1}{\|\mathbf{v}\|}\mathbf{v} = \frac{1}{3}(-2,-1,2) = \left(-\frac{2}{3}, -\frac{1}{3}, \frac{2}{3}\right)$.

4(b) The distance is $\|(2-2,\ -1-0,\ 2-1)\| = \sqrt{0^2 + (-1)^2 + 1^2} = \sqrt{2}$

(d) The distance is $\|(4-3,\ 0-2,\ -2-0)\| = \sqrt{1^2 + (-2)^2 + (-2)^2} = 3$

5(b) $\mathbf{u} \cdot \mathbf{v} = \mathbf{u} \cdot \mathbf{u} = 1^2 + 2^2 + (-1)^2 = 6$

(d) $\mathbf{u} \cdot \mathbf{v} = 3 \cdot 6 + (-1)(-7) + 5(-5) = 18 + 7 - 25 = 0$

(f) $\mathbf{u} = (0,0,0)$ so $\mathbf{u} \cdot \mathbf{v} = a \cdot 0 + b \cdot 0 + c \cdot 0 = 0$

6(b) $\cos\theta = \frac{\mathbf{u}\cdot\mathbf{v}}{\|\mathbf{u}\|\,\|\mathbf{v}\|} = \frac{-18-2+0}{\sqrt{10}\sqrt{40}} = \frac{-20}{20} = -1$. Hence $\theta = \pi$.

(d) $\cos\theta = \frac{\mathbf{u}\cdot\mathbf{v}}{\|\mathbf{u}\|\,\|\mathbf{v}\|} = \frac{6+6-3}{\sqrt{6}\cdot3\sqrt{6}} = \frac{1}{2}$. Hence $\theta = \frac{\pi}{3}$.

(f) $\cos\theta = \frac{\mathbf{u}\cdot\mathbf{v}}{\|\mathbf{u}\|\,\|\mathbf{v}\|} = \frac{0-21-4}{\sqrt{25}\sqrt{100}} = -\frac{1}{2}$. Hence $\theta = \frac{2\pi}{3}$.

7(b) Writing $\mathbf{u} = (2, -1, 1)$ and $\mathbf{v} = (1, x, 2)$, the requirement is

$$\tfrac{1}{2} = \cos \tfrac{\pi}{3} = \frac{\mathbf{u} \cdot \mathbf{v}}{\|\mathbf{u}\| \, \|\mathbf{v}\|} = \frac{2 - x + 2}{\sqrt{6}\sqrt{x^2 + 5}}.$$

Hence $6(x^2 + 5) = 4(4 - x)^2$, whence $x^2 + 16x - 17 = 0$. The roots are $x = -17$ and $x = 1$.

8(b) The conditions are $\mathbf{u}_1 \cdot \mathbf{v} = 0$ and $\mathbf{u}_2 \cdot \mathbf{v} = 0$, yielding equations

$$
\begin{array}{rcrcrcl}
3x & - & y & + & 2z & = & 0 \\
2x & & & + & z & = & 0
\end{array}
$$

The solutions are $x = -t$, $y = t$, $z = 2t$, so $\mathbf{v} = t(-1, 1, 2)$.

(d) The conditions are $\mathbf{u}_1 \cdot \mathbf{v} = 0$ and $\mathbf{u}_2 \cdot \mathbf{v} = 0$, yielding equations

$$2x - y + 3z = 0$$
$$0 = 0.$$

The solutions are $x = s$, $y = 2s + 3t$, $z = t$, so $\mathbf{v} = s(1, 2, 0) + t(0, 3, 1)$.

10(b) $\left\| \overrightarrow{PQ} \right\|^2 = \|(3, -2, 4)\|^2 = 9 + 4 + 16 = 29$

$\left\| \overrightarrow{QR} \right\|^2 = \|(2, 7, 2)\|^2 = 4 + 49 + 4 = 57$

$\left\| \overrightarrow{PR} \right\|^2 = \|(5, 5, 6)\|^2 = 25 + 25 + 36 = 86.$

Hence $\left\| \overrightarrow{PR} \right\| = \|PQ\|^2 + \|QR\|^2$. Note that this *implies* that the triangle is right angled, that PR is the hypotenuse, and hence that the angle at Q is a right angle.

12(b) We have $\overrightarrow{AB} = (2, 1, 1)$ and $\overrightarrow{AC} = (1, 2, -1)$ so the angle α at A is given by

$$\cos \alpha = \frac{\overrightarrow{AB} \cdot \overrightarrow{AC}}{\left\| \overrightarrow{AB} \right\| \left\| \overrightarrow{AC} \right\|} = \frac{2 + 2 - 1}{\sqrt{6}\sqrt{6}} = \tfrac{1}{2}.$$

Hence $\alpha = \tfrac{\pi}{3}$ or $60°$. Next $\overrightarrow{BA} = (-2, -1, -1)$ and $\overrightarrow{BC} = (-1, 1, -2)$ so the angle β at B is given by

$$\cos \beta = \frac{\overrightarrow{BA} \cdot \overrightarrow{BC}}{\left\| \overrightarrow{BA} \right\| \left\| \overrightarrow{BC} \right\|} = \frac{2 - 1 + 2}{\sqrt{6}\sqrt{6}} = \tfrac{1}{2}.$$

Hence $\beta = \tfrac{\pi}{3}$. Since the angles in any triangle add to π, the angle γ at C is $\pi - \tfrac{\pi}{3} - \tfrac{\pi}{3} = \tfrac{\pi}{3}$. Since $\overrightarrow{CA} = (-1, -2, 1)$ and $\overrightarrow{CB} = (1, -1, 2)$, this can be seen directly from

$$\cos \gamma = \frac{\overrightarrow{CA} \cdot \overrightarrow{CB}}{\left\| \overrightarrow{CA} \right\| \left\| \overrightarrow{CB} \right\|} = \frac{-1 + 2 + 2}{\sqrt{6}\sqrt{6}} = \tfrac{1}{2}.$$

14(b) $\text{proj}_{\mathbf{v}}(\mathbf{u}) = \frac{\mathbf{u}\cdot\mathbf{v}}{\|\mathbf{v}\|^2}\mathbf{v} = \frac{12-2+1}{16+1+1}(4,1,1) = \frac{11}{18}(4,1,1)$

(d) $\text{proj}_{\mathbf{v}}(\mathbf{u}) = \frac{\mathbf{u}\cdot\mathbf{v}}{\|\mathbf{v}\|^2}\mathbf{v} = \frac{-18-8-2}{36+16+4}(-6,4,2) = (-3,2,1)$

15(b) Take $u_1 = \text{proj}_{\mathbf{v}}(\mathbf{u}) = \frac{\mathbf{u}\cdot\mathbf{v}}{\|\mathbf{v}\|^2}\mathbf{v} = \frac{-6+1+0}{4+1+16}(-2,1,4) = \frac{-5}{21}(-2,1,4)$. Then $\mathbf{u}_2 = \mathbf{u}-\mathbf{u}_1 = (3,1,0) + \frac{5}{21}(-2,1,4) = \frac{1}{21}(53,26,20)$. As a check, verify that $\mathbf{u}_2 \cdot \mathbf{v} = 0$, that is $\mathbf{u}_2$ is orthogonal to $\mathbf{v}$.

(d) Take $\mathbf{u}_1 = \text{proj}_{\mathbf{v}}(\mathbf{u}) = \frac{\mathbf{u}\cdot\mathbf{v}}{\|\mathbf{v}\|^2}\mathbf{v} = \frac{-18-8-1}{36+16+1}(-6,4,-1) = \frac{27}{53}(6,-4,1)$. Then $\mathbf{u}_2$ is given by $\mathbf{u}_2 = \mathbf{u} - \mathbf{u}_1 = (3,-2,1) - \frac{27}{53}(6,-4,1) = \frac{1}{53}(-3,2,26)$. As a check, verify that $\mathbf{u}_2 \cdot \mathbf{v} = 0$, that is $\mathbf{u}_2$ is orthogonal to $\mathbf{v}$.

16(b) Write $\mathbf{p}_0 = (1,0,-1)$, $\mathbf{d} = (3,1,4)$, $\mathbf{p} = (1,-1,3)$ and write $\mathbf{u} = \overrightarrow{P_0P} = \mathbf{p} - \mathbf{p}_0 = (0,-1,4)$.

Write $\mathbf{u}_1 = \overrightarrow{P_0Q}$ and compute it as

$\mathbf{u}_1 = \text{proj}_d(\overrightarrow{P_0P}) = \frac{0-1+16}{9+1+16}(3,1,4) = \frac{15}{26}(3,1,4)$.

Then the distance from P to the line is

$\|\overrightarrow{QP}\| = \|\mathbf{u} - \mathbf{u}_1\| = \left\|\frac{1}{26}(-45,-41,44)\right\| = \frac{1}{26}\sqrt{5642}$.

To compute Q, let $\mathbf{q}$ be its position vector. Then

$\mathbf{q} = \mathbf{p}_0 + \mathbf{u}_1 = (1,0,-1) + \frac{15}{26}(3,1,4) = \frac{1}{26}(71,15,34)$.

Hence $Q = Q\left(\frac{71}{26}, \frac{15}{26}, \frac{34}{26}\right)$.

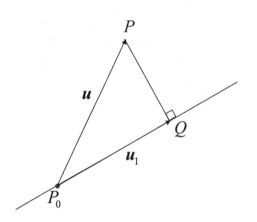

18(b) Position the cube with and vertex at the origin and sides along the positive axes. Assume each side has length a and consider the diagonal with direction $\mathbf{d} = (a,a,a)$. The face diagonals that do not meet $\mathbf{d}$ are parallel to one of

$$\mathbf{d}_1 = (0,0,a) - (0,a,0) = (0,-a,a)$$
$$\mathbf{d}_2 = (a,0,0) - (0,0,a) = (a,0,-a)$$
$$\mathbf{d}_3 = (0,a,0) - (a,0,0) = (-a,a,0)$$

Hence $\mathbf{d} \cdot \mathbf{d}_i = 0$ for each i, so each $\mathbf{d}_i$ is orthogonal to $\mathbf{d}$.

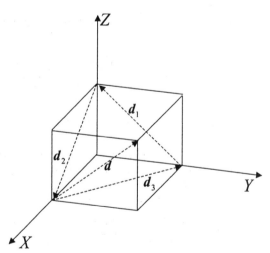

20. Position the solid with one vertex at the origin and sides, of lengths a, b, c, along the positive X-, Y- and Z-axes respectively. The diagonals are parallel to one of

$$
\begin{aligned}
(a,b,c) & \\
(-a,b,c) &= (0,b,c) - (a,0,0) \\
(a,-b,c) &= (a,0,c) - (0,b,0) \\
(a,b,-c) &= (a,b,0) - (0,0,c)
\end{aligned}
$$

The possible dot products are $\pm(-a^2 + b^2 + c^2)$, $\pm(a^2 - b^2 + c^2)$, $\pm(a^2 + b^2 - c^2)$, and one of these is zero if and only if the sum of two of a^2, b^2, c^2 equals the third.

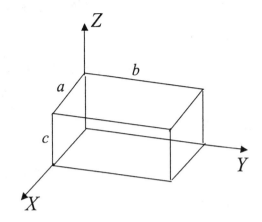

25(b) The sum of the squares of the lengths of the diagonals equals the sum of the squares of the lengths of the four sides.

29(b) The angle θ between $\mathbf{u}$ and $\mathbf{u} + \mathbf{v} + \mathbf{w}$ is given by

$$
\cos\theta = \frac{\mathbf{u}\cdot(\mathbf{u}+\mathbf{v}+\mathbf{w})}{\|\mathbf{u}\|\,\|\mathbf{u}+\mathbf{v}+\mathbf{w}\|} = \frac{\mathbf{u}\cdot\mathbf{u}+\mathbf{u}\cdot\mathbf{v}+\mathbf{u}\cdot\mathbf{w}}{\|\mathbf{u}\|\,\|\mathbf{u}+\mathbf{v}+\mathbf{w}\|} = \frac{\|\mathbf{u}\|^2+0+0}{\|\mathbf{u}\|\,\|\mathbf{u}+\mathbf{v}+\mathbf{w}\|} = \frac{\|\mathbf{u}\|}{\|\mathbf{u}+\mathbf{v}+\mathbf{w}\|}.
$$

Similarly the angles φ, ψ between $\mathbf{v}$ and $\mathbf{w}$ and $\mathbf{u} + \mathbf{v} + \mathbf{w}$ are given by

$$
\cos\varphi = \frac{\|\mathbf{v}\|}{\|\mathbf{u}+\mathbf{v}+\mathbf{w}\|} \quad \text{and} \quad \cos\psi = \frac{\|\mathbf{w}\|}{\|\mathbf{u}+\mathbf{v}+\mathbf{w}\|}.
$$

Since $\|\mathbf{u}\| = \|\mathbf{v}\| = \|\mathbf{w}\|$ we get $\cos\theta = \cos\varphi = \cos\psi$, whence $\theta = \varphi = \psi$.

NOTE: $\|\mathbf{u}+\mathbf{v}+\mathbf{w}\| = \sqrt{\|\mathbf{u}\|^2+\|\mathbf{v}\|^2+\|\mathbf{w}\|^2} = \|\mathbf{u}\|\sqrt{3}$ by part (a), so $\cos\theta = \cos\varphi = \cos\psi = \frac{1}{\sqrt{3}}$. Thus, in fact $\theta = \varphi = \psi = .955$ radians, $(54.7°)$.

30(b) If $P_1(x,y)$ is on the line then $ax + by + c = 0$. Hence $\mathbf{u} = \overrightarrow{P_1P_0} = (x_0 - x_1,\ y_0 - y_1)$ so the distance is

$$
\|\text{proj}_{\mathbf{n}}\mathbf{u}\| = \left\|\frac{\mathbf{u}\cdot\mathbf{n}}{\|\mathbf{n}\|^2}\mathbf{n}\right\| = \frac{|\mathbf{u}\cdot\mathbf{n}|}{\|\mathbf{n}\|} = \frac{|a(x_0 - x) + b(y_0 - y)|}{\sqrt{a^2+b^2}} = \frac{|ax_0 + by_0 + c|}{\sqrt{a^2+b^2}}.
$$

32(b) This follows from (a) because $\|\mathbf{v}\|^2 = a^2 + b^2 + c^2$.

34. Let $\mathbf{v} = (a,b,c)$ so that $k\mathbf{v} = (ka, kb, kc)$ by Theorem 2 §4.1. Hence Theorem 1 gives $\|k\mathbf{v}\|^2 = (ka)^2 + (kb)^2 + (kc)^2 + k^2(a^2 + b^2 + c^2) = k^2\|\mathbf{v}\|^2$. Taking positive square roots, $\|k\mathbf{v}\| = \sqrt{k^2}\|\mathbf{v}\| = |k|\,\|\mathbf{v}\|$.

36. We have $a\mathbf{v} \neq \mathbf{0}$ because $a \neq 0$ and $\mathbf{v} \neq \mathbf{0}$. Hence $\text{proj}_{a\mathbf{v}}(\mathbf{u}) = \frac{\mathbf{u}\cdot(a\mathbf{v})}{\|a\mathbf{v}\|^2}(a\mathbf{v}) = \frac{a\cdot\mathbf{u}\cdot\mathbf{v}}{a^2\|\mathbf{v}\|^2}\cdot a\mathbf{v} = \frac{\mathbf{u}\cdot\mathbf{v}}{\|\mathbf{v}\|^2}\mathbf{v} = \text{proj}_{\mathbf{v}}(\mathbf{u})$.

37(d) Take $x_1 = z_2 = x$, $y_1 = x_2 = y$ and $z_1 = y_2 = z$ in (c).

39. The three triangles in the diagram are similar so, as $p + q = h$, $\frac{a}{h} = \frac{p}{a}$ and $\frac{b}{h} = \frac{q}{b}$. Hence $a^2 = hp$ and $b^2 = hq$, so $a^2 + b^2 = h(p + q) = h^2$.

Exercises 4.3   Planes and the Cross Product

1(b) $\mathbf{u} \times \mathbf{v} = \det \begin{bmatrix} \mathbf{i} & \mathbf{j} & \mathbf{k} \\ 3 & -1 & 0 \\ -6 & 2 & 0 \end{bmatrix} = 0\mathbf{i} - 0\mathbf{j} + 0\mathbf{k} = (0,0,0) = \mathbf{0}$

(d) $\mathbf{u} \times \mathbf{v} = \det \begin{bmatrix} \mathbf{i} & \mathbf{j} & \mathbf{k} \\ 2 & 0 & -1 \\ 1 & 4 & 7 \end{bmatrix} = 4\mathbf{i} - 15\mathbf{j} + 8\mathbf{k} = (4, -15, 8)$

4(b) One vector orthogonal to $\mathbf{u}$ and $\mathbf{v}$ is $\mathbf{u} \times \mathbf{v} = \det \begin{bmatrix} \mathbf{i} & \mathbf{j} & \mathbf{k} \\ 1 & 2 & -1 \\ 3 & 1 & 2 \end{bmatrix} = (5, -5, -5)$. We have

$\|\mathbf{u} \times \mathbf{v}\| = 5\,\|(1, -1, -1)\| = 5\sqrt{3}$. Hence the unit vectors parallel to $\mathbf{u} \times \mathbf{v}$ are $\pm \frac{1}{5\sqrt{3}}(5, -5, -5) = \pm \frac{1}{\sqrt{3}(1, -1, -1)} = \pm \frac{\sqrt{3}}{3}(1, -1, -1)$.

5(b) A normal to the plane is $\mathbf{n} = \overrightarrow{AB} \times \overrightarrow{AC} = (-1, 1, -5) \times (3, 8, -17) = \det \begin{pmatrix} \mathbf{i} & \mathbf{j} & \mathbf{k} \\ -1 & 1 & -5 \\ 3 & 8 & -17 \end{pmatrix} =$

$(23, -32, -11)$. Since the plane passes through $B(0, 0, 1)$ the equation is

$$23(x - 0) - 32(y - 0) - 11(z - 1) = 0, \quad \text{that is} \ -23x + 32y + 11z = 11$$

(d) The plane with equation $2x - y + z = 3$ has normal $\mathbf{n} = (2, -1, 1)$. Since our plane is parallel to this one, $\mathbf{n}$ will serve as normal. The point $P(3, 0, -1)$ lies on our plane, the equation is $2(x - 3) - (y - 0) + (z - (-1)) = 0$, that is $2x - y + z = 5$.

(f) The plane contains $P(2, 1, 0)$ and $P_0(3, -1, 2)$, so the vector $\mathbf{u} = PP_0 = (1, -2, 2)$ is parallel to the plane. Also the direction vector $\mathbf{d} = (1, 0, -1)$ of the line is parallel to the plane. Hence

$$\mathbf{n} = \mathbf{u} \times \mathbf{d} = \det \begin{bmatrix} \mathbf{i} & \mathbf{j} & \mathbf{k} \\ 1 & -2 & 2 \\ 1 & 0 & -1 \end{bmatrix} = (2, 3, 2) \text{ is perpendicular to the plane and so serves as a}$$

normal. As $P(2, 1, 0)$ is in the plane, the equation is

$$2(x - 2) + 3(y - 1) + 2(z - 0) = 0, \quad \text{that is, } 2x + 3y + 2z = 7.$$

(h) The two direction vectors $\mathbf{d_1} = (1, -1, 3)$ and $\mathbf{d_2} = (2, 1, -1)$ are parallel to the plane, so

$$\mathbf{n} = \mathbf{d_1} \times \mathbf{d_2} = \det \begin{bmatrix} \mathbf{i} & \mathbf{j} & \mathbf{k} \\ 1 & -1 & 3 \\ 2 & 1 & -1 \end{bmatrix} = (-2, 7, 3) \text{ will serve as normal. The plane contains}$$

$P(3, 1, 0)$ so the equation is

$$-2(x - 3) + 7(y - 1) + 3(z - 0) = 0, \quad \text{that is } -2x + 7y + 3z = 1.$$

Note that this plane contains the line $(x, a, z) = (3, 1, 0) + t(1, -1, 3)$ by construction; it contains the *other* line because it contains $P(0, -2, 5)$ and is parallel to d_2. This implies that the lines intersect (both are in the same plane). In fact the point of intersection is $P(4, 0, 3)$ [$t = 1$ on the first line and $t = 2$ on the second line].

(j) The set of all points $R(x, y, z)$ equidistant from both $P(0, 1, -1)$ and $Q(2, -1, -3)$ is determined as follows: The condition is $\left\|\overrightarrow{PR}\right\| = \left\|\overrightarrow{QR}\right\|$, that is $\left\|\overrightarrow{PR}\right\|^2 = \left\|\overrightarrow{QR}\right\|^2$, that is

$$x^2 + (y - 1)^2 + (z + 1)^2 = (x - 2)^2 + (y + 1)^2 + (z + 3)^2.$$

This simplifies to $x^2 + y^2 + z^2 - 2y + 2z + 2 = x^2 + y^2 + z^2 - 4x + 2y + 6z + 14$; that is $4x - 4y - 4z = 12$; that is $x - y - z = 3$.

6(b) The normal $\mathbf{n} = (2, 1, 0)$ to the given plane will serve as direction vector for the line. Since the line passes through $P(2, -1, 3)$, the vector equation is $(x, y, z) = (2, -1, 3) + t(2, 1, 0)$.

(d) The given lines have direction vectors $\mathbf{d_1} = (1, 1, -2)$ and $\mathbf{d_2} = (1, 2, -3)$, so $\mathbf{d} = \mathbf{d_1} \times \mathbf{d_2} =$

$$\det \begin{bmatrix} \mathbf{i} & \mathbf{j} & \mathbf{k} \\ 1 & 1 & -2 \\ 1 & 2 & -3 \end{bmatrix} = (1, 1, 1)$$ is perpendicular to both lines. Hence $\mathbf{d}$ is a direction vector for

the line we seek. As $P(1, 1, -1)$ is on the line, the vector equation is $(x, y, z) = (1, 1, -1) + t(1, 1, 1)$.

(f) Each point on the given line has the form $Q(2 + t, 1 + t, t)$ for some t. So $\overrightarrow{PQ} = (1 + t, t, t - 2)$. This is perpendicular to the given line if $\overrightarrow{PQ} \cdot \mathbf{d} = 0$ (where $\mathbf{d} = (1, 1, 1)$ is the direction vector of the given line). This condition is $(1 + t) + t + (t - 2) = 0$, that is $t = \frac{1}{3}$. Hence the line we want has direction vector $\left(\frac{4}{3}, \frac{1}{3}, \frac{-5}{3}\right)$. For convenience we use $\mathbf{d} = (4, 1, -5)$. As the line we want passes through $P(1, 1, 2)$, the vector equation is $(x, y, z) = (1, 1, 2) + t(4, 1, -5)$. [Note that $Q\left(\frac{7}{3}, \frac{4}{3}, \frac{1}{3}\right)$ is the point of intersection of the two lines.]

7(b) Choose a point P_0 in the plane, say $P_0(0, 6, 0)$, and write $\mathbf{u} = \overrightarrow{P_0P} = (3, -5, -1)$. If $\mathbf{n} = (2, 1, -1)$ denotes the normal to the plane, compute

$$\mathbf{u_1} = \text{proj}_{\mathbf{n}}(\mathbf{u}) = \frac{\mathbf{u} \cdot \mathbf{n}}{\|\mathbf{n}\|^2} \mathbf{n} = \tfrac{2}{6}(2, 1, -1).$$

The distance from P to the plane is $\|\mathbf{u_1}\| = \frac{1}{3}\sqrt{6}$. If $\mathbf{p_0} = (0, 6, 0)$ and $\mathbf{q}$ are the position vectors of P_0 and Q, we get

$$\mathbf{q} = \mathbf{p_0} + (\mathbf{u} - \mathbf{u_1}) = (0, 6, 0) + (3, -5, -1) - \tfrac{1}{3}(2, 1, -1) = \tfrac{1}{3}(7, 2, -2).$$

Hence $Q = Q\left(\frac{7}{3}, \frac{2}{3}, \frac{-2}{3}\right)$.

8(b) A normal to the plane is given by

$$\mathbf{n} = \overrightarrow{PQ} \times \overrightarrow{PR} = (-2, 2, -4) \times (-3, -1, -3) = \det \begin{bmatrix} \mathbf{i} & \mathbf{j} & \mathbf{k} \\ -2 & 2 & -4 \\ -3 & -1 & -3 \end{bmatrix} = (-10, 6, 8).$$

Thus, as $P(4, 0, 5)$ is in the plane, the equation is

$$-10(x - 4) + 6(y - 0) + 8(z - 5) = 0; \quad \text{that is } 5x - 3y - 4z = 0.$$

The plane contains the origin $P(0, 0, 0)$.

10(b) The coordinates of points of intersection satisfy both equations:

$$3x + y - 2z = 1$$
$$x + y + z = 5.$$

Solve

$$\begin{bmatrix} 3 & 1 & -2 & | & 1 \\ 1 & 1 & 1 & | & 5 \end{bmatrix} \rightarrow \begin{bmatrix} 1 & 1 & 1 & | & 5 \\ 0 & -2 & -5 & | & -14 \end{bmatrix} \rightarrow \begin{bmatrix} 1 & 0 & -\frac{3}{2} & | & -2 \\ 0 & 1 & \frac{5}{2} & | & 7 \end{bmatrix}.$$

Take $z = 2t$, to eliminate fractions, whence $x = -2 + 3t$ and $y = 7 - 5t$. Thus

$$(x, y, z) = (-2, +3t, 7 - 5t, 2t) = (-2, 7, 0) + t(3, -5, 2)$$

is the line of intersection.

11(b) If $P(x, y, z)$ is the intersection point, then $x = 1 + 2t$, $y = -2 + 5t$, $z = 3 - t$ since P is on the line. Substitution in the equation of the plane gives $2(1 + 2t) - (-2 + 5t) - (3 - t) = 5$, that is $1 = 5$. Thus there is no such t, so the line does not intersect the plane.

(d) If $P(x, y, z)$ is an intersection point, then $x = 1 + 2t$, $y = -2 + 5t$ and $z = 3 - t$ since P is on the line. Substitution in the equation of the plane gives $-1(1 + 2t) + 4(-2 + 5t) + 3(3 - t) = 6$, whence $t = \frac{2}{5}$. Thus $(x, y, z) = \left(\frac{9}{5}, 0, \frac{13}{6}\right)$ so $P\left(\frac{9}{5}, 0, \frac{13}{6}\right)$ is the point of intersection.

12(b) The line has direction vector $\mathbf{d} = (3, 0, 2)$ which is a normal to all such planes. If $P_0(x_0, y_0, z_0)$ is any point, the plane $3(x - x_0) = 0(y - y_0) + 2(z - z_0) = 0$ is perpendicular to the line. This can be written $3x + 2z = 3x_0 + 2z_0$, so $3x + 2z = d$, d arbitrary.

(d) If the normal is $\mathbf{n} = (a, b, c) \neq \mathbf{0}$, the plane is $a(x - 3) + b(y - 2) + c(z + 4) = 0$, where a, b and c are not all zero.

(f) The vector $\mathbf{u} = \overrightarrow{PQ} = (-1, 1, -1)$ is parallel to these planes so the normal $\mathbf{n} = (a, b, c)$ is orthogonal to $\mathbf{u}$. Thus $0 = \mathbf{u} \cdot \mathbf{n} = -a + b - c$. Hence $c = b - a$ and $\mathbf{n} = (a, b, b - a)$. The plane passes through $Q(1, 0, 0)$ so the equation is $a(x - 1) + b(y - 0) + (b - a)(z - 0) = 0$, that is $ax + by + (b - a)z = a$. Here a and b are not both zero (as $\mathbf{n} \neq \mathbf{0}$). As a check, observe that this plane contains $P(2, -1, 1)$.

(h) Such a plane contains $P_0(3, 0, 2)$ and its normal $\mathbf{n} = (a, b, c)$ must be orthogonal to the direction vector $\mathbf{d} = (1, 2, -1)$ of the line. Thus $0 = \mathbf{d} \cdot \mathbf{n} = a + 2b - c$, whence $c = a + 2b$ and $\mathbf{n} = (a, b, a + 2b)$ (where a and b are not both zero as $\mathbf{n} \neq \mathbf{0}$). Thus the equation is

$$a(x - 3) + b(y - 0) + (a + 2b)(z - 2) = 0, \quad \text{that is } ax + by + (a + 2b)z = 5a + 4b$$

where a and b are not both zero. As a check, observe that the plane contains every point $P(3 + t, 2t, 2 - t)$ on the line.

14(b) Choose $P_1(3, 0, 2)$ on the first line. The distance in question is the distance from P_1 to the second line. Choose $P_2(-1, 2, 2)$ on the second line and let $u = \overrightarrow{P_2 P_1} = (4, -2, 0)$.

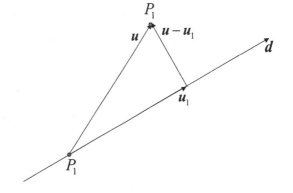

If $\mathbf{d} = (3, 1, 0)$ is the direction vector for the line, compute

$$\mathbf{u}_1 = \text{proj}_{\mathbf{d}}(\mathbf{u}) = \frac{\mathbf{u} \cdot \mathbf{d}}{\|\mathbf{d}\|^2} \mathbf{d} = \tfrac{10}{10}(3, 1, 0) = (3, 1, 0).$$

Then the required distance is

$$\|\mathbf{u} - \mathbf{u}_1\| = \|(1, -3, 0)\| = \sqrt{10}.$$

15(b) The cross product

$$\mathbf{n} = (1, 1, 1) \times (3, 1, 0) = (-1, 3, -2)$$

of the two direction vectors serves as a normal to the plane. Given $P_1(1, -1, 0)$ and $P_2(2, -1, 3)$ on the lines, let $\mathbf{u} = \overrightarrow{P_1 P_2} = (1, 0, 3)$. Compute

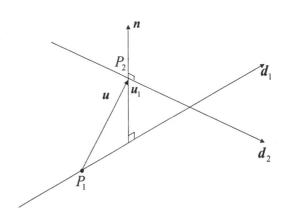

$$\begin{aligned} \mathbf{u}_1 &= \text{proj}_{\mathbf{n}}(\mathbf{u}) = \frac{\mathbf{u} \cdot \mathbf{n}}{\|\mathbf{n}\|^2} \mathbf{n} \\ &= \tfrac{-7}{14}(-1, 3, -2) = \tfrac{1}{2}(1, -3, 2). \end{aligned}$$

The required distance is $\|\mathbf{u}_1\| = \tfrac{1}{2}\sqrt{1 + 9 + 4} = \tfrac{1}{2}\sqrt{14}$.

Let $A = A(1 + s, -1 + s, s)$ and $B = B(2 + 3t, -1 + t, 3)$ be the points on the two lines that are closest together. Then $\overrightarrow{AB} = (1 + 3t - s, t - s, 3 - s)$ is orthogonal to the direction vectors $\mathbf{d}_1 = (1, 1, 1)$ and $\mathbf{d}_2 = (3, 1, 0)$. Then $\mathbf{d}_1 \cdot \overrightarrow{AB} = 0 = \mathbf{d}_2 \cdot \overrightarrow{AB}$, giving equations $4t - 3s = -4$, $10t - 4s = -3$. The solution is $t = \tfrac{1}{2}$, $s = 2$, giving $A = A(3, 1, 2)$ and $B = B\left(\tfrac{7}{2}, -\tfrac{1}{2}, 3\right)$.

(d) Analogous to (b). Answer: distance $= \frac{\sqrt{6}}{6}$, $A\left(\tfrac{19}{3}, 2, \tfrac{1}{3}\right)$, $B\left(\tfrac{37}{6}, \tfrac{13}{6}, 0\right)$.

16(b) The area of the triangle is $\tfrac{1}{2}$ the area of the parallelogram $ABCD$. By Theorem 4,

$$\begin{aligned} \text{Area of triangle} &= \tfrac{1}{2} \left\| \overrightarrow{AB} \times \overrightarrow{AC} \right\| = \tfrac{1}{2} \|(2, 1, -1) \times (4, 2, -2)\| \\ &= \tfrac{1}{2} \|(0, 0, 0)\| = 0. \end{aligned}$$

Hence $\overrightarrow{AB}$ and $\overrightarrow{AC}$ are parallel.

(d) Analogous to (b). Area $= \sqrt{5}$.

17(b) We have $\mathbf{u} \times \mathbf{v} = (-4, 5, -1)$ s $\mathbf{w} \cdot (\mathbf{u} \times \mathbf{v}) = -7$. By Theorem 5, the volume is $|\mathbf{w} \cdot (\mathbf{u} \times \mathbf{v})| = |-7| = 7$.

18(b) The line through P_0 perpendicular to the plane has direction vector $\mathbf{n}$, and so has vector equation $\mathbf{p} = \mathbf{p}_0 + t\mathbf{n}$ where $\mathbf{p} = (x, y, z)$. If $P(x, y, z)$ also lies in the plane, then $\mathbf{n} \cdot \mathbf{p} = ax + by + cz = d$. Using $\mathbf{p} = \mathbf{p}_0 + t\mathbf{n}$

$$d = \mathbf{n} \cdot \mathbf{p} = \mathbf{n} = \mathbf{n} \cdot \mathbf{p}_0 + t(\mathbf{n} \cdot \mathbf{n}) = \mathbf{n} \cdot \mathbf{p}_0 + t\|\mathbf{n}\|^2.$$

Hence $t = \frac{d - \mathbf{n} \cdot \mathbf{p}_0}{\|\mathbf{n}\|^2}$ so $\mathbf{p} = \mathbf{p}_0 + \left(\frac{d - \mathbf{n} \cdot \mathbf{p}_0}{\|\mathbf{n}\|^2} \right) \mathbf{n}$.
Finally the distance from P_0 to the plane is

$$\|\overrightarrow{PP_0}\| = \|\mathbf{p} - \mathbf{p}_0\| = \left\| \left(\frac{d - \mathbf{n} \cdot \mathbf{p}_0}{\|\mathbf{n}\|^2} \right) \mathbf{n} \right\| = \frac{|d - \mathbf{n} \cdot \mathbf{p}_0|}{\|\mathbf{n}\|}.$$

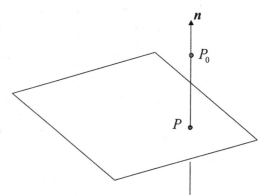

24. If $\mathbf{u}$ and $\mathbf{v}$ are perpendicular, Theorem 4 shows that $\|\mathbf{u} \times \mathbf{v}\| = \|\mathbf{u}\|\,\|\mathbf{v}\|$. Moreover, if $\mathbf{w}$ is perpendicular to both $\mathbf{u}$ and $\mathbf{v}$, it is parallel to $\mathbf{u} \times \mathbf{v}$ so $\mathbf{w} \cdot (\mathbf{u} \times \mathbf{v}) = \pm \|\mathbf{w}\|\,\|\mathbf{u} \times \mathbf{v}\|$ because the angle between them is either 0 or π. Finally, the rectangular parallelepiped has volume

$$|\mathbf{w} \cdot (\mathbf{u} \times \mathbf{v})| = \|\mathbf{w}\|\,\|\mathbf{u} \times \mathbf{v}\| = \|\mathbf{w}\|\,(\|\mathbf{u}\|\,\|\mathbf{v}\|)$$

using Theorem 5.

27(b) If $\mathbf{u} = (x, y, z)$, $\mathbf{v} = (p, q, r)$ and $\mathbf{w} = (l, m, n)$ then, by the row version of Exercise 19 §3.1, we get

$$\mathbf{u} \times (\mathbf{v} + \mathbf{w}) = \det \begin{bmatrix} \mathbf{i} & \mathbf{j} & \mathbf{k} \\ x & y & z \\ p+l & q+m & r+n \end{bmatrix}$$

$$= \det \begin{bmatrix} \mathbf{i} & \mathbf{j} & \mathbf{k} \\ x & y & z \\ p & q & r \end{bmatrix} + \det \begin{bmatrix} \mathbf{i} & \mathbf{j} & \mathbf{k} \\ x & y & z \\ l & m & n \end{bmatrix} = \mathbf{u} \times \mathbf{v} + \mathbf{u} \times \mathbf{w}.$$

28(b) Let $\mathbf{v} = (v_1, v_2, v_3)$, $\mathbf{w} = (w_1, w_2, w_3)$ and $\mathbf{u} = (u_1, u_2, u_3)$. Compute

$$\mathbf{v} \cdot [(\mathbf{u} \times \mathbf{v}) + (\mathbf{v} \times \mathbf{w}) + (\mathbf{w} \times \mathbf{u})] = \mathbf{v} \cdot (\mathbf{u} \times \mathbf{v}) + \mathbf{v} \cdot (\mathbf{v} \times \mathbf{w}) + \mathbf{v} \cdot (\mathbf{w} \times \mathbf{u})$$

$$= 0 + 0 + \det \begin{bmatrix} v_1 & v_2 & v_3 \\ w_1 & w_2 & w_3 \\ u_1 & u_2 & u_3 \end{bmatrix}$$

by Theorem 1. Similarly

$$\mathbf{w} \cdot [(\mathbf{u} \times \mathbf{v}) + (\mathbf{v} \times \mathbf{w}) + (\mathbf{w} \times \mathbf{u})] = \mathbf{w} \cdot (\mathbf{u} \times \mathbf{v}) = \det \begin{bmatrix} w_1 & w_2 & w_3 \\ u_1 & u_2 & u_3 \\ v_1 & v_2 & v_3 \end{bmatrix}.$$

These determinants are equal because each can be obtained from the other by two row interchanges. The result follows because $(\mathbf{v} - \mathbf{w}) \cdot \mathbf{x} = \mathbf{v} \cdot \mathbf{x} - \mathbf{w} \cdot \mathbf{x}$ for any vector $\mathbf{x}$.

34. Theorem 5 asserts that the volume of the parallelepiped determined by $\mathbf{u}$, $\mathbf{v}$, and $\mathbf{w}$ is

$$\left| \det \begin{bmatrix} \mathbf{u} \\ \mathbf{v} \\ \mathbf{w} \end{bmatrix} \right| \quad \text{where} \quad \begin{bmatrix} \mathbf{u} \\ \mathbf{v} \\ \mathbf{w} \end{bmatrix} \quad \text{is the matrix with these vectors as rows.} \quad \text{Since} \quad \begin{bmatrix} \mathbf{u}A \\ \mathbf{v}A \\ \mathbf{w}A \end{bmatrix} =$$

$$\begin{bmatrix} \mathbf{u} \\ \mathbf{v} \\ \mathbf{w} \end{bmatrix} A \text{ for any } 3 \times 3 \text{ matrix } A, \text{ the volume of the parallelepiped determined by } \mathbf{u}A, \mathbf{v}A,$$

and $\mathbf{w}A$ is

$$\left| \det \begin{bmatrix} \mathbf{u}A \\ \mathbf{v}A \\ \mathbf{w}A \end{bmatrix} \right| = \left| \det \begin{bmatrix} \mathbf{u} \\ \mathbf{v} \\ \mathbf{w} \end{bmatrix} A \right| = \left| \det \begin{bmatrix} \mathbf{u} \\ \mathbf{v} \\ \mathbf{w} \end{bmatrix} \det A \right| = \left| \det \begin{bmatrix} \mathbf{u} \\ \mathbf{v} \\ \mathbf{w} \end{bmatrix} \right| |\det A|$$

by the product rule for determinants (Theorem 1 §3.2).

Exercises 4.4   Least Squares Approximation

1(b) Here $M^T M = \begin{bmatrix} 1 & 1 & 1 & 1 \\ 2 & 4 & 7 & 8 \end{bmatrix} \begin{bmatrix} 1 & 2 \\ 1 & 4 \\ 1 & 7 \\ 1 & 8 \end{bmatrix} = \begin{bmatrix} 4 & 21 \\ 21 & 133 \end{bmatrix}$, $M^T Y = \begin{bmatrix} 1 & 1 & 1 & 1 \\ 2 & 4 & 7 & 8 \end{bmatrix} \begin{bmatrix} 4 \\ 3 \\ 2 \\ 1 \end{bmatrix} =$

$\begin{bmatrix} 10 \\ 42 \end{bmatrix}$. We solve the normal equation $(M^T M)A = M^T Y$ by inverting $M^T M$:

$$A = (M^T M)^{-1} M^T Y = \frac{1}{91} \begin{bmatrix} 133 & -21 \\ -21 & 4 \end{bmatrix} \begin{bmatrix} 10 \\ 42 \end{bmatrix} = \frac{1}{91} \cdot \begin{bmatrix} 448 \\ -42 \end{bmatrix} = \frac{1}{13} \begin{bmatrix} 64 \\ -6 \end{bmatrix}.$$

Hence the best fitting line has equation $y = \frac{64}{13} - \frac{6}{13}x$.

(d) Analogous to (b). The best fitting line is $y = -\frac{4}{10} - \frac{17}{10}x$.

2(b) Now $M^T M = \begin{bmatrix} 1 & 1 & 1 & 1 \\ -2 & 0 & 3 & 4 \\ 4 & 0 & 9 & 16 \end{bmatrix} \begin{bmatrix} 1 & -2 & 4 \\ 1 & 0 & 0 \\ 1 & 3 & 9 \\ 1 & 4 & 16 \end{bmatrix} = \begin{bmatrix} 4 & 5 & 29 \\ 5 & 29 & 83 \\ 29 & 83 & 353 \end{bmatrix}.$

$$M^T Y = \begin{bmatrix} 1 & 1 & 1 & 1 \\ -2 & 0 & 3 & 4 \\ 4 & 0 & 9 & 16 \end{bmatrix} \begin{bmatrix} 1 \\ 0 \\ 2 \\ 3 \end{bmatrix} = \begin{bmatrix} 6 \\ 16 \\ 70 \end{bmatrix}.$$

We use $(MM^T)^{-1}$ to solve the normal equations even though it is more efficient to solve them by Gaussian elimination.

$$A = (M^T M)^{-1} = \frac{1}{4248} \begin{bmatrix} 3348 & 642 & -426 \\ 642 & 571 & -187 \\ -426 & -187 & 91 \end{bmatrix} \begin{bmatrix} 6 \\ 16 \\ 70 \end{bmatrix} = \frac{1}{4248} \begin{bmatrix} 540 \\ -102 \\ 822 \end{bmatrix} = \begin{bmatrix} .127 \\ .024 \\ .194 \end{bmatrix}.$$

Hence the best fitting quadratic has equation $y = .127 - .024x + .194x^2$.

4. To fit $s = a + bx$ where $x = t^2$, we have

$$M^T M = \begin{bmatrix} 1 & 1 & 1 \\ 1 & 4 & 9 \end{bmatrix} \begin{bmatrix} 1 & 1 \\ 1 & 4 \\ 1 & 9 \end{bmatrix} = \begin{bmatrix} 3 & 14 \\ 14 & 98 \end{bmatrix}$$

$$M^T Y = \begin{bmatrix} 1 & 1 & 1 \\ 1 & 4 & 9 \end{bmatrix} \begin{bmatrix} 95 \\ 80 \\ 56 \end{bmatrix} = \begin{bmatrix} 231 \\ 919 \end{bmatrix}.$$

Hence $A = (M^T M)^{-1} M^T Y = \frac{1}{98} \begin{bmatrix} 98 & -14 \\ -14 & 3 \end{bmatrix} \begin{bmatrix} 231 \\ 919 \end{bmatrix} = \frac{1}{98} \begin{bmatrix} 9772 \\ -477 \end{bmatrix} = \begin{bmatrix} 99.71 \\ -4.87 \end{bmatrix}$ to two decimal places. Hence the best fitting equation is

$$y = 99.71 - 4.87x = 99.71 - 4.87t^2.$$

Hence the estimate for g comes from $-\frac{1}{2}g = -4.87$, $g = 9.74$ (the true value of g is 9.81). Now fit $s = a + bt + ct^2$. In this case

$$M^T M = \begin{bmatrix} 1 & 1 & 1 \\ 1 & 2 & 3 \\ 1 & 4 & 9 \end{bmatrix} \begin{bmatrix} 1 & 1 & 1 \\ 1 & 2 & 4 \\ 1 & 3 & 9 \end{bmatrix} = \begin{bmatrix} 3 & 6 & 14 \\ 6 & 14 & 36 \\ 14 & 36 & 98 \end{bmatrix}$$

$$M^T Y = \begin{bmatrix} 1 & 1 & 1 \\ 1 & 2 & 3 \\ 1 & 4 & 9 \end{bmatrix} \begin{bmatrix} 95 \\ 80 \\ 56 \end{bmatrix} = \begin{bmatrix} 231 \\ 423 \\ 919 \end{bmatrix}.$$

Hence

$$A = (M^T M)^{-1}(M^T Y) = \frac{1}{4} \begin{bmatrix} 76 & -84 & 20 \\ -84 & 98 & -24 \\ 20 & -24 & 6 \end{bmatrix} \begin{bmatrix} 231 \\ 423 \\ 919 \end{bmatrix} = \frac{1}{4} \begin{bmatrix} 404 \\ -6 \\ -18 \end{bmatrix} = \begin{bmatrix} 101 \\ -\frac{3}{2} \\ -\frac{9}{2} \end{bmatrix}$$

so the best quadratic is $y = 101 - \frac{3}{2}t - \frac{9}{2}t^2$. This gives $-\frac{9}{2} = -\frac{1}{2}g$ so the estimate for g is $g = 9$ in this case.

6(b) $f(x) = a_0$ here so the sum of squares is

$$s = (y_1 - a_0)^2 + (y_2 - a_0)^2 + \cdots + (y_n - a_0)^2$$

$$= \sum_{i=1}^{n}(y_i - a_0)^2$$

$$= \sum_{i=1}^{n}(a_0^2 - 2a_0 y_i + y_i^2)$$

$$= na_0^2 - \left(2\sum y_i\right)a_0 + \left(\sum y_i^2\right).$$

— a quadratic in a_0. Completing the square gives

$$s = n\left[a_0 - \frac{1}{n}\sum y_i\right]^2 - \left[\sum y_i^2 - \frac{1}{n}\left(\sum y_i\right)^2\right].$$

This is minimal when $a_0 = \frac{1}{n}\sum y_i$.

Supplementary Exercises   Chapter 4

4. Let $\mathbf{p}$ and $\mathbf{w}$ be the velocities of the airplane and the wind. Then $\|\mathbf{p}\| = 100$ knots and $\|\mathbf{w}\| = 75$ knots and the resulting actual velocity of the airplane is $\mathbf{v} = \mathbf{w} + \mathbf{p}$. Since $\mathbf{w}$ and $\mathbf{p}$ are orthogonal. Pythagoras' theorem gives $\|\mathbf{v}\|^2 = \|\mathbf{w}\|^2 + \|\mathbf{p}\|^2 = 75^2 + 100^2 = 25^2(3^2 + 4^2) = 25^2 \cdot 5^2$. Hence $\|\mathbf{v}\| = 25 \cdot 5 = 125$ knots. The angle θ satisfies $\cos\theta = \frac{\|\mathbf{w}\|}{\|\mathbf{v}\|} = \frac{75}{125} = 0.6$ so $\theta = 0.93$ radians or $53°$.

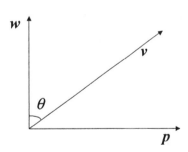

6. Let $\mathbf{v} = (x, y)$ denote the velocity of the boat in the water. If $\mathbf{c}$ is the current velocity then $\mathbf{c} = (0, -5)$ because it flows south at 5 knots. We want to choose $\mathbf{v}$ so that the resulting actual velocity $\mathbf{w}$ of the boat has easterly direction. Thus $\mathbf{w} = (z, 0)$ for some z. Now $\mathbf{w} = \mathbf{v} + \mathbf{c}$ so $(z, 0) = (x, y) + (0, -5) = (x, y - 5)$. Hence $z = x$ and $y = 5$. Finally, $13 = \|\mathbf{v}\| = \sqrt{x^2 + y^2} = \sqrt{x^2 + 25}$ gives $x^2 = 144$, $x = \pm 12$. But $x > 0$ as w heads *east*, so $x = 12$. Thus he steers $\mathbf{v} = (12, 5)$, and the resulting actual speed is $\|\mathbf{w}\| = z = 12$ knots.

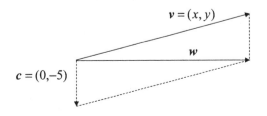

Chapter 5: The Vector Spaces $\mathbb{R}^n$

Exercises 5.1   Subspaces and Dimension

1(b) Yes. $U = \text{span}\left\{[0 \quad 1 \quad 0]^T, [0 \quad 0 \quad 1]\right\}$

(d) No. $[2 \quad 0 \quad 0]^T$ is in U but $2[2 \quad 0 \quad 0]^T = [4 \quad 0 \quad 0]^T$ is not in U.

(f) No. $[0 \quad -1 \quad 0]^T$ is in U but $(-1)[0 \quad -1 \quad 0]^T = [0 \quad 1 \quad 0]$ is not in U.

2(b) No. If $X = aT\_4Z$ equating first and third components given $a = 3$, $b = -5$. This does not satisfy the second component.

(d) Yes. $X = 3Y + 4Z$.

3(b) No. The matrix with these vectors as columns has determinant 0, so they cannot span $\mathbb{R}^4$ by Theorem 3.

5(b) Yes. The matrix with these as columns has determinant $-2 \neq 0$.

(d) No. $X_1 - X_2 + X_3 - X_4 = 0$.

6(b) Yes. If $a(X + Y) + b(Y + Z) + c(Z + X) = 0$ then $(a + c)X + (a + b)Y + (b + c)Z = 0$ so $a + c = 0$, $a + b = 0$, $b + c = 0$ by hypothesis. So $a = b = c = 0$.

(d) No. $(X + Y) - (Y + Z) + (Z + W) - (W + X) = 0$.

7(b) $X_3 = 3X_1 + 4X_2$ so the space is $\text{span}\{X_1, X_2\}$. This is a basis so the dimension is 2.

(d) $X_3 = 3X_1 + 4X_2$ and $X_4 = X_1 + X_2$ so the space is $\text{span}[X_1, X_2]$. This is a basis so the dimension is 2.

8(b) $[a + b \quad a - b \quad b \quad a]^T = a[1 \quad 1 \quad 0 \quad 1]^T + b[1 \quad -1 \quad 1 \quad 0]^T$ so $U = \text{span}\left\{[1 \quad 1 \quad 0 \quad 1]^T, [1 \quad -1 \quad 1 \quad 0]^T\right\}$. This is a basis so $\dim U = 2$.

(d) $[a - b \quad b + c \quad a \quad b + c]^T = a[1 \quad 0 \quad 1 \quad 0]^T + b[-1 \quad 1 \quad 0 \quad 1]^T + c[0 \quad 1 \quad 0 \quad 1]^T$. Hence $U = \text{span}\left\{[1 \quad 0 \quad 1 \quad 0]^T, [-1 \quad 1 \quad 0 \quad 1]^T, [0 \quad 1 \quad 0 \quad 1]^T\right\}$. This is a basis so $\dim U = 3$.

(f) $U = \{[-b + c - d \quad b \quad c \quad d]^T \mid b, c, d \text{ in } \mathbb{R}\}$ so $U = \text{span}\left\{[-1 \quad 1 \quad 0 \quad 0]^T, [1 \quad 0 \quad 1 \quad 0]^T, [-1 \quad 0 \quad 0 \quad 1]^T\right\}$. This is a basis so $\dim U = 3$.

9(b) Let $a(X + W) + b(Y + W) + c(Z + W) + dW = 0$, that is $aX + bY + cZ + (a + b + c + d)W = 0$. As $\{X, Y, Z, W\}$ is independent, $a = 0$, $b = 0$, $c = 0$ and $a + b + c + d = 0$. (So $d = 0$.)

10(b) The system is $A^T X^T = 0$ and the general solution is $X^T = [2s - 7t \quad -2s + t \quad s \quad t]$, so the basic solutions are $\{[-2 \quad -2 \quad 1 \quad 0], [-2 \quad 1 \quad 0 \quad 1]\}$. This is a basis so $\dim U = 2$.

12. $\text{span}\{0\} = \{t0 \mid t \text{ in } \mathbb{R}\} = \{0\}$ is the zero subspace.

14. No. $\mathbb{R}^2$ consists of ordered pairs and $\mathbb{R}^3$ consists of ordered tuples, so $\mathbb{R}^2$ is not contained in $\mathbb{R}^3$.

16. If Y is in U write $Y = t_1X_1 + \cdots + t_kX_k, t_i$ in $\mathbb{R}$. Then $AY = t_1AX_1 + \cdots + t_1AX_k = t_10 + \cdots + t_k0 = 0$.

18. If $rX_2 + sX_3 + tX_5 = 0$ then $0X_1 + rX_2 + sX_3 + 0X_4 + tX_5 + 0X_6 = 0$. Since the larger set is independent, this implies $r = s = t = a$.

20. If $t_1X_1 + t_2(X_1+X_2) + \cdots + t_k(X_1+X_2+\cdots+X_k) = 0$ then, collecting terms in $X_1, X_2, \ldots$,
$$(t_1 + t_2 + \cdots + t_k)X_1 + (t_2 + \cdots + t_k)X_2 + \cdots + (t_{k-1} + t_k)X_{k-1} + t_kX_k = 0.$$

Since $[X_1, X_2, \ldots, X_k\}$ is independent we get
$$t_1 + t_2 + \cdots + t_k = 0$$
$$t_2 + \cdots + t_k = 0$$
$$\vdots$$
$$t_{k-1} + t_k = 0$$
$$t_k = 0.$$

The solution (from the bottom up) is $t_k = 0, t_{k-1} = 0, \ldots, t_2 = 0, t_1 = 0$.

22. Let $tY + t_1X_1 + \cdots + t_kX_k = 0$. We claim first that $t = 0$. For if $t \neq 0$ then $Y = (-t^{-1}t_1)X_1 + (-t^{-1}t_2)X_2 + \cdots + (-t^{-1}t_k)X_k$ is in $\text{span}\{X_1, X_2, \ldots, X_k\}$, contrary to our hypothesis. So $t = 0$ and we get $t_1X_1 + \cdots + t_kX_k = 0$. Thus each $t_i = 0$ because $\{X_1, \ldots, X_k\}$ is independent.

23(b) We show that A^T is invertible. Suppose $A^TX = 0$, X in $\mathbb{R}^2$. By Theorem 5 §2.3, we must show that $X = 0$. If $X = [s \quad t]^T$ then $A^TX = 0$ gives $as + ct = 0$, $bs + dt = 0$. But then $s[aX + bY) + t(cX + dY) = (sa + tc)X + (sb + td)Y = 0$. Hence $s = t = 0$ because $\{aX + bY, cX + dY\}$ is independent.

24(b) We have $Y = (-1)X + 1(X + Y)$, so Y is in U because X and $X + Y$ are in U.

26(b) Note first that each $V^{-1}X_i$ is in $\text{null}(AV)$ because $(AV)(V^{-1}X_i) = AX_i = 0$. If $t_1V^{-1}X_1 + \cdots + t_kV^{-1}X_k = 0$ then $V^{-1}(t_1X_1 + \cdots + t_kX_k) = 0$ so $t_1X_1 + \cdots + t_kX_k = 0$ (by multiplication by V). Thus $t_1 = \cdots = t_k = 0$ because $\{X_1, \ldots, X_k\}$ is independent. So $\{V^{-1}X_1, \ldots, V^{-1}X_k\}$ is independent. To see that it spans $\text{null}(AV)$, let Y be in $\text{null}(AV)$, so that $AVY = 0$. Then VY is in $\text{null} A$ so $VY = s_1X_1 + \cdots + s_nX_n$ because $\{X_1, \ldots, X_n\}$ spans $\text{null} A$. Hence $Y = s_1V^{-1}X_1 + \cdots + s_kV^{-1}X_k$, as required.

28. We have $U \subseteq W \subseteq \mathbb{R}^n$, $\dim U = n - 1$, and $\dim \mathbb{R}^n = n$. So $\dim W = n - 1$ or $\dim W = n$ by Theorem 5(2). But then $W = U$ or $W = \mathbb{R}^n$ by Theorem 5(3).

Exercises 5.2   Rank of a Matrix

1(b) $\begin{bmatrix} 2 & -1 & 1 \\ -2 & 1 & 1 \\ 4 & -2 & 3 \\ -6 & 3 & 0 \end{bmatrix} \rightarrow \begin{bmatrix} 2 & -1 & 1 \\ 0 & 0 & 2 \\ 0 & 0 & 1 \\ 0 & 0 & 3 \end{bmatrix} \rightarrow \begin{bmatrix} 1 & -\frac{1}{2} & \frac{1}{2} \\ 0 & 0 & 1 \\ 0 & 0 & 0 \\ 0 & 0 & 0 \end{bmatrix}$

Hence, rank $A = 2$ and $\{(1, -\frac{1}{2}, \frac{1}{2}), (0, 0, 1)\}$ is a basis of row A. Thus $\{(2, -1, 1)(0, 0, 1)\}$ is also a basis of row A. Since the leading 1's are in columns 1 and 3, columns 1 and 3 of A are a basis of col A.

(d) $\begin{bmatrix} 1 & 2 & -1 & 3 \\ -3 & -6 & 3 & -2 \end{bmatrix} \rightarrow \begin{bmatrix} 1 & 2 & -1 & 3 \\ 0 & 0 & 0 & 7 \end{bmatrix} \rightarrow \begin{bmatrix} 1 & 2 & -1 & 3 \\ 0 & 0 & 0 & 1 \end{bmatrix}$

Hence, rank $A = 2$ and $\{(1, 2, -1, 3), (0, 0, 0, 1)\}$ is a basis of row A. Since the leading 1's are in columns 1 and 4, columns 1 and 4 of A are a basis of col A.

2(b) Apply the Gaussian Algorithm to the matrix with these vectors as rows:

$\begin{bmatrix} 1 & -1 & 2 & 5 & 1 \\ 3 & 1 & 4 & 2 & 7 \\ 1 & 1 & 0 & 0 & 0 \\ 5 & 1 & 6 & 7 & 8 \end{bmatrix} \rightarrow \begin{bmatrix} 1 & 1 & 0 & 0 & 0 \\ 0 & -2 & 2 & 5 & 1 \\ 0 & -2 & 4 & 2 & 7 \\ 0 & -4 & 6 & 7 & 8 \end{bmatrix} \rightarrow \begin{bmatrix} 1 & 1 & 0 & 0 & 0 \\ 0 & 1 & -1 & -\frac{5}{2} & -\frac{1}{2} \\ 0 & 0 & 1 & -\frac{3}{2} & 3 \\ 0 & 0 & 0 & 0 & 0 \end{bmatrix}$

Hence, $\{(1, 1, 0, 0, 0), (0, 2, -2, -5, -1), (0, 0, 2, -3, 6)\}$ is a basis of U (where we have cleared fractions using scalar multiples).

(d) Write these columns as rows:

$\begin{bmatrix} 1 & 5 & -6 \\ 2 & 6 & -8 \\ 3 & 7 & -10 \\ 4 & 8 & 12 \end{bmatrix} \rightarrow \begin{bmatrix} 1 & 5 & -6 \\ 0 & 1 & -1 \\ 0 & 0 & 24 \\ 0 & 0 & 0 \end{bmatrix} \rightarrow \begin{bmatrix} 1 & 5 & -6 \\ 0 & 1 & -1 \\ 0 & 0 & 24 \\ 0 & 0 & 0 \end{bmatrix} \rightarrow \begin{bmatrix} 1 & 5 & -6 \\ 0 & 1 & -1 \\ 0 & 0 & 1 \\ 0 & 0 & 0 \end{bmatrix}$

Hence, $\left\{ \begin{bmatrix} 1 \\ 5 \\ -6 \end{bmatrix}, \begin{bmatrix} 0 \\ 1 \\ -1 \end{bmatrix}, \begin{bmatrix} 0 \\ 0 \\ 1 \end{bmatrix} \right\}$ is a basis of U.

3(b) No. If the 3 columns were independent, the rank would be 3.
No. If the 4 rows were independent, the rank would be 4, a contradiction here as the rank cannot exceed the number of columns.

(d) No. If the rows (or columns) were independent, the rank equals the number of rows (or columns). Thus the number of rows would equal the number of columns; that is A would be square.

(f) If both the rows and the columns were independent, the matrix would be square by (d).

4(b) Theorem 2 gives $\text{col}(AV) \subseteq \text{col}(A)$, so $\dim[\text{col}(AV)] \leq \dim[\text{col}(A)]$; that is rank $AV \leq$ rank A.

7(b) $\left[\begin{array}{ccccc|c} 3 & 5 & 5 & 2 & 0 & 0 \\ 1 & 0 & 2 & 2 & 1 & 0 \\ 1 & 1 & 1 & -2 & -2 & 0 \\ -2 & 0 & -4 & -4 & -2 & 0 \end{array} \right] \rightarrow \left[\begin{array}{ccccc|c} 1 & 0 & 2 & 2 & 1 & 0 \\ 0 & 5 & -1 & -4 & -3 & 0 \\ 0 & 1 & -1 & -4 & -3 & 0 \\ 0 & 0 & 0 & 0 & 0 & 0 \end{array} \right]$

$\rightarrow \left[\begin{array}{ccccc|c} 1 & 0 & 2 & 2 & 1 & 0 \\ 0 & 1 & -1 & -4 & -3 & 0 \\ 0 & 0 & 4 & 16 & 12 & 0 \\ 0 & 0 & 0 & 0 & 0 & 0 \end{array} \right] \rightarrow \left[\begin{array}{ccccc|c} 1 & 0 & 0 & -6 & -5 & 0 \\ 0 & 1 & 0 & 0 & 0 & 0 \\ 0 & 0 & 1 & 4 & 3 & 0 \\ 0 & 0 & 0 & 0 & 0 & 0 \end{array} \right]$

Hence, the set of solutions is null $A = \left\{ \begin{bmatrix} 6s+5t \\ 0 \\ -4s+3t \\ s \\ t \end{bmatrix} \mid s,t \text{ in } \mathbb{R} \right\} = \text{span } B$ where

$B = \left\{ \begin{bmatrix} 6 \\ 0 \\ -4 \\ 1 \\ 0 \end{bmatrix}, \begin{bmatrix} 5 \\ 0 \\ 3 \\ 0 \\ 1 \end{bmatrix} \right\}$. Since B is independent, it is the required basis of null A. We have $r = \text{rank } A = 3$ by the above reduction, so $n - r = 5 - 3 = 2$. This is the dimension of null A, as Theorem 3 asserts.

8(b) The r columns containing leading 1's are independent because the leading 1's are in different rows. If R is $m \times n$, col R is contained in the subspace of all columns in $\mathbb{R}^m$ with the last $m - r$ entries zero. This space has dimension r so the r (independent) columns containing leading 1's are a basis.

9(b) $[A \mid I] = \begin{bmatrix} 3 & 2 & | & 1 & 0 \\ 2 & 1 & | & 0 & 1 \end{bmatrix} \rightarrow \begin{bmatrix} 1 & 1 & | & 1 & -1 \\ 2 & 1 & | & 0 & 1 \end{bmatrix} \rightarrow \begin{bmatrix} 1 & 1 & | & 1 & -1 \\ 0 & -1 & | & -2 & 3 \end{bmatrix}$

$\rightarrow \begin{bmatrix} 1 & 0 & | & -1 & 2 \\ 0 & 1 & | & 2 & -3 \end{bmatrix}$ so $U = \begin{bmatrix} -1 & 2 \\ 2 & -3 \end{bmatrix}$. Hence, $UA = R = I_2$ in this case so $U = A^{-1}$.

Thus, $r = \text{rank } A = 2$ and, taking $V = I_2$, $\tilde{U}AV = UA = I_2$.

(d) $[A \mid I] = \begin{bmatrix} 1 & 1 & 0 & -1 & | & 1 & 0 & 0 \\ 3 & 2 & 1 & 1 & | & 0 & 1 & 0 \\ 1 & 0 & 1 & 3 & | & 0 & 0 & 1 \end{bmatrix} \rightarrow \begin{bmatrix} 1 & 1 & 0 & -1 & | & 1 & 0 & 0 \\ 0 & -1 & 1 & 4 & | & -3 & 1 & 0 \\ 0 & -1 & 1 & 4 & | & -1 & 0 & 1 \end{bmatrix} \rightarrow$

$\begin{bmatrix} 1 & 0 & 1 & 3 & | & -2 & 1 & 0 \\ 0 & 1 & -1 & -4 & | & 3 & -1 & 0 \\ 0 & 0 & 0 & 0 & | & 2 & -1 & 1 \end{bmatrix}$. Hence, $UA = R$ where $U = \begin{bmatrix} -2 & 1 & 0 \\ 3 & -1 & 0 \\ 2 & -1 & 1 \end{bmatrix}$ and $R =$

$\begin{bmatrix} 1 & 0 & 1 & 3 \\ 0 & 1 & -1 & -4 \\ 0 & 0 & 0 & 0 \end{bmatrix}$. Note that rank $A = 2$. Next, $[R^T \mid I] = \begin{bmatrix} 1 & 0 & 0 & | & 1 & 0 & 0 & 0 \\ 0 & 1 & 0 & | & 0 & 1 & 0 & 0 \\ 1 & -1 & 0 & | & 0 & 0 & 1 & 0 \\ 3 & -4 & 0 & | & 0 & 0 & 0 & 1 \end{bmatrix} \rightarrow$

$\begin{bmatrix} 1 & 0 & 0 & | & 1 & 0 & 0 & 0 \\ 0 & 1 & 0 & | & 0 & 1 & 0 & 0 \\ 0 & 0 & 0 & | & -1 & 1 & 1 & 0 \\ 0 & 0 & 0 & | & -3 & 4 & 0 & 1 \end{bmatrix}$ so $V^T = \begin{bmatrix} 1 & 0 & 0 & 0 \\ 0 & 1 & 0 & 0 \\ -1 & 1 & 1 & 0 \\ -3 & 4 & 0 & 1 \end{bmatrix}$.

Hence, $(UAV)^T = (RV)^T = V^T R^T = \begin{bmatrix} 1 & 0 & 0 \\ 0 & 1 & 0 \\ 0 & 0 & 0 \\ 0 & 0 & 0 \end{bmatrix}$, so $UAV = \begin{bmatrix} 1 & 0 & 0 & 0 \\ 0 & 1 & 0 & 0 \\ 0 & 0 & 0 & 0 \end{bmatrix}$.

10(b) Since A is $m \times n$, dim(null A) $= n-$ rank $A = n - 1$ using (a).

12(b) Write $r = $ rank A. Then, dim(col A) $= r$ and dim(null A) $= n - r$. As col $A \subseteq$ null A by (a), this shows that $r \leq n - r$; that is $r \leq \frac{n}{2}$.

14. Let $C_1, \ldots, C_n$ denote the columns of A. If $X = [x_1 \ \ x_2 \ \ \ldots \ \ x_n]^T$ then

$$AX = [C_1 \ \ \ldots \ \ C_n] \begin{bmatrix} x_1 \\ \vdots \\ x_n \end{bmatrix} = x_1 C_1 + \cdots + x_n C_n.$$

Thus, $\{AX \mid X \text{ in } \mathbb{R}^n\} \subseteq \text{span}\{C_1, \ldots, C_n\} = \text{col } A$. Conversely, every element of span$\{C_1, \ldots, C_n\}$ is a linear combination $x_1 C_1 + \cdots + x_n C_n$, and so equals AX for some X in $\mathbb{R}^n$.

17(b) Let $\{U_1, \ldots, U_r\}$ be a basis of col A where $r = $ rank A. Since $AX = B$ has no solution, B is not in col A by Exercise 14, so $\{U_1, \ldots, U_r, B\}$ is independent. Hence, it suffices to show that col$[A, B] = \text{span}\{U_1, \ldots, U_r, B\}$. It is clear that B is in col$[A, B]$, and each U_j is in col$[A, B]$ because it is a linear combination of columns of A (and so of $[A, B]$). Hence

$$\text{span}\{U_1, \ldots, U_r, B\} \subseteq \text{col}[A, B].$$

On the other hand, each column X in col$[A, B]$ is a linear combination of B and the columns C_j of A. These C_j are themselves linear combinations of the U_j, so X is a linear combination of B and the U_j. That is, X is in span$\{U_1, \ldots, U_r, B\}$.

19. If $BA = I_n$ for some $n \times m$ matrix B, we show that rank $A = n$ by verifying that $AX = 0$, X in $\mathbb{R}^n$, implies $X = 0$ (Theorem 4). But $AX = 0$ implies that $X = IX = BAX = BO = 0$. Conversely, if rank $A = n$ then $A^T A$ is invertible (Theorem 4 again) so $\left[(A^T A)^{-1} A^T\right] A = I_n$. Hence take $B = (A^T A)^{-1} A^T$.

23(b) Let $A = \begin{bmatrix} 3 & 2 \\ 2 & 1 \end{bmatrix}$. Then $UAV = I_2$ where $U = A^{-1} = \begin{bmatrix} -1 & 2 \\ 2 & -3 \end{bmatrix}$ and $V = I_2$. Since rank $A = 2$ here, take $P = U^{-1}$, $Q = V = I_2$.

Now consider $A = \begin{bmatrix} 1 & 1 & 0 & -1 \\ 3 & 2 & 1 & 1 \\ 1 & 0 & 1 & 3 \end{bmatrix}$. By Exercise 9(d), we have $UAV = \begin{bmatrix} 1 & 0 & 0 & 0 \\ 0 & 1 & 0 & 0 \\ 0 & 0 & 0 & 0 \end{bmatrix}$

where $U = \begin{bmatrix} -2 & 1 & 0 \\ 3 & -1 & 0 \\ \hline 2 & -1 & 1 \end{bmatrix}$ and $V = \begin{bmatrix} 1 & 0 & -1 & -3 \\ 0 & 1 & 1 & 4 \\ \hline 0 & 0 & 1 & 0 \\ 0 & 0 & 0 & 1 \end{bmatrix}$.

Then $U^{-1} = \begin{bmatrix} 1 & 1 & 0 \\ 3 & 2 & 0 \\ \hline 1 & 0 & 1 \end{bmatrix} = \begin{bmatrix} U_1 & U_2 \\ U_3 & U_4 \end{bmatrix}$ and $V^{-1} = \begin{bmatrix} 1 & 0 & 1 & 3 \\ 0 & 1 & -1 & -4 \\ \hline 0 & 0 & 1 & 0 \\ 0 & 0 & 0 & 1 \end{bmatrix} = \begin{bmatrix} V_1 & V_2 \\ V_3 & V_4 \end{bmatrix}$

as in the hint. Then block multiplication gives

$$A = U^{-1} \begin{bmatrix} I_2 & 0 \\ 0 & 0 \end{bmatrix} V^{-1} = \begin{bmatrix} U_1 & U_2 \\ U_3 & U_4 \end{bmatrix} \begin{bmatrix} I_2 & 0 \\ 0 & 0 \end{bmatrix} \begin{bmatrix} V_1 & V_2 \\ V_3 & V_3 \end{bmatrix}$$

$$= \begin{bmatrix} U_1 V_1 & U_1 V_2 \\ U_3 V_1 & U_3 V_2 \end{bmatrix} = \begin{bmatrix} U_1 \\ U_3 \end{bmatrix} [V_1 \quad V_2]. \text{ Thus } P = \begin{bmatrix} U_1 \\ U_2 \end{bmatrix} = \begin{bmatrix} 1 & 1 \\ 3 & 2 \\ 1 & 0 \end{bmatrix} \text{ and } Q = [V_1 \quad V_2] =$$

$$\begin{bmatrix} 1 & 0 & 1 & 3 \\ 0 & 1 & -1 & -4 \end{bmatrix} \text{ satisfy our requirements.}$$

Exercises 5.3   Similarity and Diagonalization

1(b) $\det A = -5$, $\det B = -1$ (so A and B are not similar). However, $\operatorname{tr} A = 2 = \operatorname{tr} B$, and rank $A = 2 = \operatorname{rank} B$ (both are invertible).

(d) tr $A = 5$, tr $B = 4$ (so A and B are not similar). However, $\det A = 7 = \det B$, so rank $A = 2 = \operatorname{rank} B$ (both are invertible).

(f) tr $A = -5 = \operatorname{tr} B$; $\det A = 0 = \det B$; however rank $A = 2$, rank $B = 1$ (so A and B are not similar).

3(b) We have $A \sim B$, say $B = P^{-1}AP$. Hence $B^{-1} = (P^{-1}AP)^{-1} = P^{-1}A^{-1}(P^{-1})^{-1}$, so $A^{-1} \sim B^{-1}$ because P^{-1} is invertible.

4(b) $c_A(x) = \begin{vmatrix} x-3 & 0 & -6 \\ 0 & x+3 & 0 \\ -5 & 0 & x-2 \end{vmatrix} = (x+3)(x^2 - 5x - 24) = (x+3)^2(x-8)$; eigenvalues $\lambda_1 = -3$, $\lambda_2 = 8$.

$\lambda_1 = -3$: $\begin{bmatrix} -6 & 0 & -6 \\ 0 & 0 & 0 \\ -5 & 0 & -5 \end{bmatrix} \rightarrow \begin{bmatrix} 1 & 0 & 1 \\ 0 & 0 & 0 \\ 0 & 0 & 0 \end{bmatrix}$; basic eigenvectors $\begin{bmatrix} -1 \\ 0 \\ 1 \end{bmatrix}$, $\begin{bmatrix} 0 \\ 1 \\ 0 \end{bmatrix}$.

$\lambda_2 = 8$: $\begin{bmatrix} 5 & 0 & -6 \\ 0 & 11 & 0 \\ -5 & 0 & 6 \end{bmatrix} \rightarrow \begin{bmatrix} 1 & 0 & -\frac{6}{5} \\ 0 & 1 & 0 \\ 0 & 0 & 0 \end{bmatrix}$; basic eigenvector $\begin{bmatrix} 6 \\ 0 \\ 5 \end{bmatrix}$.

Since $\left\{ \begin{bmatrix} -1 \\ 0 \\ 1 \end{bmatrix}, \begin{bmatrix} 0 \\ 1 \\ 0 \end{bmatrix}, \begin{bmatrix} 6 \\ 0 \\ 5 \end{bmatrix} \right\}$ is a basis of eigenvectors, A is diagonalizable and

$$P = \begin{bmatrix} -1 & 0 & 6 \\ 0 & 1 & 0 \\ 1 & 0 & 5 \end{bmatrix} \text{ will satisfy } P^{-1}AF = \begin{bmatrix} -3 & 0 & 0 \\ 0 & -3 & 0 \\ 0 & 0 & 8 \end{bmatrix}.$$

(d) $c_A(x) = \begin{vmatrix} x-4 & 0 & 0 \\ 0 & x-2 & -2 \\ 2 & -3 & x-1 \end{vmatrix} = (x-4)^2(x+1).$ For $\lambda = 4$ $\begin{bmatrix} 0 & 0 & 0 \\ 0 & 2 & -2 \\ 2 & -3 & 3 \end{bmatrix} \rightarrow \begin{bmatrix} 1 & 0 & 0 \\ 0 & 1 & -1 \\ 0 & 0 & 0 \end{bmatrix};$

$E_1 = \begin{bmatrix} 0 \\ 1 \\ 1 \end{bmatrix}$. Hence A is not diagonalizable.

8(b) If $B = P^{-1}AF$ and $A^k = 0$ then $B^k = (P^{-1}AF)^k = P^{-1}A^kP = P^{-1}OF = 0$.

9(b) Let the diagonal entries of A all equal λ. If A is diagonalizable then $P^{-1}AF = \lambda I$ by Theorem 3 for some invertible matrix P. Hence $A = P(I)P^{-1} = \lambda(PIF^{-1}) = \lambda I$.

10(b) Let $P^{-1}AF = D = \text{diag}\{\lambda_1, \lambda_2, \ldots, \lambda_n\}$. Since A and D are similar matrices, they have the same trace by Theorem 1. That is

$$\text{tr } A = \text{ tr } (P^{-1}AF) = \text{ tr } D = \lambda_1 + \lambda_2 + \cdots + \lambda_n.$$

12(b) $T_P(A)T_P(B) = (P^{-1}AF)(P^{-1}BF) = P^{-1}AIBF = P^{-1}ABF = T_P(AB)$.

13(b) A and A^T have the same eigenvalues (Example 5 §3.3). If A is diagonalization, say $A \sim D$ where D is diagonal, then $A^T \sim D^T$ is also diagonal. So (a) applies.

16. Let $P^{-1}AF = D$ be diagonal and write $E = P^{-1}CF$. It suffices to show that E is diagonal. Since $CA = AC$, we get

$$ED = (P^{-1}CF)(P^{-1}AF) = P^{-1}CAF = P^{-1}ACF = (P^{-1}AF)(P^{-1}CF) = DE.$$

If $E = [e_{ij}]$, we show that $e_{ij} = 0$ whenever $i \neq j$. Write $D = \text{diag}\{\lambda_1, \lambda_2, \ldots, \lambda_n\}$. Observe:

the (i,j)-entry ED is $\lambda_j e_{ij}$

the (i,j)-entry DE is $\lambda_i e_{ij}$

These are equal (as $ED = DE$) so $(\lambda_i - \lambda_j)e_{ij} = 0$. As $\lambda_i \neq \lambda_j$ when $i \neq j$, it follows that $e_{ij} = 0$ when $i \neq j$.

17(b) We use Theorem 7. The characteristic polynomial of B is computed by first adding rows 2 and 3 to row 1. For convenience, write $s = a + b + c$, $k = a^2 + b^2 + c^2 - (ab + ac + bc)$.

$$c_B(x) = \begin{vmatrix} x-c & -a & -b \\ -a & x-b & -c \\ -b & -c & x-a \end{vmatrix} = \begin{vmatrix} x-s & x-s & x-s \\ -a & x-b & -c \\ -b & -c & x-a \end{vmatrix} = \begin{vmatrix} x-s & 0 & 0 \\ -a & x+(a-b) & a-c \\ -b & b-c & x-(a-b) \end{vmatrix}$$

$$= (x-s)\left[x^2 - (a-b)^2 - (a-c)(b-c)\right]$$
$$= (x-s)(x^2 - k).$$

Hence, the eigenvalues of B are s, $\sqrt{k}$ and $-\sqrt{k}$. These must be real by Theorem 7, so $k \geq 0$. Thus $a^2 + b^2 + c^2 \geq ab + ac + bc$.

20(b) To compute $c_A(x) = \det(xI - A)$, add x times column 2 to column 1, and expand along row 1. The result is another matrix of the same type, but one size smaller. So repeat the procedure. It leads to the given expression.

Exercises 5.4 Linear Transformations

1(b) $\frac{1}{5} \begin{bmatrix} -3 & 4 \\ 4 & 3 \end{bmatrix}$ by Example 4 with $m = 2$.

(d) $\begin{bmatrix} 0 & 1 \\ -1 & 0 \end{bmatrix}$ by Example 9 with $\theta = -\frac{\pi}{2}$.

(f) $\frac{1}{2} \begin{bmatrix} 1 & -1 \\ -1 & 1 \end{bmatrix}$ by Example 3 with $m = -1$.

2(b) If $X = [0 \ \ 1]^T$ then $T(2X) = [0 \ \ 4]^T$ while $2T(X) = [0 \ \ 2]^T$.

3(b) $A = \frac{1}{5} \begin{bmatrix} -4 & 3 \\ 3 & 4 \end{bmatrix}$ by Example 4 so $T \begin{bmatrix} 1 \\ 1 \end{bmatrix} = A \begin{bmatrix} 1 \\ 1 \end{bmatrix} = \frac{1}{5} \begin{bmatrix} -1 \\ 7 \end{bmatrix}$

and $T \begin{bmatrix} 2 \\ -1 \end{bmatrix} = A \begin{bmatrix} 2 \\ -1 \end{bmatrix} = \frac{1}{5} \begin{bmatrix} -11 \\ 2 \end{bmatrix}$.

(d) $A = \frac{\sqrt{2}}{2} \begin{bmatrix} 1 & 1 \\ -1 & 1 \end{bmatrix}$ by Example 9 so $T \begin{bmatrix} 1 \\ 1 \end{bmatrix} = A \begin{bmatrix} 1 \\ 1 \end{bmatrix} = \begin{bmatrix} \sqrt{2} \\ 0 \end{bmatrix}$

and $T \begin{bmatrix} 1 \\ -1 \end{bmatrix} = A \begin{bmatrix} 2 \\ -1 \end{bmatrix} = \frac{\sqrt{2}}{2} \begin{bmatrix} 1 \\ -3 \end{bmatrix}$.

(f) $A = \frac{1}{5} \begin{bmatrix} 4 & -2 \\ -2 & 1 \end{bmatrix}$ by Example 3 so $T \begin{bmatrix} 1 \\ 1 \end{bmatrix} = A \begin{bmatrix} 1 \\ 1 \end{bmatrix} = \frac{1}{5} \begin{bmatrix} 2 \\ -1 \end{bmatrix}$

and $T \begin{bmatrix} 2 \\ -1 \end{bmatrix} = A \begin{bmatrix} 2 \\ -1 \end{bmatrix} = \begin{bmatrix} 2 \\ -1 \end{bmatrix}$.

4(b) $A = \frac{1}{2} \begin{bmatrix} 1 & -1 \\ -1 & 1 \end{bmatrix}$ so T is projection on the line $y = -x$ by Example 3.

(d) $A = \frac{1}{5} \begin{bmatrix} -3 & 4 \\ 4 & 3 \end{bmatrix}$ so T is reflection in the line $y = 2x$ by Example 4.

(f) $A = \frac{1}{2}\begin{bmatrix} 1 & -\sqrt{3} \\ \sqrt{3} & 1 \end{bmatrix}$ so T is rotation through the angle $\frac{\pi}{6}$ by Example 9.

6. $T(X) = aX = (aI)X$, so T is the matrix transformation induced by aI.

8(b) $T(X) = aR(X) = a(AX) = (aA)X$ for all X in $\mathbb{R}^n$. So T is the matrix transformation induced by aA.

9(b) $T(-X) = T[(-1)X] = (-1)T(X) = -T(X)$.

10. We must show $T(X) = 0$ for every X in $\mathbb{R}^n$. Since $\mathbb{R}^n = \text{span}\{X_1, \ldots, X_k\}$, write $X = t_1X_1 + \cdots + t_kX_k$, t_k in $\mathbb{R}$. Then Theorem 2 gives $T(X) = t_1T(X_1) = \cdots + t_kT(X_k) = t_10 + \cdots + t_k0 = 0$.
Alternatively $T(X_i) = 0 = 0(X_i)$ for each i so $T = 0$ by Theorem 2.

11(b) Observe: $X = 2[1 \quad 1 \quad 1 \quad 1]^T - 3[-1 \quad 1 \quad 0 \quad 2]^T$.
Hence $T(X) = 2[5 \quad 1 \quad -3]^T - 3[2 \quad 0 \quad 1]^T = [4 \quad 2 \quad -9]^T$ using Theorem 1.

12(b) If $a = 0$ the line is the Y axis, so the reflection is $\begin{bmatrix} x \\ y \end{bmatrix} \to \begin{bmatrix} -x \\ y \end{bmatrix}$ with matrix $\begin{bmatrix} -1 & 0 \\ 0 & 1 \end{bmatrix}$,
as required. If $a \neq 0$ then $\mathbf{0}$ and $\mathbf{d}$ are in the line so the slope is $m = \frac{b}{a}$. Hence Example 4
gives the matrix $\frac{1}{1+m^2}\begin{bmatrix} 1 - m^2 & 2m \\ 2m & m^2 - 1 \end{bmatrix} = \frac{1}{a^2+b^2}\begin{bmatrix} a^2 - b^2 & 2ab \\ 2ab & b^2 - a^2 \end{bmatrix}$.

13(b) $T\begin{bmatrix} x \\ y \end{bmatrix} = \begin{bmatrix} x + 5y \\ y \end{bmatrix}$ so the "reverse" transformation is $T^{-1}\begin{bmatrix} x \\ y \end{bmatrix} = \begin{bmatrix} x - 5y \\ y \end{bmatrix}$. So
$A^{-1} = \begin{bmatrix} 1 & -5 \\ 0 & 1 \end{bmatrix}$.

(d) T is rotation through $\frac{\pi}{4}$, so the "reverse" transformation T^{-1} is rotation through $-\frac{\pi}{4}$. Hence
$A^{-1} = \frac{1}{\sqrt{2}}\begin{bmatrix} 1 & 1 \\ -1 & 1 \end{bmatrix}$.

(f) T is reflection in $y = mx$, so the "reverse" transformation is $T^{-1} = T$ itself. So $A^{-1} = A$.

14. In each case let T and S be the first and second transformation, with matrices A and B respectively.

(b) T is rotation through π, $A = \begin{bmatrix} -1 & 0 \\ 0 & -1 \end{bmatrix}$; S is reflection in the X-axis, $B = \begin{bmatrix} -1 & 0 \\ 0 & 1 \end{bmatrix}$.

So $S \circ T$ has matrix $BA = \begin{bmatrix} -1 & 0 \\ 0 & 1 \end{bmatrix}$; that is $S \circ T$ is reflection in the Y-axis.

(d) T is reflection in the X-axis, $A = \begin{bmatrix} 1 & 0 \\ 0 & -1 \end{bmatrix}$; S is rotation through $\frac{\pi}{2}$, $B = \begin{bmatrix} 0 & -1 \\ 1 & 0 \end{bmatrix}$.

So $S \circ T$ has matrix $BA = \begin{bmatrix} 0 & -1 \\ -1 & 0 \end{bmatrix}$, that is $S \circ T$ is reflection in the line $y = -x$.

16(b) Let T and S be projection on $y = x$ and projection on $y = -x$, with matrices $A = \frac{1}{2}\begin{bmatrix} 1 & 1 \\ 1 & 1 \end{bmatrix}$

and $B = \frac{1}{2}\begin{bmatrix} 1 & -1 \\ -1 & 1 \end{bmatrix}$. So the matrix of $S \circ T$ is $BA = 0$; that is, $S \circ T$ is the zero transformation.

18. Let T denote reflection in a line L through the origin, with matrix A.

(a) Given X in $\mathbb{R}^2$, let $Y = T(X)$ be the reflection in L. Then the reflection of Y is X; that is $T(Y) = X$, that is $(T \circ T)(X) = X$ for all X. Thus $T \circ T = 1_{\mathbb{R}^2}$; that is $T^{-1} = T$.

(b) If L is the Y-axis then $A = \begin{bmatrix} -1 & 0 \\ 0 & 1 \end{bmatrix}$ so $A^2 = I$. Thus $(T \circ T)(X) = A^2 X = X$ for all

X in $\mathbb{R}^n$, so $T \circ T = 1_{\mathbb{R}^2}$ in this case. In general $A = \frac{1}{1+m^2}\begin{bmatrix} 1 - m^2 & 2m \\ 2m & m^2 - 1 \end{bmatrix}$ so, after

some computation $A^2 = I$, and again $T \circ T = 1_{\mathbb{R}^2}$ and $T^{-1} = T$. [Note that $A^{-1} = A$ too.]

20. We have $AE_j = BE_j$ for all j so, as $I = [E_1 \ldots E_n]$, $A = AI = [AE_1, \ldots, AE_n] = [BE_1, \ldots, BE_n] = BI = B$.

21(b) If $B^2 = I$ then $T^2(X) = B^2 X = IX = X$ for all X in $\mathbb{R}^n$. Thus $T^2 = 1_{\mathbb{R}^n}$. Conversely, if $T^2 = 1_{\mathbb{R}^n}$ then $B^2 X = T^2(X) = 1_{\mathbb{R}^n}(X) = X = IX$ for all X in $\mathbb{R}^n$. So $B^2 = I$ by Exercise 20.

22. First the argument in the solution to Example 9 shows that T is linear. Thus

$$T\begin{bmatrix} x \\ y \\ z \end{bmatrix} = T\left\{\begin{bmatrix} x \\ y \\ 0 \end{bmatrix} + \begin{bmatrix} 0 \\ 0 \\ z \end{bmatrix}\right\} = T\begin{bmatrix} x \\ y \\ 0 \end{bmatrix} + T\begin{bmatrix} 0 \\ 0 \\ z \end{bmatrix}.$$

Since T fixes the Z-axis, $T[0 \ 0 \ z]^T = [0 \ 0 \ z]^T$. Also points in the X-Y plane remain

there so $T[x \ y \ 0]^T = [p \ q \ 0]^T$ where $\begin{bmatrix} p \\ q \end{bmatrix} = R_\theta \begin{bmatrix} x \\ y \end{bmatrix} = \begin{bmatrix} \cos\theta & -\sin\theta \\ \sin\theta & \cos\theta \end{bmatrix}\begin{bmatrix} x \\ y \end{bmatrix}$.

Hence

$$T\begin{bmatrix} x \\ y \\ z \end{bmatrix} = \begin{bmatrix} x\cos\theta - y\sin\theta \\ x\sin\theta + y\cos\theta \\ z \end{bmatrix} = \begin{bmatrix} \cos\theta & -\sin\theta & 0 \\ \sin\theta & \cos\theta & 0 \\ 0 & 0 & 1 \end{bmatrix}\begin{bmatrix} x \\ y \\ z \end{bmatrix}.$$

Supplementary Exercises for Chapter 5

1(b) F   (d) T   (f) T   (h) F   (j) F   (l) T   (n) F   (p) F   (r) F

Chapter 6: Vector Spaces

Exercises 6.1   Examples and Basic Properties

1(b) No: S5 fails $1(x, y, z) = (1x, 0, 1z) = (x, 0, z) \neq (x, y, z)$ for all (x, y, z) in V. Note that the other nine axioms do hold.

(d) No: S4 and S5 fail, S5 because $1(x, y, z) = (2x, 2y, 2z) \neq (x, y, z)$, and S5 because $a[b(x, y, z)] = a(2bx, 2by, 2bz) = (4abx, 4aby, 4abz) \neq (2abx, 2aby, 2abz) = ab(x, y, z)$. Note that the eight other axioms hold.

2(b) No: A1 fails — for example $(x^3 + x + 1) + (-x^3 + x + 1) = 2x + 2$ is not in the set.

(d) A1, S1. Suppose $A = \begin{bmatrix} a & b \\ c & d \end{bmatrix}$ and $B = \begin{bmatrix} x & y \\ z & w \end{bmatrix}$ are in V, so $a + c = b + d$ and $x + z = y + w$.

Then $A + B = \begin{bmatrix} a + x & b + y \\ c + z & d + w \end{bmatrix}$ is in V because

$$(a + x) + (c + z) = (a + c) + (x + z) = (b + d) + (y + w) = (b + y) + (d + w).$$

Also $rA = \begin{bmatrix} ra & rb \\ rc & rd \end{bmatrix}$ is in V for all r in $\mathbb{R}$ because $ra + rc = r(a + c) = r(b + d) = rb + rd$.

A2, A3, S2, S3, S4, S5. These hold for matrices in general.

A4. $\begin{bmatrix} 0 & 0 \\ 0 & 0 \end{bmatrix}$ is in V and so serves as the zero of V.

A5. Given $A = \begin{bmatrix} a & b \\ c & d \end{bmatrix}$ with $a + c = b + d$, then $-A = \begin{bmatrix} -a & -b \\ -c & -d \end{bmatrix}$ is also in V because

$-a - c = -(a + c) = -(b + d) = -b - d$. So $-A$ is the negative of A in V.

(f) Yes. The vector space axioms are the basic laws of arithmetic.

(h) Yes.
A1. $\mathbf{0} + \mathbf{0} = \mathbf{0}$ is in V.
A2. $\mathbf{0} + \mathbf{0} = \mathbf{0}$
A3. $\mathbf{0} + (\mathbf{0} + \mathbf{0}) = \mathbf{0} + \mathbf{0} = (\mathbf{0} + \mathbf{0}) + \mathbf{0}$
A4. $\mathbf{0} + \mathbf{0} = \mathbf{0}$
A5. $\mathbf{0} + \mathbf{0} = \mathbf{0}$ so the negative of $\mathbf{0}$ is $\mathbf{0}$ (actually true in every vector space).
S1. $0 \cdot \mathbf{0} = \mathbf{0}$ is in V
S2. $a(\mathbf{0} + \mathbf{0}) = \mathbf{0} = \mathbf{0} + \mathbf{0} = a\mathbf{0} + a\mathbf{0}$
S3. $(a + b)\mathbf{0} = \mathbf{0} = \mathbf{0} + \mathbf{0} = a\mathbf{0} + b\mathbf{0}$
S4. $a(b\mathbf{0}) = a\mathbf{0} = \mathbf{0} = (ab)\mathbf{0}$
S5. $1 \cdot \mathbf{0} = \mathbf{0}$

(j) No. S3 fails: Given $f : \mathbb{R} \to \mathbb{R}$ and a, b in $\mathbb{R}$, we have

$$((a + b)f)(x) = f((a + b)x) = f(ax + bx)$$
$$(af + bf)(x) = (af)(x) + (bf)(x) = f(ax) + f(bx).$$

These need not be equal: for example, if f is the function defined by $f(x) = x^2$;
Then $f(ax + bx) = (ax + bx)^2$ need not equal $(ax)^2 + (bx)^2 = f(ax) + f(bx)$.
Note that the other axioms hold. A1-A4 hold by Example 6 as we are using pointwise addition.

S2. $\begin{aligned}[t] a(f + g)(x) &= (f + g)(ax) && \text{definition of scalar multiplication in } V \\ &= f(ax) + g(ax) && \text{definition of pointwise addition} \\ &= (af)(x) + (ag)(x) && \text{definition of scalar multiplication in } V \\ &= (af + ag)(x) && \text{definition of pointwise addition} \end{aligned}$

As this is true for all x, $a(f + g) = af + ag$.

S4. $[a(bf)](x) = (bf)(ax) = f[b(ax)] = f[(ba)x] = [(ba)f](x) = [abf](x)$ for all x,
so $a(bf) = (ab)f$.

S5. $(1f)(x) = f(1x) = f(x)$ for all x, so $1f = f$.

(1) No. S4, S5 fail: $a * (b * X) = a * (bX^T) = a(bX^T)^T = abX^{TT} = abX$, while $(ab) * X = abX^T$.
These need not be equal. Similarly: $1 * X = 1X^T = X^T$ need not equal X. Note that the
other axioms hold:

A1-A5. These hold for matrix addition generally.

S1. $a * X = aX^T$ is in V.

S2. $a * (X + Y) = a(X + Y)^T = a(X^T + Y^T) = aX^T + aY^T = a * X + a * Y$.

S3 $(a + b) * X = (a + b)X^T = aX^T + bX^T = a * X + b * X$.

4. A1. $(x, y) + (x_1, y_1) = (x + x_1, y + y_1 + 1)$ is in V for all (x, y) and (x_1, y_1) in V.

A2. $(x, y) + (x_1, y_1) = (x + x_1, y + y_1 + 1) = (x_1 + x, y_1 + y + 1) = (x_1, y_1) + (x_1, y)$.

A3. $\begin{aligned}[t] (x, y) + ((x_1, y_1) + (x_2, y_2)) &= (x, y) + (x_1 + x_2, y_1 + y_2 + 1) \\ &= (x + (x_1 + x_2), y + (y_1 + y_2 + 1) + 1) \\ &= (x + x_1 + x_2, y + y_1 + y_2 + 2) \\ ((x, y) + (x_1, y_1)) + (x_2, y_2) &= (x + x_1, y + y_1 + 1) + (x_2, y_2) \\ &= ((x + x_1) + x_2, (y + y_1 + 1) + y_2 + 1) \\ &= (x + x_1 + x_2, y + y_1 + y_2 + 2). \end{aligned}$

These are equal for all (x, y), (x_1, y_1) and (x_2, y_2) in V.

A4. $(x, y) + (0, -1) = (x + 0, y + (-1) + 1) = (x, y)$ for all (x, y), so $(0, -1)$ is the zero of V.

A5. $(x, y) + (-x, -y - 2) = (x + (-x), y + (-y - 2) + 1) = (0, -1)$ is the zero of V (from A4)
so the negative of (x, y) is $(-x, -y - 2)$.

S1. $a(x, y) = (ax, ay + a - 1)$ is in V for all (x, y) in V and a in $\mathbb{R}$.

S2. $\begin{aligned}[t] a[(x, y) + (x_1, y_1)] &= a(x + x_1, y + y_1 + 1) &&= (a(x + x_1), a(y + y_1 + 1) + a - 1) \\ & &&= (ax + ax_1, ay + ay_1 + 2a - 1) \\ a(x, y) + a(x_1, y_1) &= (ax, ay + a - 1) + (ax_1, ay_1 - a - 1) \\ &= ((ax + ax_1), (ay + a - 1) + (ay_1 + a - 1) + 1) \\ &= (ax + ax_1, ay + ay_1 + 2a - 1). \end{aligned}$

These are equal.

S4. $a[b(x, y)] = a(bx, by + b - 1) = (a(bx), a(by + b - 1) + a - 1) = (abx, aby + ab - 1) = (ab)(x, y)$.

S5. $1(x, y) = (1x, 1y + 1 - 1) = (x, y)$ for all (x, y) in V.

5(b) Subtract the first equation from the second to get $\mathbf{x} - 3\mathbf{y} = \mathbf{v} - \mathbf{u}$, whence $\mathbf{x} = 3\mathbf{y} + \mathbf{v} - \mathbf{u}$.

Substitute in the first equation to get

$$3(3 + \mathbf{v} - \mathbf{u}) - 2\mathbf{y} = \mathbf{u}$$
$$7\mathbf{y} = 4\mathbf{u} - 3\mathbf{v}$$
$$\mathbf{y} = \tfrac{4}{7}\mathbf{u} - \tfrac{3}{7}\mathbf{v}.$$

Substitute this in the first equation to get $\mathbf{x} = \tfrac{5}{7}\mathbf{u} - \tfrac{2}{7}\mathbf{v}$.

6(b) $a\mathbf{u} + b\mathbf{v} + c\mathbf{w} = \mathbf{0}$ becomes
$$\begin{bmatrix} a & 0 \\ 0 & a \end{bmatrix} + \begin{bmatrix} 0 & b \\ b & 0 \end{bmatrix} + \begin{bmatrix} c & c \\ c & -c \end{bmatrix} = \begin{bmatrix} 0 & 0 \\ 0 & 0 \end{bmatrix}.$$
Equating corresponding entries gives equations for a and b.

$$a + c = 0, \quad b + c = 0, \quad b + c = 0, \quad a - c = 0.$$

The only solution is $a = b = c = 0$.

(d) $a\mathbf{u} + b\mathbf{v} + c\mathbf{w} = \mathbf{0}$ means $a \sin x + b \cos x + c1 = 0$ for all choices of x. If $x = 0, \frac{\pi}{2}, \pi$, we get equations $b + c = 0$, $a + c = 0$, and $-b + c = 0$. The only solution is $a = b = c = 0$.

7(b) $4(3\mathbf{u} - \mathbf{v} + \mathbf{w}) - 2[(3\mathbf{u} - 2\mathbf{v}) - 3(\mathbf{v} - \mathbf{w})] + 6(\mathbf{w} - \mathbf{u} - \mathbf{v})$
$= (12\mathbf{u} - 4\mathbf{v} + 4\mathbf{w}) - 2[3\mathbf{u} - 2\mathbf{v} - 3\mathbf{v} + 3\mathbf{w}] + (6\mathbf{w} - 6\mathbf{u} - 6\mathbf{v})$
$= (12\mathbf{u} - 4\mathbf{v} + 4\mathbf{w}) - (6\mathbf{u} - 10\mathbf{v} + 6\mathbf{w}) + (6\mathbf{w} - 6\mathbf{u} - 6\mathbf{v})$
$= 4\mathbf{w}$

11. We have $\mathbf{v} + (-\mathbf{v}) = \mathbf{0}$ by A5. Suppose $\mathbf{v} + \mathbf{x} = \mathbf{0}$ for some other vector $\mathbf{x}$. Then $\mathbf{v} + \mathbf{x} = \mathbf{v} + (-\mathbf{v})$ so $\mathbf{x} = -\mathbf{v}$ by cancellation.

12(b) $(-a)\mathbf{v} + a\mathbf{v} = (-a + a)\mathbf{v} = 0\mathbf{v} = \mathbf{0}$. Since also $-(a\mathbf{v}) + a\mathbf{v} = \mathbf{0}$ we get $(-a)\mathbf{v} + a\mathbf{v} = -(a\mathbf{v}) + a\mathbf{v}$. Thus $(-a)\mathbf{v} = -(a\mathbf{v})$ by cancellation.
Alternatively: $(-a)\mathbf{v} = [(-1)a]\mathbf{v} = (-1)(a\mathbf{v}) = -a\mathbf{v}$ using Theorem 3.

15(b) Since $a \neq 0$, a^{-1} exists in $\mathbb{R}$. Hence $a\mathbf{v} = a\mathbf{w}$ gives $a^{-1}a\mathbf{v} = a^{-1}a\mathbf{w}$; that is $1\mathbf{v} = 1\mathbf{w}$, that is $\mathbf{v} = \mathbf{w}$.
Alternatively: $a\mathbf{v} = a\mathbf{w}$ gives $a\mathbf{v} - a\mathbf{w} = \mathbf{0}$, so $a(\mathbf{v} - \mathbf{w}) = \mathbf{0}$. As $a \neq 0$, it follows that $\mathbf{v} - \mathbf{w} = \mathbf{0}$ by Theorem 3, that is $\mathbf{v} = \mathbf{w}$.

Exercises 6.2   Subspaces and Spanning Sets

1(b) Yes. U is a subset of $\mathbf{P}_3$ because $xg(x)$ has degree one more than the degree of $g(x)$. Clearly $0 = x \cdot 0$ is in U. Given $\mathbf{u} = xg(x)$ and $\mathbf{v} = xh(x)$ in U (where $g(x)$ and $b(x)$ are in $\mathbf{P}_2$) we have

$$\mathbf{u} + \mathbf{v} = x(g(x) + h(x)) \text{ is in } U \text{ because } g(x) + h(x) \text{ is in } \mathbf{P}_2$$
$$k\mathbf{u} = x(kg(x)) \text{ is in } U \text{ for all } k \text{ in } \mathbb{R} \text{ because } kg(x) \text{ is in } \mathbf{P}_2$$

(d) Yes. As in (b), U is a subset of $\mathbf{P}_3$. Clearly $0 = x \cdot 0 + (1 - x) \cdot 0$ is in U. If $\mathbf{u} = xg(x) + (1 - x)h(x)$ and $\mathbf{v} = xg_1(x) + (1 - x)h_1(x)$ are in U then

$$\mathbf{u} + \mathbf{v} = x[g(x) + g_1(x)] + (1 - x)[h(x) + h_1(x)]$$
$$k\mathbf{u} = x[kg(x)] + (1 - x)[kh(x)]$$

both lie in U because $g(x) + g_1(x)$ and $h(x) + h_1(x)$ are in $\mathbf{P}_2$.

(f) No. U is not closed under addition (for example $\mathbf{u} = 1 + x^3$ and $\mathbf{v} = x - x^3$ are in U but $\mathbf{u} + \mathbf{v} = 1 + x$ is not in U). Also, the zero polynomial is not in U.

2(b) Yes. Clearly $\mathbf{0} = \begin{bmatrix} 0 & 0 \\ 0 & 0 \end{bmatrix}$ is in U. If $\mathbf{u} = \begin{bmatrix} a & b \\ c & d \end{bmatrix}$ and $\mathbf{u}_1 = \begin{bmatrix} a_1 & b_1 \\ c_1 & d_1 \end{bmatrix}$ are in U then

$$\mathbf{u} + \mathbf{u}_1 = \begin{bmatrix} a + a_1 & b + b_1 \\ c + c_1 & d + d_1 \end{bmatrix} \text{ is in } U \text{ because } \begin{aligned} (a + a_1) + (b + b_1) &= (a + b) + (a_1 + b_1) \\ &= (c + d) + (c_1 + d1) \\ &= (c + c_1) + (d + d_1). \end{aligned}$$

$ku = \begin{bmatrix} ka & kb \\ kc & kd \end{bmatrix}$ is in U because $ka + kb = k(a + b) = k(c + d) = kc + kd$.

(d) Yes. Here 0 is in U as $0B = 0$. If A and A_1 are in U then $AB = 0 = A_1B$, so $(A + A_1)B = AB + A_1B = 0 + 0 = 0$ and $(kA)B = k(AB) = k0 = 0$ for all k in $\mathbb{R}$, so $A + A_1$ and kA are also in U.

(f) No. U is not closed under addition. In fact, $A = \begin{bmatrix} 1 & 0 \\ 0 & 0 \end{bmatrix}$ and $A_1 \begin{bmatrix} 0 & 0 \\ 0 & 1 \end{bmatrix}$ are both in U,

but $A + A_1 = \begin{bmatrix} 1 & 0 \\ 0 & 1 \end{bmatrix}$ is $\underline{\text{not}}$ in U.

3(b) No. U is not closed under addition. For example if f and g are defined by $f(x) = x + 1$ and $g(x) = x^2 + 1$, then f and g are in U but $f + g$ is not in U because $(f + g)(0) = f(0) + g(0) = 1 + 1 = 2$.

(d) No. U is not closed under scalar multiplication. For example, if f is defined by $f(x) = x$, then f is in U but $(-1)f$ is not in U.

(f) Yes. 0 is in U because $0(x + y) = 0 = 0 + 0 = 0(x) + 0(y)$ for all x and y in $[0, 1]$. If f and g are in U then, for all k in $\mathbb{R}$:

$$\begin{aligned} (f + g)(x + y) &= f(x + y) + g(x + y) \\ &= (f(x) + f(y)) + (g(x) + g(y)) \\ &= (f(x) + g(x)) + (f(y) + g(y)) \\ &= (f + g)(x) + (f + g)(y) \end{aligned}$$

$$\begin{aligned} (kf)(x + y) = k[f(x + y)] = k[f(x) + f(y)] &= k[f(x)] + k[f(y)] \\ &= (kf)(x) + (kf)(y) \end{aligned}$$

Hence $f + g$ and kf are in U.

5(b) Suppose $X = \begin{bmatrix} x_1 \\ \vdots \\ x_n \end{bmatrix} \neq 0$, let $x_k \neq 0$. Given $Y = \begin{bmatrix} y_1 \\ \vdots \\ y_n \end{bmatrix}$ let A be the $m \times n$ matrix with k^{th} column $x_k^{-1}Y$ and the other columns zero. Then $Y = AX$ by matrix multiplication, so Y is in U. Hence $U = \mathbb{R}^m$.

6(b) Want $2x^2 - 3x + 1 = r(x+1) + s(x^2 + x) + t(x^2 + 2)$. Equating coefficients of x^2, x and 1 gives $s + t = 2$, $r + s = -3$, $r + 2t = 1$. The unique solution is $r = -3$, $s = 0$, $t = 2$.

(d) As in (b), $x = \frac{2}{3}(x+1) + \frac{1}{3}(x^2 + x) - \frac{1}{3}(x^2 + 2)$.

7(b) If $\mathbf{v} = s\mathbf{u} + t\mathbf{w}$ then $(3, 1, -3) = s(1, 1, 1) + t(0, 1, 3)$. Then $s = 3$, $s + t = 1$, $s + 3t = -3$, whence $s = 3$, $t = -2$. Hence $\mathbf{v}$ does lie in span$\{\mathbf{u}, \mathbf{w}\}$.

(d) If $\mathbf{v} = s\mathbf{u} + t\mathbf{w}$ then $x = s(x^2 + 1) + t(x + 2)$. Thus $s = 0$, $t = 1$, $s + 2t = 0$. These equations have *no* solution, so $\mathbf{v}$ is *not* in span$\{\mathbf{u}, \mathbf{w}\}$.

(f) If $\mathbf{v} = s\mathbf{u} + t\mathbf{w}$, then $\begin{bmatrix} 1 & -4 \\ 5 & 3 \end{bmatrix} = s \begin{bmatrix} 1 & -1 \\ 2 & 1 \end{bmatrix} + t \begin{bmatrix} 2 & 1 \\ 1 & 0 \end{bmatrix}$. Thus $s + 2t = 1$, $-s + t = -4$, $2s + t = 5$ and $s = 3$. The solution is $s = 3$, $t = -1$, so $\mathbf{v}$ is in span$\{\mathbf{u}, \mathbf{w}\}$

8(b) Yes. The trigonometry identity $1 = \sin^2 x + \cos^2 x$ for all x means that 1 is in span$\{\sin^2 x, \cos^2 x\}$.

(d) Suppose $1 + x^2 = s \sin^2 x + t \cos^2 x$ for some s and t. This must hold for all x. Taking $x = 0$ gives $1 = t$; taking $x = \pi$ gives $1 + \pi^2 = -t$. Thus $2 + \pi^2 = 0$, a contradiction. So no such s and t exist, that is $1 + x^2$ is *not* in span$\{\sin^2 x, \cos^2 x\}$.

9(b) Write $U = \text{span}\{1 + 2x^2, 3x, 1 + x\}$, then successively

$$x = \tfrac{1}{3}(3x) \text{ is in } U$$
$$1 = (1 + x) - x \text{ is in } U$$
$$x^2 = \tfrac{1}{2}[(1 + 2x^2) - 1] \text{ is in } U.$$

Since $\mathbf{P}_2 = \text{span}\{1, x, x^2\}$, this shows that $\mathbf{P}_2 \subseteq U$. Clearly $U \subseteq \mathbf{P}_2$ so $U = \mathbf{P}_2$.

11(b) The vectors $\mathbf{u} - \mathbf{v} = 1\mathbf{u} + (-1)\mathbf{v}$, $\mathbf{u} + \mathbf{v}$, and $\mathbf{w}$ are all in span$\{\mathbf{u}, \mathbf{v}, \mathbf{w}\}$ so span$\{\mathbf{u} - \mathbf{v}, \mathbf{u} + \mathbf{w}, \mathbf{w}\} \subseteq$ span$\{\mathbf{u}, \mathbf{v}, \mathbf{w}\}$. The other inclusion follows by Theorem 2 because

$$\mathbf{u} = (\mathbf{u} + \mathbf{w}) - \mathbf{w}$$
$$\mathbf{v} = -(\mathbf{u} - \mathbf{v}) + (\mathbf{u} + \mathbf{w}) - \mathbf{w}$$
$$\mathbf{w} = \mathbf{w}$$

show that $\mathbf{u}$, $\mathbf{v}$ and $\mathbf{w}$ are in span$\{\mathbf{u} - \mathbf{v}, \mathbf{u} + \mathbf{v}, \mathbf{w}\}$.

14. No. For example $(1, 1, 0)$ is not in span$\{(1, 2, 0), (1, 1, 1)\}$. Indeed $(1, 1, 0) = s(1, 2, 0) + t(1, 1, 1)$ gives $s + t = 1$, $2s + t = 1$, $t = 0$, and this has no solution.

18. Write $W = \text{span}\{\mathbf{u}, \mathbf{v}_2, \dots, \mathbf{v}_n\}$. Since $\mathbf{u}$ is in V we have $W \subseteq V$. But the fact that $a_1 \neq 0$ means

$$\mathbf{v}_1 = \tfrac{1}{a_1}\mathbf{u} - \tfrac{a_2}{a_1}\mathbf{v}_2 - \cdots - \tfrac{a_n}{a_1}\mathbf{v}_n$$

so $\mathbf{v}_1$ is in W. Since $\mathbf{v}_2, \dots, \mathbf{v}_n$ are all in W, this shows that $V \subseteq W$. Hence $V = W$.

21(b) If $\mathbf{u}$ and $\mathbf{u} + \mathbf{v}$ are in U then $\mathbf{v} = (\mathbf{u} + \mathbf{v}) - \mathbf{u} = (\mathbf{u} + \mathbf{v}) + (-1)\mathbf{u}$ is in U because U is closed under addition and scalar multiplication.

22. If (2) and (3) hold in Theorem 2, and if **u** is in U, then $\mathbf{0} = 0 \cdot \mathbf{u}$ is in U by (3), so U is a subspace by Theorem 1. Conversely, if U is a subspace then U is nonempty because **0** is in U, and (2) and (3) hold by Theorem 2.

Exercises 6.3   Linear Independence and Dimension

1(b) Independent. If $rx^2 + s(x+1) + t(1 - x - x^2) = 0$ then $r - t = 0$, $s - t = 0$, $s + t = 0$. The only solution is $r = s = t = 0$.

(d) Independent. If $r \begin{bmatrix} 1 & 1 \\ 1 & 0 \end{bmatrix} + s \begin{bmatrix} 0 & 1 \\ 1 & 1 \end{bmatrix} + t \begin{bmatrix} 1 & 0 \\ 1 & 1 \end{bmatrix} + u \begin{bmatrix} 1 & 1 \\ 0 & 1 \end{bmatrix} = \begin{bmatrix} 0 & 0 \\ 0 & 0 \end{bmatrix}$, then $r + t + u = 0$, $r + s + u = 0$, $r + s + t = 0$, $s + t + u = 0$. The ony solution is $r = s = t = u = 0$.

2(b) Dependent. $3(x^2 - x + 3) - 2(2x^2 + x + 5) + (x^2 + 5x + 1) = 0$

(d) Dependent. $2 \begin{bmatrix} -1 & 0 \\ 0 & -1 \end{bmatrix} + \begin{bmatrix} 1 & -1 \\ -1 & 1 \end{bmatrix} + \begin{bmatrix} 1 & 1 \\ 1 & 1 \end{bmatrix} + 0 \begin{bmatrix} 0 & -1 \\ -1 & 0 \end{bmatrix} = \begin{bmatrix} 0 & 0 \\ 0 & 0 \end{bmatrix}$.

(f) Dependent. $\frac{5}{x^2 + x - 6} + \frac{1}{x^2 - 5x + 6} - \frac{6}{x^2 - 9} = 0$.

3(b) Dependent. $1 - \sin^2 x - \cos^2 x = 0$.

4(b) If $r(2, x, 1) + s(1, 0, 1) + t(0, 1, 3) = (0, 0, 0)$ then

$$
\begin{aligned}
2r &+ s &&= 0 \\
xr &&+ t &= 0 \\
r &+ s &+ 3t &= 0.
\end{aligned}
$$

Gaussian elimination gives

$$
\begin{bmatrix} 2 & 1 & 0 & | & 0 \\ x & 0 & 1 & | & 0 \\ 1 & 1 & 3 & | & 0 \end{bmatrix} \rightarrow \begin{bmatrix} 1 & 1 & 3 & | & 0 \\ 2 & 1 & 0 & | & 0 \\ x & 0 & 1 & | & 0 \end{bmatrix} \rightarrow \begin{bmatrix} 1 & 1 & 3 & | & 0 \\ 0 & 1 & 6 & | & 0 \\ 0 & -x & 1-3x & | & 0 \end{bmatrix} \rightarrow \begin{bmatrix} 1 & 1 & 3 & | & 0 \\ 0 & 1 & 6 & | & 0 \\ 0 & 0 & 1+3x & | & 0 \end{bmatrix}.
$$

This has only the trivial solution $r = s = t = 0$ if and only if $x \neq \frac{-1}{3}$. Alternatively, the coefficient matrix has determinant

$$
\det \begin{bmatrix} 2 & 1 & 0 \\ x & 0 & 1 \\ 1 & 1 & 3 \end{bmatrix} = \det \begin{bmatrix} 2 & 1 & 0 \\ x & 0 & 1 \\ -1 & 0 & 3 \end{bmatrix} = -\det \begin{bmatrix} x & 1 \\ -1 & 3 \end{bmatrix} = -(1 + 3x).
$$

This is nonzero if and only if $x \neq -\frac{1}{3}$.

5(b) **Independence:** If $r(-1, 1, 1) + s(1, -1, 1) + t(1, 1, -1) = (0, 0, 0)$ then $-r + s + t = 0$, $r - s + t = 0$, $r + s - t = 0$. The only solution is $r = s = t = 0$.

Spanning: Write $U = \text{span}\{(-1, 1, 1), (1, -1, 1), (1, 1, -1)\}$. Then $(1, 0, 0) = \frac{1}{2}[(1, 1, -1) + (1, -1, 1)]$ is in U; similarly $(0, 1, 0)$ and $(0, 0, 1)$ are in U. Thus $\mathbb{R}^3 = \text{span}\{(1, 0, 0), (0, 1, 0), (0, 0, 1)\}$ is contained in U. As $U \subseteq \mathbb{R}^3$, we have $\mathbb{R}^3 = U$.

(d) **Independence**: If $r(1+x)+s(x+x^2)+t(x^2+x^3)+ux^3 = 0$ then $r = 0$, $r+s = 0$, $s+t = 0$, $t+u = 0$. The only solution is $r = s = t = u = 0$.

Spanning: Write $U = \text{span}\{1+x, \quad x+x^2, \quad x^2+x^3, \quad x^3\}$. Then x^3 is in U; whence $x^2 = (x^2+x^3) - x^3$ is in U; whence $x = (x+x^2) - x^2$ is in U; whence $1 = (1+x) - x$ is in U. Hence $\mathbf{P}_3 = \text{span}\{1, x, x^2, x^3\}$ is contained in U. As $U \subseteq \mathbf{P}_3$, we have $U = \mathbf{P}_3$.

6(b) Write $U = \{a + b(x+x^2) \mid a, b \text{ in } \mathbb{R}\} = \text{span } B$ where $B = \{1, x+x^2\}$. But B is independent ($s + t(x+x^2) = 0$ implies $s = t = 0$). Hence B is a basis of U, so $\dim U = 2$.

(d) Write $U = \{p(x) \mid p(x) = p(-x)\}$. As $U \subseteq \mathbf{P}_2$, write $p(x) = a + bx + cx^2$. The condition $p(x) = p(-x)$ becomes $a+bx+cx^2 = a-bx+cx^2$, so $b = 0$. Thus $U = \{a + bx^2 \mid a, b \text{ in } \mathbb{R}\} = \text{span}\{1, x^2\}$. As $\{1, x^2\}$ is independent ($s + tx^2 = 0$ implies $s = 0 = t$), it is a basis of U, so $\dim U = 2$.

7(b) Write $U = \left\{ A \mid A \begin{bmatrix} 1 & 1 \\ -1 & 0 \end{bmatrix} = \begin{bmatrix} 1 & 1 \\ -1 & 0 \end{bmatrix} A \right\}$. If $A = \begin{bmatrix} x & y \\ z & w \end{bmatrix}$, A is in U if and only

if $\begin{bmatrix} x & y \\ z & w \end{bmatrix} \begin{bmatrix} 1 & 1 \\ -1 & 0 \end{bmatrix} = \begin{bmatrix} 1 & 1 \\ -1 & 0 \end{bmatrix} \begin{bmatrix} x & y \\ z & w \end{bmatrix}$, that is $\begin{bmatrix} x-y & x \\ z-w & z \end{bmatrix} = \begin{bmatrix} x+z & y+w \\ -x & -y \end{bmatrix}$.

This holds if and only if $x = y + w$ and $z = -y$, that is

$$A = \begin{bmatrix} y+w & y \\ -y & w \end{bmatrix} = y \begin{bmatrix} 1 & 1 \\ -1 & 0 \end{bmatrix} + w \begin{bmatrix} 1 & 0 \\ 0 & 1 \end{bmatrix}.$$

Hence $U = \text{span } B$ where $B = \left\{ \begin{bmatrix} 1 & 1 \\ -1 & 0 \end{bmatrix}, \begin{bmatrix} 1 & 0 \\ 0 & 1 \end{bmatrix} \right\}$. But B is independent because

$s \begin{bmatrix} 1 & 1 \\ -1 & 0 \end{bmatrix} + t \begin{bmatrix} 1 & 0 \\ 0 & 1 \end{bmatrix} = \begin{bmatrix} 0 & 0 \\ 0 & 0 \end{bmatrix}$ means $s + t = 0$, $s = 0$, $-s = 0$, $t = 0$, so $s = t = 0$.

Thus B is a basis of U, so $\dim U = 2$.

(d) Write $U = \left\{ A \mid A \begin{bmatrix} 1 & 1 \\ -1 & 0 \end{bmatrix} = \begin{bmatrix} 0 & 1 \\ -1 & 1 \end{bmatrix} A \right\}$. If $A = \begin{bmatrix} x & y \\ z & w \end{bmatrix}$ then A is in U if and only

if $\begin{bmatrix} x & y \\ z & w \end{bmatrix} \begin{bmatrix} 1 & 1 \\ -1 & 0 \end{bmatrix} = \begin{bmatrix} 0 & 1 \\ -1 & 1 \end{bmatrix} \begin{bmatrix} x & y \\ z & w \end{bmatrix}$; that is $\begin{bmatrix} x-y & x \\ z-w & z \end{bmatrix} = \begin{bmatrix} z & w \\ z-x & w-y \end{bmatrix}$.

This holds if and only if $z = x - y$ and $x = w$; that is

$$A = \begin{bmatrix} x & y \\ x-y & x \end{bmatrix} = x \begin{bmatrix} 1 & 0 \\ 1 & 1 \end{bmatrix} + y \begin{bmatrix} 0 & 1 \\ -1 & 0 \end{bmatrix}.$$

Thus $U = \text{span } B$ where $B = \left\{ \begin{bmatrix} 1 & 0 \\ 1 & 1 \end{bmatrix}, \begin{bmatrix} 0 & 1 \\ -1 & 0 \end{bmatrix} \right\}$. But B is independent because

$$s \begin{bmatrix} 1 & 0 \\ 1 & 1 \end{bmatrix} + t \begin{bmatrix} 0 & 1 \\ -1 & 0 \end{bmatrix} = \begin{bmatrix} 0 & 0 \\ 0 & 0 \end{bmatrix}$$ implies $s = t = 0$. Hence B is a basis of U,

so $\dim U = 2$.

8(b) If $X = \begin{bmatrix} x & y \\ z & w \end{bmatrix}$ the condition $AX = X$ is $\begin{bmatrix} x+z & y+w \\ 0 & 0 \end{bmatrix} = \begin{bmatrix} x & y \\ z & w \end{bmatrix}$ and this holds if

and only if $z = w = 0$. Hence $X = \begin{bmatrix} x & y \\ 0 & 0 \end{bmatrix} = x \begin{bmatrix} 1 & 0 \\ 0 & 0 \end{bmatrix} + y \begin{bmatrix} 0 & 1 \\ 0 & 0 \end{bmatrix}$. So $U = \text{span } B$

where $B = \left\{ \begin{bmatrix} 1 & 0 \\ 0 & 0 \end{bmatrix}, \begin{bmatrix} 0 & 1 \\ 0 & 0 \end{bmatrix} \right\}$. As B is independent, it is a basis of U, so $\dim U = 2$.

10(b) If the common column sum is m, V has the form

$$V = \left\{ \begin{bmatrix} a & q & r \\ b & p & s \\ m-a-b & m-p-q & m-r-s \end{bmatrix} \;\middle|\; a,b,p,q,e,r,s,m \text{ in } \mathbb{R} \right\} = \text{span } B \text{ where}$$

$$B = \left\{ \begin{bmatrix} 0 & 0 & 0 \\ 0 & 0 & 0 \\ 1 & 1 & 1 \end{bmatrix}, \begin{bmatrix} 1 & 0 & 0 \\ 0 & 0 & 0 \\ -1 & 0 & 0 \end{bmatrix}, \begin{bmatrix} 0 & 0 & 0 \\ 1 & 0 & 0 \\ -1 & 0 & 0 \end{bmatrix}, \begin{bmatrix} 0 & 1 & 0 \\ 0 & 0 & 0 \\ 0 & -1 & 0 \end{bmatrix}, \right.$$

$$\left. \begin{bmatrix} 0 & 0 & 0 \\ 0 & 1 & 0 \\ 0 & -1 & 0 \end{bmatrix}, \begin{bmatrix} 0 & 0 & 1 \\ 0 & 0 & 0 \\ 0 & 0 & -1 \end{bmatrix}, \begin{bmatrix} 0 & 0 & 0 \\ 0 & 0 & 1 \\ 0 & 0 & -1 \end{bmatrix} \right\}.$$

The set B is independent (a bilinear combination using coefficients $a, b, p, q, r, s,$ and m yields the matrix in V, and this is 0 if and only if $a = b = p = q = r = s = m = 0$.) Hence B is a basis of B, so $\dim V = 7$.

11(b) A general polynomial in $\mathbf{P}_3$ has the form $p(x) = a + bx + cx^2 + dx^3$, so
$$V = \left\{ (x^2 - x)(a + bx + cx^2 + dx^3) \mid a, b, c, d \text{ in } \mathbb{R} \right\}$$
$$= \left\{ a(x^2 - x) + bx(x^2 - x) + cx^2(x^2 - x) + dx^3(x^2 - x) \mid a, b, c, d \text{ in } \mathbb{R} \right\}$$
$$= \text{span } B$$

where $B = \left\{ (x^2 - x), x(x^2 - x), x^2(x^2 - x), x^3(x^2 - x) \right\}$. We claim that B is independent: for if $a(x^2 - x) + bx(x^2 - x) + cx^2(x^2 - x) + dx^3(x^2 - x) = 0$ then $(a + bx + cx^2 + dx^3)(x^2 - x) = 0$, whence $a + bx + cx^2 + dx^3 = 0$ by the hint in (a). Thus $a = b = c = d = 0$. [This also follows by comparing coefficients.] Thus B is a basis of V, so $\dim V = 4$.

12(b) No. If $\mathbf{P}_3 = \text{span}\{f_1(x), f_2(x), f_3(x), f_4(x)\}$ where $f_i(0) = 0$ for each i, then each polynomial $p(x)$ in $\mathbf{P}_3$ is a linear combination
$$p(x) = a_1 f_1(x) + a_2 f_2(x) + a_3 f_3(x) + a_4 f_4(x)$$

when the a_i are in $\mathbb{R}$. But then

$$p(0) = a_1 f_1(0) + a_2 f_2(0) + a_3 f_3(0) + a_4 f_4(0) = 0$$

for every $p(x)$ in $\mathbf{P}_3$. This is not the case, so no such basis of $\mathbf{P}_3$ can exist. [Indeed, no such spanning set of $\mathbf{P}_3$ can exist.]

(d) No. $B = \left\{ \begin{bmatrix} 1 & 0 \\ 0 & 1 \end{bmatrix}, \begin{bmatrix} 1 & 1 \\ 0 & 1 \end{bmatrix}, \begin{bmatrix} 1 & 0 \\ 1 & 1 \end{bmatrix}, \begin{bmatrix} 0 & 1 \\ 1 & 1 \end{bmatrix} \right\}$ is a basis of invertible matrices.

Independent: $r \begin{bmatrix} 1 & 0 \\ 0 & 1 \end{bmatrix} + s \begin{bmatrix} 1 & 1 \\ 0 & 1 \end{bmatrix} + t \begin{bmatrix} 1 & 0 \\ 1 & 1 \end{bmatrix} + u \begin{bmatrix} 0 & 1 \\ 1 & 1 \end{bmatrix} = \begin{bmatrix} 0 & 0 \\ 0 & 0 \end{bmatrix}$ gives

$r + s + t = 0,\ s + u = 0,\ t + u = 0,\ r + s + t + u = 0$. The only solution is $r = s = t = u = 0$.

Spanning:
$\begin{bmatrix} 0 & 1 \\ 0 & 0 \end{bmatrix} = \begin{bmatrix} 1 & 1 \\ 0 & 1 \end{bmatrix} - \begin{bmatrix} 1 & 0 \\ 0 & 1 \end{bmatrix}$ is in span B

$\begin{bmatrix} 0 & 0 \\ 1 & 0 \end{bmatrix} = \begin{bmatrix} 1 & 0 \\ 1 & 1 \end{bmatrix} - \begin{bmatrix} 1 & 0 \\ 0 & 1 \end{bmatrix}$ is in span B

$\begin{bmatrix} 0 & 0 \\ 0 & 1 \end{bmatrix} = \begin{bmatrix} 0 & 1 \\ 1 & 1 \end{bmatrix} - \begin{bmatrix} 0 & 1 \\ 0 & 0 \end{bmatrix} - \begin{bmatrix} 0 & 0 \\ 1 & 0 \end{bmatrix}$ is in span B

$\begin{bmatrix} 1 & 0 \\ 0 & 0 \end{bmatrix} = \begin{bmatrix} 1 & 0 \\ 0 & 1 \end{bmatrix} - \begin{bmatrix} 0 & 0 \\ 0 & 1 \end{bmatrix}$ is in span B

Hence $\mathbf{M}_{22} = \text{span} \left\{ \begin{bmatrix} 0 & 1 \\ 0 & 0 \end{bmatrix}, \begin{bmatrix} 0 & 0 \\ 1 & 0 \end{bmatrix}, \begin{bmatrix} 0 & 0 \\ 0 & 1 \end{bmatrix}, \begin{bmatrix} 1 & 0 \\ 0 & 0 \end{bmatrix} \right\} \subseteq \text{span } B$. Clearly span $B \subseteq \mathbf{M}_{22}$.

(f) Yes. Indeed, $0\mathbf{u} + 0\mathbf{v} + 0\mathbf{w} = \mathbf{0}$ for any $\mathbf{u}, \mathbf{v}, \mathbf{w}$.

(h) Yes. If $s\mathbf{u} + t(\mathbf{u} + \mathbf{v}) = \mathbf{0}$ then $(s + t)\mathbf{u} + t\mathbf{v} = \mathbf{0}$, so $s + t = 0$ and $t = 0$ (because $\{\mathbf{u}, \mathbf{v}\}$ is independent). Thus $s = t = 0$.

(j) Yes. If $s\mathbf{u} + t\mathbf{v} = \mathbf{0}$ then $s\mathbf{u} + t\mathbf{v} + 0\mathbf{w} = \mathbf{0}$, so $s = t = 0$ (because $\{\mathbf{u}, \mathbf{v}, \mathbf{w}\}$ is independent).

15. If a linear combination of the vectors in the subset vanishes, it is a linear combination of the vectors in the larger set (coefficients outside the subset are zero). Since it still vanishes, all the coefficients are zero because the larger set is independent.

20. We have $s\mathbf{u}' + t\mathbf{v}' = s(a\mathbf{u} + b\mathbf{v}) + t(c\mathbf{u} + d\mathbf{v}) = (sa + tc)\mathbf{u} + (sb + td)\mathbf{v}$. Since $\{\mathbf{u}, \mathbf{v}\}$ is independent, we have

$$s\mathbf{u}' + t\mathbf{v}' = \mathbf{0} \quad \text{if and only if} \quad sa + tc = 0 \text{ and } sb + td = 0$$

$$\text{if and only if} \quad \begin{bmatrix} a & c \\ b & d \end{bmatrix} \begin{bmatrix} s \\ t \end{bmatrix} = \begin{bmatrix} 0 \\ 0 \end{bmatrix}.$$

Hence $\{\mathbf{u}', \mathbf{v}'\}$ is independent if and only if $\begin{bmatrix} a & c \\ b & d \end{bmatrix} \begin{bmatrix} s \\ t \end{bmatrix} = \begin{bmatrix} 0 \\ 0 \end{bmatrix}$ implies $\begin{bmatrix} s \\ t \end{bmatrix} = \begin{bmatrix} 0 \\ 0 \end{bmatrix}$.

By Theorem 5 §2.3, this is equivalent to A being invertible.

23(b) **Independent:** If $r(\mathbf{u}+\mathbf{v})+s(\mathbf{v}+\mathbf{w})+t(\mathbf{w}+\mathbf{u}) = \mathbf{0}$ then $(r+t)\mathbf{u}+(r+s)\mathbf{v}+(s+t)\mathbf{w} = \mathbf{0}$. Thus $r+t = 0$, $r+s = 0$, $s+t = 0$ (because $\{\mathbf{u}, \mathbf{v}, \mathbf{w}, \mathbf{z}\}$ is independent). Hence $r = s = t = 0$.

(d) **Dependent:** $(\mathbf{u}+\mathbf{v}) - (\mathbf{v}+\mathbf{w}) + (\mathbf{w}+\mathbf{z}) - (\mathbf{z}+\mathbf{u}) = \mathbf{0}$.

26. If $rz + sz^2 = 0$, r, s in $\mathbb{R}$, then $z(r + sz) = 0$. If z is not real then $z \neq 0$ so $r + sz = 0$. Thus $s = 0$ (otherwise $z = \frac{-r}{s}$ is real), whence $r = 0$. Conversely, if z is real then $rz + sz^2 = 0$ when $r = z$, $s = -1$, so $\{z, z^2\}$ is not independent.

28(b) If U is not invertible, let $UX = 0$ where $X \neq 0$ in $\mathbb{R}^n$ (Theorem 5, §2.3). We claim that no set $\{A_1U, A_2U, \dots\}$ can span $\mathbf{M}_{mn}$ (let along be a basis). For if it did, we could write any matrix B in $\mathbf{M}_{mn}$ as a linear combaintion

$$B = a_1 A_1 U + a_2 A_2 U + \cdots$$

Then $BX = a_1 AUX + a_2 A_2 UX + \cdots = 0 + 0 + \cdots = 0$, a contradiction. In fact, if entry k of X is nonzero, then $BX \neq 0$ where all entries of B are zero except column k, which consists of 1's.

32(b) Suppose $U \cap W = 0$. If $s\mathbf{u} + t\mathbf{w} = \mathbf{0}$ with $\mathbf{u}$ and $\mathbf{w}$ nonzero in U and W, then $s\mathbf{u} = -t\mathbf{w}$ is in $U \cap W$. Hence $s\mathbf{u} = \mathbf{0} = t\mathbf{w}$. So $s = 0 = t$ (as $\mathbf{u} \neq \mathbf{0}$ and $\mathbf{w} \neq \mathbf{0}$). Thus $\{\mathbf{u}, \mathbf{v}\}$ is independent. Conversely, assume that the condition holds. If $\mathbf{v} \neq \mathbf{0}$ lies in $U \cap W$, then $\{\mathbf{v}, -\mathbf{v}\}$ is independent by the hypothesis, a contradiction because $1\mathbf{v} + 1(-\mathbf{v}) = \mathbf{0}$.

35(b) If $p(x) = a_0 + a_1 x + \cdots + a_n x^n$ is in 0_n, then $p(-x) = -p(x)$, so

$$a_0 - a_1 x + a_2 x^2 + a_3 x^3 + a_4 x^4 - \cdots = -a_0 - a_1 x - a_2 x^2 - a_3 x^3 - a_4 x^4 - \cdots .$$

Hence $a_0 = a_2 = a_4 = \cdots = 0$ and $p(x) = a_1 x + a_3 x^3 + a_5 x^5 + \cdots$. Thus $0_n = \text{span}\{x, x^3, x^5, \dots\}$ is spanned by the odd powers of x in $\mathbf{P}_n$. The set $B = \{x, x^3, x^5, \dots\}$ is independent (because $\{1, x, x^2, x^3, x^4, \dots\}$ is independent) so it is a basis of 0_n. If n is even, $B = \{x, x^3, x^5, \dots, x^{n-1}\}$ has $\frac{n}{2}$ members, so $\dim 0_n = \frac{n}{2}$. If n is odd, $B = \{x, x^3, x^5, \dots, x^n\}$ has $\frac{n+1}{2}$ members, so $\dim 0_n = \frac{n+1}{2}$.

Exercises 6.4   Existence of Bases

1(b) $B = \{(1,0,0), (0,1,0), (0,1,1)\}$ is independent because $r(1,0,0) + s(0,1,0) + t(0,1,1) = (0,0,0)$ implies $r = 0$, $s+t = 0$, $t = 0$, whence $r = s = t = 0$. Hence B is a basis by Theorem 3 because $\dim \mathbb{R}^3 = 3$.

(d) $B = \{1, x, x^2 - x + 1, x^3\}$ is independent because $r \cdot 1 + sx + t(x^2 - x - 1) + ux^3 = 0$ implies $r - t = 0$, $s - t = 0$, $t = 0$, $u = 0$; whence $r = s = t = u = 0$. Hence B is a basis by Theorem 3 because $\dim \mathbf{P}_3 = 4$.

2(b) As $\mathbf{P}_2 = 3$, any independent set of three vectors is a basis by Theorem 3. Since $-(x^2 + 3) + 2(x+2) + (x^2 - 2x - 1) = 0$, $\{x^2 + 3, x + 2, x^2 - 2x - 1\}$ is dependent. However any other subset of three vectors from $\{x^2 + 3, x + 2, x^2 - 2x - 1, x^2 + x\}$ is independent. (Verify).

3(b) $B = \{(0,1,0,0),(0,0,1,0),(0,0,1,1),(1,1,1,1)\}$ spans $\mathbb{R}^4$ because

$$(1,0,0,0) = (1,1,1,1) - (0,1,0,0) - (0,0,1,1) \quad \text{is in span } B$$
$$(0,0,0,1) = (0,0,1,1) - (0,0,1,0) \quad \text{is in span } B$$

and, of course, $(0,1,0,0)$ and $(0,0,1,0)$ are in span B. Hence B is a basis of $\mathbb{R}^4$ by Theorem 3 because $\dim \mathbb{R}^4 = 4$.

(d) $B = \{1, x^2 + x, x^2 + 1, x^3\}$ spans $\mathbf{P}_3$ because $x^2 = (x^2 + 1) - 1$ and $x = (x^2 + x) - x^2$ are in span B (together with 1 and x^3). So B is a basis of $\mathbf{P}_3$ by Theorem 3 because $\dim \mathbf{P}_3 = 4$.

4(b) Let $z = a + bi$; a, b in $\mathbb{R}$. Then $b \neq 0$ as z is not real and $a \neq 0$ as z is not pure imaginary. Since $\dim \mathbb{C} = 2$, it suffices (by Theorem 3) to show that $\{z, \bar{z}\}$ is independent. If $rz + s\bar{z} = 0$ then $0 = r(a+bi) + s(a-bi) = (r+s)a + (r-s)bi$. Hence $(r+s)a = 0 = (r-s)b$ so (because $a \neq 0 \neq b$) $r + s = 0 = r - s$. Thus $r = s = 0$.

7(b) Not a basis because $(2\mathbf{u} + \mathbf{v} + 3\mathbf{w}) - (3\mathbf{u} + \mathbf{v} - \mathbf{w}) + (\mathbf{u} - 4\mathbf{w}) = \mathbf{0}$.

(d) Not a basis because $2\mathbf{u} - (\mathbf{u} + \mathbf{w}) - (\mathbf{u} - \mathbf{w}) + 0(\mathbf{v} + \mathbf{w}) = \mathbf{0}$.

8(b) Yes, four vectors can span $\mathbb{R}^3$ — say any basis together with any other vector.
No, four vectors in $\mathbb{R}^3$ cannot be independent by Theorem 3 §6.3.

10. If A is the matrix then rank $A = m$ because the m rows are a basis of row A. Hence $m = \text{rank } A \leq n$ by Corollary 3, Theorem 2 §5.2. Hence the m rows of A are independent in $\mathbb{R}^n$, they are part of a basis of $\mathbb{R}^n$. Let U be the $n \times n$ matrix with the rows of A in order at the top and the rest of the basis below. Then U is invertible and fills our requirements.

17(b) The two-dimensional subspaces of $\mathbb{R}^3$ are the planes through the origin, and the one-dimensional subspaces are the lines through the origin. Hence part (a) asserts that if U and W are distinct planes through the origin, then $U \cap W$ is a line through the origin.

20(b) Let $\mathbf{v}_n$ denote the sequence with 1 in the nth coordinate and zeros elsewhere thus $\mathbf{v}_0 = (1,0,0,\dots)$, $\mathbf{v}_1 = (0,1,0,\dots)$ etc. Then $a_0\mathbf{v}_0 + a_1\mathbf{v}_1 + \cdots + a_n\mathbf{v}_n = (a_0, a_1, \dots, a_n, 0, 0, \dots)$ so $a_0\mathbf{v}_0 + a_1\mathbf{v}_1 + \cdots + a_n\mathbf{v}_n$ implies $a_0 = a_1 = \cdots = a_n = 0$. Thus $\{\mathbf{v}_0, \mathbf{v}_1, \dots, \mathbf{v}_n\}$ is an independent set of $n + 1$ vectors. Since n is arbitrary, $\dim V$ cannot be finite by Theorem 5 §6.3.

22(b) $\mathbb{R}\mathbf{u} + \mathbb{R}\mathbf{v} = \{s\mathbf{u} + t\mathbf{v} \mid s\mathbf{u} \text{ in } \mathbb{R}\mathbf{u}, \quad t\mathbf{v} \text{ in } \mathbb{R}\mathbf{v}\} = \text{span}\{\mathbf{u}, \mathbf{v}\}$

Exercises 6.5   An Application to Polynomials

2(b) $f^{(0)}(x) = f(x) = x^3 + x + 1$, so $f^{(1)}(x) = 3x^2 + 1$, $f^{(2)}(x) = 6x$, $f^{(3)}(x) = 6$. Hence, Taylor's theorem gives

$$f(x) = f^{(0)}(1) + f^{(1)}(1)(x - 1) + \frac{f^{(2)}(1)}{2!}(x - 1)^2 + \frac{f^{(3)}(1)}{3!}(x - 1)^3$$
$$= 3 + 4(x - 1) + 3(x - 1)^2 + (x - 1)^3.$$

(d) $f^{(0)}(x) = f(x) = x^3 - 3x^2 + 3x$, $f^{(1)}(x) = 3x^2 - 6x + 3$, $f^{(2)}(x) = 6x - 6$, $f^{(3)}(x) = 6$. Hence, Taylor's theorem gives

$$f(x) = f^{(0)}(1) + f^{(1)}(1)(x - 1) + \frac{f^{(2)}(1)}{2!}(x - 1)^2 + \frac{f^{(3)}(1)}{3!}(x - 1)^3$$
$$= 1 + 0(x - 1) + \frac{0}{2!}(x - 1)^2 + 1(x - 1)^3$$
$$= 1 + (x - 1)^3.$$

6(b) The three polynomials are $x^2 - 3x + 2 = (x - 1)(x - 2)$, $x^2 - 4x + 3 = (x - 1)(x - 3)$ and $x^2 - 5x + 6 = (x - 2)(x - 3)$, so use $a_0 = 3$, $a_1 = 2$, $a_3 = 1$, in Theorem 2.

7(b) The Lagrange polynomials for $a_0 = 1$, $a_1 = 2$, $a_2 = 3$, are

$$\delta_0(x) = \frac{(x - 2)(x - 3)}{(1 - 2)(1 - 3)} = \tfrac{1}{2}(x - 2)(x - 3)$$

$$\delta_1(x) = \frac{(x - 1)(x - 3)}{(2 - 1)(2 - 3)} = -(x - 1)(x - 3)$$

$$\delta_2(x) = \frac{(x - 1)(x - 2)}{(3 - 1)(3 - 2)} = \tfrac{1}{2}(x - 1)(x - 2).$$

Given $f(x) = x^2 + x + 1$:

$$f(x) = f(1)\delta_0(x) + f(2)\delta_1(x) + f(3)\delta_2(x)$$
$$= \tfrac{3}{2}(x - 2)(x - 3) - 7(x - 1)(x - 3) + \tfrac{13}{2}(x - 1)(x - 2).$$

10(b) If $r(x - a)^2 + s(x - a)(x - b) + t(x - b)^2 = 0$, then taking $x = a$ gives $t(a - b)^2 = 0$, so $t = 0$ because $a \neq b$; and taking $x = b$ gives $r(b - a)^2 = 0$, so $r = 0$. Thus, we are left with $s(x - a)(x - b) = 0$. If x is any number except, a, b, this implies $s = 0$. Thus $B = \{(x - a)^2, (x - a)(x - b), (x - b)^2\}$ is independent in $\mathbf{P}_2$; since $\dim \mathbf{P}_2 = 3$, B is a basis.

11(b) Have $U_n = \{f(x) \text{ in } \mathbf{P}_n \mid f(a) = 0 = f(b)\}$. Let $\{p_1(x), \dots, p_{n-1}(x)\}$ be a basis of $\mathbf{P}_{n-2}$; it suffices to show that

$$B = \{(x - a)(x - b)p_1(x), \dots, (x - a)(x - b)p_{n-1}(x)\}$$

is a basis of U_n. Clearly $B \subseteq U_n$.

Independent: Let $s_1(x - a)(x - b)p_1(x) + \cdots + s_{n-1}(x - a)(x - b)p_{n-1}(x) = 0$. Then $(x - a)(x - b)[s_1 p_1(x) + \cdots + s_{n-1} p_{n-1}(x)] = 0$, so (by the hint) $s_1 p_1(x) + \cdots + s_{n-1} p_{n-1}(x) = 0$. Thus $s_1 = s_2 = \cdots = s_{n-1} = 0$.

Spanning: Given $f(x)$ in $\mathbf{P}_n$ with $f(a) = 0$, we have $f(x) = (x-a)g(x)$ for some polynomial $g(x)$ in $\mathbf{P}_{n-1}$ by the factor theorem. But $0 = f(b) = (b-a)g(b)$ so (as $b \neq a$) $g(b) = 0$. Then $g(x) = (x-b)h(x)$ with $h(x) = r_1p_1(x) + \cdots + r_{n-1}p_{n-1}(x)$, r_i in $\mathbb{R}$, whence

$$
\begin{aligned}
f(x) &= (x-a)g(x) \\
&= (x-a)(x-b)g(x) \\
&= (x-a)(x-b)(r_1p_1(x) + \cdots + r_{n-1}p_{n-1}(x)) \\
&= r_1(x-a)(x-b)p_1(x) + \cdots + r_{n-1}(x-a)(x-b)p_{n-1}(x).
\end{aligned}
$$

Exercises 6.6   An Application to Differential Equations

1(b) By Theorem 1, $f(x) = ce^{-x}$ for some constant c. We have $1 = f(1) = ce^{-1}$, so $c = e$. Thus $f(x) = e^{1-x}$.

(d) The characteristic polynomial is $x^2 + x - 6 = (x-2)(x+3)$. Hence $f(x) = ce^{2x} + de^{-3x}$ for some c, d. We have $0 = f(0) = c + d$ and $1 = f(1) = ce^2 + de^{-3}$. Hence, $d = -c$ and $c = \frac{1}{e^2 - e^{-3}}$ so $f(x) = \frac{e^{2x} - e^{-3x}}{e^2 - e^{-3}}$.

(f) The characteristic polynomial is $x^2 - 4x + 4 = (x-2)^2$. Hence, $f(x) = ce^{2x} + dxe^{2x} = (c + dx)e^{2x}$ for some c, d. We have $2 = f(0) = c$ and $0 = f(-1) = (c - d)e^{-2}$. Thus $c = d = 2$ and $f(x) = 2(1 + x)e^{2x}$.

(h) The characteristic polynomial is $x^2 - a^2 = (x-a)(x+a)$, so (as $a \neq -a$) $f(x) = ce^{ax} + de^{-ax}$ for some c, d. We have $1 = f(0) = c + d$ and $0 = f(1) = ce^a + de^{-a}$. Thus $d = 1 - c$ and $c = \frac{1}{1 - e^{2a}}$ whence

$$
f(x) = c^{ax} + (1 - c)e^{-ax} = \frac{e^{ax} - e^{a(2-x)}}{1 - e^{2a}}.
$$

(j) The characteristic polynomial is $x^2 + 4x + 5$. The roots are $\lambda = -2 \pm i$, so $f(x) = e^{-2x}(c\sin x + d\cos x)$ for some real c and d. We have $0 = f(0) = d$ and $1 = f\left(\frac{\pi}{2}\right) = e^{-\pi}(c)$. Hence $f(x) = e^{\pi - 2x}\sin x$.

5(b) If $f(x) = g(x) + 2$ then $f' + f = 2$ becomes $g' + g = 0$, whence $g(x) = ce^{-x}$ for some c. Thus $f(x) = ce^{-x} + 2$ for some c.

6(b) If $f(x) = -\frac{x^3}{3}$ then $f'(x) = -x^2, f''(x) = -2x$, so

$$
f''(x) + f'(x) - 6f(x) = -2x - x^2 + 2x^3.
$$

Hence, $f(x) = \frac{-x^3}{3}$ is a particular solution. Now, if $h = h(x)$ is any solution, write $g(x) = h(x) - f(x) = h(x) + \frac{x^3}{3}$, then

$$
g'' + g' - 6g = (h'' + h' - 6h) - (f'' + f' - 6f) = 0.
$$

So, to find g, the characteristic polynomial is $x^2 + x - 6 = (x-2)(x+3)$. Hence, $g(x) = ce^{-3x} + de^{2x}$, where c and d are constants, so

$$
h(x) = ce^{-3x} + de^{2x} - \frac{x^3}{3}.
$$

7(b) If $m = m(t)$ is the mass at time t, then the rate $m'(t)$ of decay b is proportional to $m(t)$: $m'(t) = km(t)$ for some k. Thus, $m' - km = 0$ so $m = ce^{kt}$ for some constant c. Since $m(0) = 10$, we obtain $c = 10$, whence $m(t) = 10e^{kt}$. Also, $8 = m(3) = 10e^{3k}$ so $e^{3k} = \frac{4}{5}$, $e^k = \left(\frac{4}{5}\right)^{1/3}$, $m(t) = 10(e^k)^t = 10\left(\frac{4}{5}\right)^{t/3}$.

9. In Example 5, we found that the period of oscillation is $\frac{2\pi}{\sqrt{k}}$. Hence $\frac{2\pi}{\sqrt{k}} = 30$ so we obtain $k = \left(\frac{\pi}{15}\right)^2 = 0.044$.

Supplementary Exercises   Chapter 6

2(b) Suppose $\{Ax_1, \ldots, Ax_n\}$ is a basis of $\mathbb{R}^n$. To show that A is invertible, we show that $YA = 0$ implies $Y = 0$. (This shows A^T is invertible by Theorem 5 §2.3, so A is invertible). So assume that $YA = 0$. Let $C_1, \ldots, C_m$ denote the columns of I_m, so $I_m = [C_1, C_2, \ldots, C_m]$. Then $Y = YI_m = Y[C_1 \quad C_2 \quad \ldots \quad C_m] = [YC_1 \quad YC_2 \quad \ldots \quad YC_m]$, so it suffices to show that $YC_j = 0$ for each j. But C_j is in $\mathbb{R}^n$ so our hypothesis shows that $C_j = r_1 A\mathbf{v}_1 + \cdots + r_n A\mathbf{v}_n$ for some r_j in $\mathbb{R}$. Hence,

$$C_j = A(r_1\mathbf{v}_1 + \cdots + r_n\mathbf{v}_n)$$

so $YC_j = YA(r_1\mathbf{v}_1 + \cdots + r_n\mathbf{v}_n) = 0$, as required.

4. Assume that A is $m \times n$. If X is in null A, then $AX = 0$ so $(A^TA)X = A^T0 = 0$. Thus X is n null A^TA, so null $A \subseteq$ null A^TA. Conversely, let X be in A^TA; that is $A^TAX = 0$. Write

$$AX = Y = \begin{bmatrix} y_1 \\ \vdots \\ y_m \end{bmatrix}.$$

Then $y_1^2 + y_2^2 + \cdots + y_m^2 = Y^TY = (AX)^T(AX) = X^TA^TAX = X^T0 = 0$. Since the y_i are real numbers, this implies that $y_1 = y_2 = \cdots = y_m = 0$; that is $Y = 0$, that is $AX = 0$, that is X is in null A.

Chapter 7: Orthogonality

Exercises 7.1   Orthogonality in $\mathbb{R}^n$

1(b) $\left\{ \frac{1}{\sqrt{3}} \begin{bmatrix} 1 & 1 & 1 \end{bmatrix}, \frac{1}{\sqrt{42}} \begin{bmatrix} 4 & 1 & -5 \end{bmatrix}, \frac{1}{\sqrt{14}} \begin{bmatrix} 2 & -3 & 1 \end{bmatrix} \right\}$

3(b) Write $E_1 = \begin{bmatrix} 1 & 0 & -1 \end{bmatrix}$, $E_2 = \begin{bmatrix} 1 & 4 & 1 \end{bmatrix}$, $E_3 = \begin{bmatrix} 2 & -1 & 2 \end{bmatrix}$. Then $E_1 \cdot E_2 = 1 + 0 - 1 = 0$, $E_1 \cdot E_3 = 2 + 0 - 2 = 0$, $E_2 \cdot E_3 = 2 - 4 + 2 = 0$, so $\{E_1, E_2, E_3\}$ is orthogonal and hence a basis of $\mathbb{R}^3$. If $X = \begin{bmatrix} a & b & c \end{bmatrix}$, Theorem 4 gives

$$\begin{bmatrix} a & b & c \end{bmatrix} = X = \frac{X \cdot E_1}{\|E_1\|^2} E_1 + \frac{X \cdot E_2}{\|E_2\|^2} E_2 + \frac{X \cdot E_3}{\|E_3\|^2} E_3$$
$$= \left(\frac{a - c}{2} \right) E_1 + \left(\frac{a + 4b + c}{6} \right) E_2 + \left(\frac{2a - b + 2c}{9} \right) E_3.$$

(d) Analogous to (b).

4(b) If $E_1 = \begin{bmatrix} 2 & -1 & 0 & 3 \end{bmatrix}$ and $E_2 = \begin{bmatrix} 2 & 1 & -2 & -1 \end{bmatrix}$ then $\{E_1, E_2\}$ is orthogonal, and so is an orthogonal basis of the space U it spans. If $X = \begin{bmatrix} 14 & 1 & -8 & 5 \end{bmatrix}$ is in U, Theorem 4 gives

$$X = \frac{X \cdot E_1}{\|E_1\|^2} E_1 + \frac{X \cdot E_2}{\|E_2\|^2} E_2 = \tfrac{42}{14} E_1 + \tfrac{40}{10} E_2 = 3E_1 + 4E_2.$$

[The fact that these are equal confirms that X is in U; in any case, $X - \left(\dfrac{X \cdot E_1}{\|E_1\|^2} E_1 + \dfrac{X \cdot E_2}{\|E_2\|^2} E_2 \right)$ is orthogonal to U by Theorem 7.]

5(b) The condition that $\begin{bmatrix} a & b & c & d \end{bmatrix}$ is orthogonal to each of the other rows gives equations for a, b, c, and d.

$$\begin{array}{rrrrrl} a & & - c & + d & = & 0 \\ 2a & + b & + c & - d & = & 0 \\ a & - 3b & + c & & = & 0 \end{array}$$

$$\begin{bmatrix} 1 & 0 & -1 & 1 & | & 0 \\ 2 & 1 & 1 & -1 & | & 0 \\ 1 & -3 & 1 & 0 & | & 0 \end{bmatrix} \rightarrow \begin{bmatrix} 1 & 0 & -1 & 1 & | & 0 \\ 0 & 1 & 3 & -3 & | & 0 \\ 1 & -3 & 2 & -1 & | & 0 \end{bmatrix}$$

$$\rightarrow \begin{bmatrix} 1 & 0 & -1 & 1 & | & 0 \\ 0 & 1 & 3 & -3 & | & 0 \\ 0 & 0 & 11 & -10 & | & 0 \end{bmatrix} \rightarrow \begin{bmatrix} 1 & 0 & 0 & \frac{1}{11} & | & 0 \\ 0 & 1 & 0 & -\frac{3}{11} & | & 0 \\ 0 & 0 & 1 & -\frac{10}{11} & | & 0 \end{bmatrix}.$$

The solution is $\begin{bmatrix} a & b & c & d \end{bmatrix} = t \begin{bmatrix} -1 & 3 & 10 & 11 \end{bmatrix}$, t in $\mathbb{R}$.

7(b) $$\begin{aligned} \|2X + 7Y\|^2 &= (2X + 6Y) \cdot (2X + 7Y) \\ &= 4(X \cdot X) + 14(X \cdot Y) + 14(Y \cdot X) + 49(Y \cdot Y) \\ &= 4\|X\|^2 + 28(X \cdot Y) + 49\|Y\|^2 \\ &= 36 - 56 + 49 \\ &= 29. \end{aligned}$$

(d) $\begin{aligned}(X - 2Y) \cdot (3X + 5Y) &= 3(X \cdot X) + 5(X \cdot Y) - 6(Y \cdot X) - 10(Y \cdot Y) \\ &= 3\|X\|^2 - (X \cdot Y) - 10\|Y\|^2 \\ &= 27 + 2 - 10 \\ &= 19.\end{aligned}$

8(b) Write $X_1 = [2 \quad 1]$ and $X_2 = [1 \quad 2]$. The Gram-Schmidt algorithm gives

$$\begin{aligned} E_1 &= X_1 = [2 \quad 1] \\ E_2 &= X_2 - \frac{X_2 \cdot E_1}{\|E_1\|^2} E_1 \\ &= [1 \quad 2] - \tfrac{4}{5}[2 \quad 1] \\ &= \tfrac{1}{5}\{[5 \quad 10] - [8 \quad 4]\} \\ &= \tfrac{3}{5}[-1 \quad 2]. \end{aligned}$$

In hand calculations, $\{[2 \quad 1], [-1 \quad 2]\}$ may be a more convenient orthogonal basis.

(d) If $X_1 = [0 \quad 1 \quad 1]$, $X_2 = [1 \quad 1 \quad 1]$, $X_3 = [1 \quad -2 \quad 2]$ then
$E_1 = X_1 = 0 \quad 1 \quad 1]$
$E_2 = X_2 - \dfrac{X_2 \cdot E_1}{\|E_1\|^2} E_1 = [1 \quad 1 \quad 1] - \tfrac{2}{2}[0 \quad 1 \quad 1] = [1 \quad 0 \quad 0]$
$E_3 = X_3 - \dfrac{X_3 \cdot E_1}{\|E_1\|^2} E_1 - \dfrac{X_3 \cdot E_2}{\|E_2\|^2} E_2 = [1 \quad -2 \quad 2] - \tfrac{0}{2}[0 \quad 1 \quad 1] - \tfrac{1}{1}[1 \quad 0 \quad 0] = [0 \quad -2 \quad 2].$

9(b) Write $E_1 = [3 \quad -1 \quad 2]$ and $E_2 = [2 \quad 0 \quad -3]$. Then $\{E_1, E_2\}$ is orthogonal and so is an orthogonal basis of $U = \text{span}\{E_1, E_2\}$. Now $X = [2 \quad 1 \quad 6]$ so take

$$\begin{aligned} X_1 = \text{proj}_U(X) &= \frac{X \cdot E_1}{\|E_1\|^2} E_1 + \frac{X \cdot E_2}{\|E_2\|^2} E_2 \\ &= \tfrac{17}{14}[3 \quad -1 \quad 2] - \tfrac{14}{13}[2 \quad 0 \quad -3] \\ &= \tfrac{1}{182}[271 \quad -221 \quad 1030]. \end{aligned}$$

Then $X_2 = X - X_1 = \tfrac{1}{182}[93 \quad 402 \quad 62]$. As a check: X_2 is orthogonal to both E_1 and E_2 (and so is in $U^\perp$).

(d) If $E_1 = [1 \quad 1 \quad 1 \quad 1]$, $E_2 = [1 \quad 1 \quad -1 \quad -1]$, $E_3 = [1 \quad -1 \quad 1 \quad -1]$ and $X = [2 \quad 0 \quad 1 \quad 6]$, then $\{E_1, E_2, E_3\}$ is orthogonal so take

$$\begin{aligned} X_1 = \text{proj}_U(X) &= \frac{X \cdot E_1}{\|E_1\|^2} E_1 + \frac{X \cdot E_2}{\|E_2\|^2} E_2 + \frac{X \cdot E_3}{\|E_3\|^2} E_3 \\ &= \tfrac{9}{4}[1 \quad 1 \quad 1 \quad 1] - \tfrac{5}{4}[1 \quad 1 \quad -1 \quad -1] - \tfrac{3}{4}[1 \quad -1 \quad 1 \quad -1] \\ &= \tfrac{1}{4}[1 \quad 7 \quad 11 \quad 17]. \end{aligned}$$

Then, $X_2 = X - X_1 = \tfrac{1}{4}[7 \quad -7 \quad -7 \quad 7] = \tfrac{7}{4}[1 \quad -1 \quad -1 \quad 1]$. Check: X_2 is orthogonal to each E_i, hence X_2 is in $U^\perp$.

(f) If $E_1 = [1 \quad -1 \quad 2 \quad 0]$ and $E_2 = [-1 \quad 1 \quad 1 \quad 1]$ then (as $X = [a \quad b \quad c \quad d]$)

$$X_1 = \text{proj}_U(X) = \frac{a-b+2c}{6}[1 \quad -1 \quad 2 \quad 0] + \frac{-a+b+c+d}{4}[-1 \quad 1 \quad 1 \quad 1]$$

$$= \left[\frac{5a-5b+c-3d}{12} \quad \frac{-5a+5b-c+3d}{12} \quad \frac{a-b+11c+3d}{12} \quad \frac{-3a+3b+3c+3d}{12}\right]$$

$$X_2 = X - X_1 = \left[\frac{7a+5b-c+3d}{12} \quad \frac{5a+7b+c-3d}{12} \quad \frac{-a+b+c-3d}{12} \quad \frac{3a-3b-3c+9d}{12}\right].$$

10(a) Write $E_1 = [2 \quad 1 \quad 3 \quad -4]$ and $E_2 = [1 \quad 2 \quad 0 \quad 1]$, so $\{E_1, E_2\}$ is orthogonal.
As $X = [1 \quad -2 \quad 1 \quad 6]$

$$\text{proj}_U(X) = \frac{X \cdot E_1}{\|E_1\|^2}E_1 + \frac{X \cdot E_2}{\|E_2\|^2}E_2$$

$$= -\tfrac{21}{30}[2 \quad 1 \quad 3 \quad -4] + \tfrac{3}{6}[1 \quad 2 \quad 0 \quad 1] = \tfrac{3}{10}[-3 \quad 1 \quad -7 \quad 11].$$

(c) $\text{proj}_U(X) = -\tfrac{15}{14}[1 \quad 0 \quad 2 \quad -3] + \tfrac{3}{70}[4 \quad 7 \quad 1 \quad 2] = \tfrac{3}{10}[-3 \quad 1 \quad -7 \quad 11].$

11(b) $U = \text{span}\{[1 \quad -1 \quad 0], [-1 \quad 0 \quad 1]\}$ but this basis is not orthogonal. By Gram-Schmidt:

$$E_1 = [1 \quad -1 \quad 0]$$

$$E_2 = [-1 \quad 0 \quad 1] - \frac{[-1 \quad 0 \quad 1] \cdot [1 \quad -1 \quad 0]}{\|[1 \quad -1 \quad 0]\|^2}[1 \quad -1 \quad 0] = -\tfrac{1}{2}[1 \quad 1 \quad -2].$$

So we use $U = \text{span}\{[1 \quad -1 \quad 0], [1 \quad 1 \quad -2]\}$. Then the vector X_1 in U closest to $X = [2 \quad 1 \quad 0]$ is

$$X_1 = \text{proj}_U(X) = \frac{2-1+0}{2}[1 \quad -1 \quad 0] + \frac{2+1+0}{6}[1 \quad 1 \quad -2] = [1 \quad 0 \quad -1].$$

(d) The given basis of U is not orthogonal. The Gram-Schmidt algorithm gives

$$E_1 = [1 \quad -1 \quad 0 \quad 1]$$
$$E_2 = [1 \quad 1 \quad 0 \quad 0] = \tfrac{0}{3}E_1 = [1 \quad 1 \quad 0 \quad 0]$$
$$E_3 = [1 \quad 1 \quad 0 \quad 1] - \tfrac{1}{3}[1 \quad -1 \quad 0 \quad 1] - \tfrac{2}{2}[1 \quad 1 \quad 0 \quad 0] = \tfrac{1}{3}[-1 \quad 1 \quad 0 \quad 2].$$

Given $X = [2 \quad 0 \quad 3 \quad 1]$, we get (using $E_3' = [-1 \quad 1 \quad 0 \quad 2]$ for convenience)
$\text{proj}_U(X) = \tfrac{3}{3}[1 \quad -1 \quad 0 \quad 1] + \tfrac{2}{2}[1 \quad 1 \quad 0 \quad 0] + \tfrac{0}{6}[-1 \quad 1 \quad 0 \quad 2] = [2 \quad 0 \quad 0 \quad 1].$

12(b) Here $A = \begin{bmatrix} 1 & -1 & 2 & 1 \\ 1 & 0 & -1 & 1 \end{bmatrix} \to \begin{bmatrix} 1 & -1 & 2 & 1 \\ 0 & 1 & -3 & 0 \end{bmatrix} \to \begin{bmatrix} 1 & 0 & -1 & 1 \\ 0 & 1 & -3 & 0 \end{bmatrix}$. Hence, $AX^T = 0$ has solution $X = [s-t \quad 3s \quad s \quad t] = s[1 \quad 3 \quad 1 \quad 0] + t[-1 \quad 0 \quad 0 \quad 1]$. Thus $U^\perp = \text{span}\{[1 \quad 3 \quad 1 \quad 0], [-1 \quad 0 \quad 0 \quad 1]\}$.

13(b) $(X+Y) \cdot (X-Y) = X \cdot X - X \cdot Y + Y \cdot X - Y \cdot Y = \|X\|^2 - \|Y\|^2$. Hence, $X+Y$ and $X-Y$ are orthogonal (that is $(X+Y) \cdot (X-Y) = 0$) if and only if $\|X\|^2 = \|Y\|^2$, that is (as $\|X\|$ and $\|Y\|$ are non-negative) if and only if $\|X\| = \|Y\|$.

20. If X is in $U^\perp$ then $X \cdot Y = 0$ for all Y in U. In particular, $X \cdot X_i = 0$ for all i because each X_i is in U. Conversely, if $X \cdot X_i = 0$ for all i, let Y be any vector in $U = \text{span}\{X_1, X_2, \dots, X_m\}$. Then $Y = r_1X_1 + r_2X_2 + \cdots + r_mX_m$ for some r_i in $\mathbb{R}$, so

$$X \cdot Y = X \cdot (r_1X_1 + r_2X_2 + \cdots + r_mX_m)$$
$$= r_1X \cdot X_1 + r_2X \cdot X_2 + \cdots + r_mX \cdot X_m$$
$$= r_1 \cdot 0 + r_2 \cdot 0 + \cdots + r_m \cdot 0$$
$$= 0.$$

Thus X is in $U^\perp$.

24. If $\text{proj}_U(X) = 0$ then $X = X - \text{proj}_U(X)$ so X is in $U^\perp$ by the projection theorem. Conversely, if X is in $U^\perp$ write $P = \text{proj}_U(X)$. Then both X and $X - P$ are in $U^\perp$, so $P = X - (X - P)$ is in $U^\perp$. But P is also in U (projection theorem) so P is in both U and $U^\perp$. This means that P is orthogonal to itself:

$$0 = P \cdot P = \|P\|^2.$$

So $\|P\| = 0$, whence $P = 0$.

31(d) If AA^T is invertible and $E = A^T(AA^T)^{-1}A$, then

$$E^2 = A^T(AA^T)^{-1}A \cdot A^T(AA^T)^{-1}A = A^TI(AA^T)^{-1}A = E$$
$$E^T = \left[A^T(AA^T)^{-1}A\right]^T = A^T\left[(AA^T)^{-1}\right]^T(A^T)^T$$
$$= A^T\left[(AA^T)^T\right]^{-1}A = A^T\left[(A^T)^TA^T\right]^{-1}A$$
$$= A^T\left[AA^T\right]^{-1}A = E.$$

Thus, $E^2 = E = E^T$.

Exercises 7.2   Orthogonal Diagonalization

1(b) Since $3^2 + 4^2 = 5^2$, each row has length 5. So $\begin{bmatrix} \frac{3}{5} & \frac{-4}{5} \\ \frac{4}{5} & \frac{3}{5} \end{bmatrix} = \frac{1}{5}\begin{bmatrix} 3 & -4 \\ 4 & 3 \end{bmatrix}$ is orthogonal.

(d) Each row has length $\sqrt{a^2 + b^2}$, so $\frac{1}{\sqrt{a^2+b^2}}\begin{bmatrix} a & b \\ -b & a \end{bmatrix}$ is orthogonal.

(f) The rows have length $\sqrt{6}$, $\sqrt{3}$, $\sqrt{2}$ respectively, so

$$\begin{bmatrix} \frac{2}{\sqrt{6}} & \frac{1}{\sqrt{6}} & -\frac{1}{\sqrt{6}} \\ \frac{1}{\sqrt{3}} & -\frac{1}{\sqrt{3}} & \frac{1}{\sqrt{3}} \\ 0 & \frac{1}{\sqrt{2}} & \frac{1}{\sqrt{2}} \end{bmatrix} = \frac{1}{\sqrt{6}}\begin{bmatrix} 2 & 1 & -1 \\ \sqrt{2} & -\sqrt{2} & \sqrt{2} \\ 0 & \sqrt{3} & \sqrt{3} \end{bmatrix}$$

is orthogonal.

(h) Each row has length $\sqrt{4 + 36 + 9} = \sqrt{49} = 7$. Hence

$$\begin{bmatrix} \frac{2}{7} & \frac{6}{7} & -\frac{3}{7} \\ \frac{3}{7} & \frac{2}{7} & \frac{6}{7} \\ -\frac{6}{7} & \frac{3}{7} & \frac{2}{7} \end{bmatrix} = \frac{1}{7} \begin{bmatrix} 2 & 6 & -3 \\ 3 & 2 & 6 \\ -6 & 3 & 2 \end{bmatrix}$$

is orthogonal.

2. Let P be orthogonal, so $P^{-1} = P^T$. If P is upper triangular, so also is P^{-1}, so $P^{-1} = P^T$ is both upper triangular (P^{-1}) and lower triangular P^T). Hence, $P^{-1} = P^T$ is diagonal, whence $P = (P^{-1})^{-1}$ is diagonal. In particular, P is symmetric so $P^{-1} = P^T = P$. Thus $P^2 = I$. Since P is diagonal, this implies that all diagonal entries are ± 1.

5(b) $c_A(x) = \begin{vmatrix} x - 3 & 0 & -7 \\ 0 & x - 5 & 0 \\ -7 & 0 & x - 3 \end{vmatrix} = (x - 5)(x^2 - 6x - 40) = (x - 5)(x + 4)(x - 10)$. Hence the

eigenvalues are $\lambda_1 = 5$, $\lambda_2 = 10$, $\lambda_3 = -4$.

$\lambda_1 = 5 : \begin{bmatrix} 2 & 0 & -7 \\ 0 & 0 & 0 \\ -7 & 0 & 2 \end{bmatrix} \rightarrow \begin{bmatrix} 1 & 0 & 0 \\ 0 & 0 & 1 \\ 0 & 0 & 0 \end{bmatrix}$; $E_5(A) = \text{span} \left\{ \begin{bmatrix} 0 \\ 1 \\ 0 \end{bmatrix} \right\}$.

$\lambda_2 = 10 : \begin{bmatrix} 7 & 0 & -7 \\ 0 & 5 & 0 \\ -7 & 0 & 7 \end{bmatrix} \rightarrow \begin{bmatrix} 1 & 0 & -1 \\ 0 & 1 & 0 \\ 0 & 0 & 0 \end{bmatrix}$; $E_{10}(A) = \text{span} \left\{ \begin{bmatrix} 1 \\ 0 \\ 1 \end{bmatrix} \right\}$.

$\lambda_3 = -4 : \begin{bmatrix} -7 & 0 & -7 \\ 0 & -9 & 0 \\ -7 & 0 & -7 \end{bmatrix} \rightarrow \begin{bmatrix} 1 & 0 & 1 \\ 0 & 1 & 0 \\ 0 & 0 & 0 \end{bmatrix}$; $E_{-4}(A) = \text{span} \left\{ \begin{bmatrix} 1 \\ 0 \\ -1 \end{bmatrix} \right\}$.

Note that the three eigenvectors are pairwise orthogonal (as Theorem 4 asserts). Normalizing them gives an orthogonal matrix

$$P = \begin{bmatrix} 0 & \frac{1}{\sqrt{2}} & \frac{1}{\sqrt{2}} \\ 1 & 0 & 0 \\ 0 & \frac{1}{\sqrt{2}} & -\frac{1}{\sqrt{2}} \end{bmatrix} = \frac{1}{\sqrt{2}} \begin{bmatrix} 0 & 1 & 1 \\ \sqrt{2} & 0 & 0 \\ 0 & 1 & -1 \end{bmatrix}.$$

Then $P^{-1} = P^T$ and $P^T A P = \begin{bmatrix} 5 & 0 & 0 \\ 0 & 10 & 0 \\ 0 & 0 & -4 \end{bmatrix}$.

(d) $c_A(x) = \begin{vmatrix} x - 5 & 2 & 4 \\ 2 & x - 8 & 2 \\ 4 & 2 & x - 5 \end{vmatrix} = \begin{vmatrix} x - 9 & 0 & 9 - x \\ 2 & x - 8 & 2 \\ 4 & 2 & x - 5 \end{vmatrix} = \begin{vmatrix} x - 9 & 0 & 0 \\ 2 & x - 8 & 4 \\ 4 & 2 & x - 1 \end{vmatrix}$

$$= (x-9)\begin{vmatrix} x-8 & 4 \\ 2 & x-1 \end{vmatrix} = (x-9)(x^2-9x) = x(x-9)^2. \text{ The eigenvalues are } \lambda_1 = 0, \lambda_2 = 9.$$

$$\lambda_1 = 0: \begin{bmatrix} -5 & 2 & 4 \\ 2 & -8 & 2 \\ 4 & 2 & -5 \end{bmatrix} \rightarrow \begin{bmatrix} 1 & -4 & 1 \\ 0 & -18 & 9 \\ 0 & 18 & -9 \end{bmatrix} \rightarrow \begin{bmatrix} 1 & -4 & 1 \\ 0 & 1 & -\frac{1}{2} \\ 0 & 0 & 0 \end{bmatrix} \rightarrow \begin{bmatrix} 1 & 0 & -1 \\ 0 & 1 & -\frac{1}{2} \\ 0 & 0 & 0 \end{bmatrix};$$

$$E_0(A) = \text{span}\left\{ \begin{bmatrix} 2 \\ 1 \\ 2 \end{bmatrix} \right\}.$$

$$\lambda_2 = 9: \begin{bmatrix} 4 & 2 & 4 \\ 2 & 1 & 2 \\ 4 & 2 & 4 \end{bmatrix} \rightarrow \begin{bmatrix} 1 & \frac{1}{2} & 1 \\ 0 & 0 & 0 \\ 0 & 0 & 0 \end{bmatrix}; E_9(A) = \text{span}\left\{ \begin{bmatrix} -1 \\ 0 \\ 1 \end{bmatrix}, \begin{bmatrix} -1 \\ 2 \\ 0 \end{bmatrix} \right\}.$$

However, these are not orthogonal and the Gram-Schmidt algorithm replaces $\begin{bmatrix} -1 \\ 2 \\ 0 \end{bmatrix}$ with

$$Z_2 = \begin{bmatrix} 1 \\ -4 \\ 1 \end{bmatrix}. \text{ Hence } P = \begin{bmatrix} \frac{2}{3} & \frac{-1}{\sqrt{2}} & \frac{1}{3\sqrt{2}} \\ \frac{1}{3} & 0 & \frac{-4}{3\sqrt{2}} \\ \frac{2}{3} & \frac{1}{\sqrt{2}} & \frac{1}{3\sqrt{2}} \end{bmatrix} = \frac{1}{3\sqrt{2}}\begin{bmatrix} 2\sqrt{2} & -3 & 1 \\ \sqrt{2} & 0 & -4 \\ 2\sqrt{2} & 3 & 1 \end{bmatrix} \text{ is orthogonal and}$$

satisfies $P^T A P = \begin{bmatrix} 0 & 0 & 0 \\ 0 & 9 & 0 \\ 0 & 0 & 9 \end{bmatrix}$. Note that $\begin{bmatrix} -2 \\ 2 \\ 1 \end{bmatrix}$ and $\begin{bmatrix} 1 \\ 2 \\ -2 \end{bmatrix}$ are another orthogonal

basis of $E_9(A)$, so $Q = \frac{1}{3}\begin{bmatrix} 2 & -2 & 1 \\ 1 & 2 & 2 \\ 2 & 1 & -2 \end{bmatrix}$ also satisfies $Q^T A Q = \begin{bmatrix} 0 & 0 & 0 \\ 0 & 9 & 0 \\ 0 & 0 & 9 \end{bmatrix}$.

(f) To evaluate $c_A(x)$, we begin adding rows 2, 3 and 4 to row 1.

$$c_A(x) = \begin{vmatrix} x-3 & -5 & 1 & -1 \\ -5 & x-3 & -1 & 1 \\ 1 & -1 & x-3 & -5 \\ -1 & 1 & -5 & x-3 \end{vmatrix} = \begin{vmatrix} x-8 & x-8 & x-8 & x-8 \\ -5 & x-3 & -1 & 1 \\ 1 & -1 & x-3 & -5 \\ -1 & 1 & -5 & x-3 \end{vmatrix}$$

$$= \begin{vmatrix} x-8 & 0 & 0 & 0 \\ -5 & x-2 & 4 & 6 \\ 1 & -2 & x-4 & -6 \\ -1 & 2 & -4 & x-2 \end{vmatrix} = (x-8)\begin{vmatrix} x+2 & 4 & 6 \\ -2 & x-4 & -6 \\ 2 & -4 & x-2 \end{vmatrix}$$

$$= (x-8) \begin{vmatrix} x+2 & 4 & 6 \\ x & x & 0 \\ 2 & -4 & x-2 \end{vmatrix} = (x-8) \begin{vmatrix} x-2 & 4 & 6 \\ 0 & x & 0 \\ 6 & -4 & x-2 \end{vmatrix}$$

$$= x(x-8) \begin{vmatrix} x-2 & 6 \\ 6 & x-2 \end{vmatrix} = x(x-8)(x^2 - 4x - 32) = x(x+4)(x-8)^2.$$

$\lambda_1 = 0$:
$$\begin{bmatrix} -3 & -5 & 1 & -1 \\ -5 & -3 & -1 & 1 \\ 1 & -1 & -3 & -5 \\ -1 & 1 & -5 & -3 \end{bmatrix} \rightarrow \begin{bmatrix} -3 & -5 & 1 & -1 \\ -8 & -8 & 0 & 0 \\ 1 & -1 & -3 & -5 \\ 0 & 0 & -8 & -8 \end{bmatrix} \rightarrow \begin{bmatrix} 1 & -1 & -3 & -5 \\ 0 & -8 & -8 & -16 \\ 0 & -16 & -24 & -40 \\ 0 & 0 & 1 & 1 \end{bmatrix}$$

$$\rightarrow \begin{bmatrix} 1 & 0 & -2 & -3 \\ 0 & 1 & 1 & 2 \\ 0 & 0 & 1 & 1 \\ 0 & 0 & 1 & 1 \end{bmatrix} \rightarrow \begin{bmatrix} 1 & 0 & 0 & -1 \\ 0 & 1 & 0 & 1 \\ 0 & 0 & 1 & 1 \\ 0 & 0 & 0 & 0 \end{bmatrix}; \; E_0(A) = \text{span} \left\{ \begin{bmatrix} 1 \\ -1 \\ -1 \\ 1 \end{bmatrix} \right\}.$$

$\lambda_2 = -4$:
$$\begin{bmatrix} -7 & -5 & 1 & -1 \\ -5 & -7 & -1 & 1 \\ 1 & -1 & -7 & -5 \\ -1 & 1 & -5 & -7 \end{bmatrix} \rightarrow \begin{bmatrix} 1 & -1 & -7 & -5 \\ 0 & -12 & -48 & -36 \\ 0 & -12 & -36 & -24 \\ 0 & 0 & -12 & -12 \end{bmatrix} \rightarrow \begin{bmatrix} 1 & 0 & -3 & -2 \\ 0 & 1 & 4 & 3 \\ 0 & 0 & -1 & -1 \\ 0 & 0 & 1 & 1 \end{bmatrix}$$

$$\rightarrow \begin{bmatrix} 1 & 0 & 0 & 1 \\ 0 & 1 & 0 & -1 \\ 0 & 0 & 1 & 1 \\ 0 & 0 & 0 & 0 \end{bmatrix}; \; E_{-4}(A) = \text{span} \left\{ \begin{bmatrix} -1 \\ 1 \\ -1 \\ 1 \end{bmatrix} \right\}.$$

$\lambda_3 = 8$:
$$\begin{bmatrix} 5 & -5 & 1 & -1 \\ -5 & 5 & -1 & 1 \\ 1 & -1 & 5 & -5 \\ -1 & 1 & -5 & 5 \end{bmatrix} \rightarrow \begin{bmatrix} 1 & -1 & 5 & -5 \\ 0 & 0 & -24 & 24 \\ 0 & 0 & 24 & -24 \\ 0 & 0 & 0 & 0 \end{bmatrix} \rightarrow \begin{bmatrix} 1 & -1 & 0 & 0 \\ 0 & 0 & 1 & -1 \\ 0 & 0 & 0 & 0 \\ 0 & 0 & 0 & 0 \end{bmatrix};$$

$$E_8(A) = \text{span} \left\{ \begin{bmatrix} 1 \\ 1 \\ 0 \\ 0 \end{bmatrix}, \begin{bmatrix} 0 \\ 0 \\ 1 \\ 1 \end{bmatrix} \right\}.$$

Hence, $P = \begin{bmatrix} \frac{1}{2} & -\frac{1}{2} & \frac{1}{\sqrt{2}} & 0 \\ -\frac{1}{2} & \frac{1}{2} & \frac{1}{\sqrt{2}} & 0 \\ -\frac{1}{2} & -\frac{1}{2} & 0 & \frac{1}{\sqrt{2}} \\ \frac{1}{2} & \frac{1}{2} & 0 & \frac{1}{\sqrt{2}} \end{bmatrix} = \frac{1}{2} \begin{bmatrix} 1 & -1 & \sqrt{2} & 0 \\ -1 & 1 & \sqrt{2} & 0 \\ -1 & -1 & 0 & \sqrt{2} \\ 1 & 1 & 0 & \sqrt{2} \end{bmatrix}$ gives $P^T A P = \begin{bmatrix} 0 & 0 & 0 & 0 \\ 0 & -4 & 0 & 0 \\ 0 & 0 & 8 & 0 \\ 0 & 0 & 0 & 8 \end{bmatrix}.$

6. $c_A(x) = \begin{vmatrix} x & -a & 0 \\ -a & x & -c \\ 0 & -c & x \end{vmatrix} = x \begin{vmatrix} x & -c \\ -c & x \end{vmatrix} + a \begin{vmatrix} -a & 0 \\ -c & x \end{vmatrix} = x(x^2 - c^2) - a^2 x = x(x^2 - k^2) =$

$x(x-k)(x+k)$, where $k^2 = a^2 + c^2$. So the eigenvalues are $\lambda_1 = 0$, $\lambda_2 = k$, $\lambda_3 = -k$. They are all distinct ($k \neq 0$, and $a \neq 0$ or $c \neq 0$) so the eigenspaces are all one dimensional.

$\lambda_1 = 0 : \begin{bmatrix} 0 & -a & 0 \\ -a & 0 & -c \\ 0 & -c & 0 \end{bmatrix} \begin{bmatrix} c \\ 0 \\ -a \end{bmatrix} = \begin{bmatrix} 0 \\ 0 \\ 0 \end{bmatrix}$; $E_0(A) = \text{span} \left\{ \begin{bmatrix} c \\ 0 \\ -a \end{bmatrix} \right\}$.

$\lambda_2 = k : \begin{bmatrix} k & -a & 0 \\ -a & k & -c \\ 0 & -c & k \end{bmatrix} \begin{bmatrix} a \\ k \\ c \end{bmatrix} = \begin{bmatrix} 0 \\ 0 \\ 0 \end{bmatrix}$; $E_k(A) = \text{span} \left\{ \begin{bmatrix} a \\ k \\ c \end{bmatrix} \right\}$.

$\lambda_3 = -k : \begin{bmatrix} -k & -a & 0 \\ -a & -k & -c \\ 0 & -c & -k \end{bmatrix} \begin{bmatrix} a \\ -k \\ c \end{bmatrix} = \begin{bmatrix} 0 \\ 0 \\ 0 \end{bmatrix}$; $E_{-k}(A) = \text{span} \left\{ \begin{bmatrix} a \\ -k \\ c \end{bmatrix} \right\}$,

These eigenvalues are orthogonal and have length, k, $\sqrt{2}k$, $\sqrt{2}k$ respectively. Hence, $P =$

$\frac{1}{\sqrt{2}k} \begin{bmatrix} c\sqrt{2} & a & a \\ 0 & k & -k \\ -a\sqrt{2} & c & c \end{bmatrix}$ is orthogonal and $P^T A P = \begin{bmatrix} 0 & 0 & 0 \\ 0 & k & 0 \\ 0 & 0 & -k \end{bmatrix}$.

12(b) Let A and B be orthogonally similar, say $B = P^T A P$ where $P^T = P^{-1}$. Then $B^2 = P^T A P P^T A P = P^T A I A P = P^T A^2 P$. Hence A^2 and B^2 are orthogonally similar.

14. Suppose $(AX) \cdot Y = X \cdot AY$ for all columns X and Y. Thus $(AX)^T Y = X^T(AY)$; $X^T A^T Y = X^T A Y$. If X_i denotes column i of the identity matrix, then writing $A = [a_{ij}]$ we have

$$X_i^T A X_j = [0 \quad \dots \quad 1 \quad \dots \quad 0] \begin{bmatrix} \vdots & \dots & \vdots & \dots & \vdots \\ a_{i1} & \dots & a_{ij} & \dots & a_{in} \\ \vdots & \dots & \vdots & \dots & \vdots \end{bmatrix} \begin{bmatrix} 0 \\ \vdots \\ 1 \\ \vdots \\ 0 \end{bmatrix}$$

$$= [a_{i1} \quad \dots \quad a_{ij} \quad \dots \quad a_{in}] \begin{bmatrix} 0 \\ \vdots \\ 1 \\ \vdots \\ 0 \end{bmatrix} = a_{ij}.$$

Similarly, $X_i^T A^T X_j$ is the (i,j)-entry of A^T, so the condition $X^T A^T Y = X^T A Y$ for all X, Y means each entry of A^T equals the corresponding entry of A. Thus $A^T = A$.

17(b) If $P = \begin{bmatrix} \cos\theta & \sin\theta \\ -\sin\theta & \cos\theta \end{bmatrix}$ and $Q = \begin{bmatrix} \cos\theta & \sin\theta \\ \sin\theta & -\cos\theta \end{bmatrix}$ then P and Q are orthogonal matrices, $\det P = 1$ and $\det Q = -1$. (We note that every 2×2 orthogonal matrix has the form of P or Q for some θ.)

(d) Since P is orthogonal, $P^T = P^{-1}$. Hence

$$P^T(I - P) = P^T - P^T P = P^T - I = -(I - P^T) = -(I - P)^T.$$

Since P is $n \times n$, taking determinants gives

$$\det P^T \det(I - P) = (-1)^n \det[(I - P)^T] = (-1)^n \det(I - P).$$

Hence, if $I - P$ is invertible, then $\det(I - P) \neq 0$ so this gives $\det P^T = (-1)^n$; that is $\det P = (-1)^n$, contrary to assumption.

20. By the definition of matrix multiplication, the $[i, j]$-entry of AA^T is $R_i \cdot R_j$. This is zero if $i \neq j$, and equals $\|R_i\|^2$ if $i = j$. Hence, $AA^T = D = \text{diag}(\|R_1\|^2, \|R_2\|^2, \dots, \|R_n\|^2)$. Since D is invertible ($\|R_i\|^2 \neq 0$ for each i), it follows that A is invertible and, since row i of A^T is $\begin{bmatrix} a_{1i} & a_{2i} & \dots & a_{ji} & \dots & a_{ni} \end{bmatrix}$

$$A^{-1} = A^T D^{-1} = \begin{bmatrix} \vdots & \dots & \vdots & \dots & \vdots \\ a_{1i} & \dots & a_{ji} & \dots & a_{ni} \\ \vdots & \dots & \vdots & \dots & \vdots \end{bmatrix} \begin{bmatrix} \frac{1}{\|R_1\|^2} & 0 & \dots & 0 \\ 0 & \frac{1}{\|R_2\|^2} & \dots & 0 \\ \vdots & \vdots & \dots & \vdots \\ 0 & 0 & \dots & \frac{1}{\|R_n\|^2} \end{bmatrix}.$$

Thus, the (i, j)-entry of A^{-1} is $\dfrac{a_{ji}}{\|R_j\|^2}$.

22(b) Observe first that $I - A$ and $I + A$ commute, whence $I - A$ and $(I + A)^{-1}$ commute. Moreover, $\left[(I + A)^{-1}\right]^T = \left[(I + A)^T\right]^{-1} = (I^T + A^T)^{-1} = (I - A)^{-1}$. Hence,

$$\begin{aligned} PP^T &= (I - A)(I + A)^{-1}[(I - A)(I + A)^{-1}]^T \\ &= (I - A)(I + A)^{-1}[(I + A)^{-1}]^T(I - A)^T \\ &= (I - A)(I + A)^{-1}(I - A)^{-1}(I + A) \\ &= (I + A)^{-1}(I - A)(I - A)^{-1}(I + A) \\ &= (I + A)^{-1}I(I + A) \\ &= I. \end{aligned}$$

Exercises 7.3   Positive Definite Matrices

1(b) $\begin{bmatrix} 2 & -1 \\ -1 & 1 \end{bmatrix} \rightarrow \begin{bmatrix} 2 & -1 \\ 0 & \frac{1}{2} \end{bmatrix}$. Hence $U = \begin{bmatrix} \sqrt{2} & -\frac{1}{\sqrt{2}} \\ 0 & \frac{1}{\sqrt{2}} \end{bmatrix} = \frac{\sqrt{2}}{2} \begin{bmatrix} 2 & -1 \\ 0 & 1 \end{bmatrix}$. Then $A = U^T U$.

(d) $\begin{bmatrix} 20 & 4 & 5 \\ 4 & 2 & 3 \\ 5 & 3 & 5 \end{bmatrix} \rightarrow \begin{bmatrix} 20 & 4 & 5 \\ 0 & \frac{6}{5} & 2 \\ 0 & 2 & \frac{15}{4} \end{bmatrix} \rightarrow \begin{bmatrix} 20 & 4 & 5 \\ 0 & \frac{6}{5} & 2 \\ 0 & 0 & \frac{5}{12} \end{bmatrix}$. Hence, $U = \begin{bmatrix} \frac{2}{\sqrt{5}} & \frac{1}{\sqrt{5}} & \frac{5}{2\sqrt{5}} \\ 0 & \frac{6}{\sqrt{30}} & \frac{10}{\sqrt{30}} \\ 0 & 0 & \frac{5}{2\sqrt{15}} \end{bmatrix} =$

$\frac{1}{30} \begin{bmatrix} 60\sqrt{5} & 12\sqrt{5} & 15\sqrt{5} \\ 0 & 6\sqrt{30} & 10\sqrt{30} \\ 0 & 0 & 5\sqrt{15} \end{bmatrix}$, and $A = U^T U$.

4. If $X \neq 0$ is a column then $X^T(A+B)X = X^T AX + X^T BX > 0$ because $X^T AX > 0$ and $X^T BX > 0$ by hypothesis.

10. Since A is symmetric, the principal axis theorem asserts that an orthogonal matrix P exists such that $P^T AP = D = \mathrm{diag}(\lambda_1, \lambda_2, \ldots, \lambda_n)$ where the λ_i are the eigenvalues of A. Since each $\lambda_i > 0$, $\sqrt{\lambda_i}$ is real and positive, so define $B = \mathrm{diag}(\sqrt{\lambda_1}, \sqrt{\lambda_2}, \ldots, \sqrt{\lambda_n})$. Then $B^2 = D$. As $A = PDP^T$, take $C = PBP^T$. Then

$$C^2 = PBP^T PBP^T = PB^2 P^T = PDP^T = A.$$

Finally, C is symmetric because B is $\left(C^T = P^{TT} B^T P^T = PBP^T = C\right)$ and C has eigenvalues $\sqrt{\lambda_i} > 0$ (C is similar to B). Hence C is positive definite.

12(b) Suppose that A is positive definite so $A = U_0^T U_0$ where U_0 is upper triangular with positive diagonal entries $d_1, d_2, \ldots, d_n$. Put $D_0 = \mathrm{diag}(d_1, d_2, \ldots, d_n)$. Then $L = U_0^T D_0^{-1}$ is lower triangular with 1's on the diagonal, $U = D_0^{-1} U_0$ is upper triangular with 1's on the diagonal, and $A = LD_0^2 U$. Take $D = D_0^2$.

Conversely, if $A = LDU$ as in (a), then $A^T = U^T DL^T$. Hence, $A^T = A$ implies $U^T DL^T = LDU$ so $U^T = L$ and $L^T = U$ by (a). Hence, $A = U^T DU$. If $D = \mathrm{diag}(d_1, d_2, \ldots, d_n)$, let $D_1 = \mathrm{diag}(\sqrt{d_1}, \sqrt{d_2}, \ldots, \sqrt{d_n})$. Then $D = D_1^2$ so $A = U^T D_1^2 U = (D_1 U)^T (D_1 U)$. Hence, A is positive definite.

Exercises 7.4   QR-Factorization

1(b) The columns of A and $C_1 = \begin{bmatrix} 2 \\ 1 \end{bmatrix}$ and $C_2 = \begin{bmatrix} 1 \\ 1 \end{bmatrix}$. First apply the Gram-Schmidt algorithm

$$F_1 = C_1 = \begin{bmatrix} 2 \\ 1 \end{bmatrix}$$

$$F_2 = C_2 - \frac{C_2 \cdot F_1}{\|F_1\|^2} F_1 = \begin{bmatrix} 1 \\ 1 \end{bmatrix} - \frac{3}{5} \begin{bmatrix} 2 \\ 1 \end{bmatrix} = \begin{bmatrix} -\frac{1}{5} \\ \frac{2}{5} \end{bmatrix} .$$

Now normalize to obtain

$$Q_1 = \frac{1}{\|F_1\|} F_1 = \frac{1}{\sqrt{5}} \begin{bmatrix} 2 \\ 1 \end{bmatrix}$$

$$Q_2 = \frac{1}{\|F_2\|} F_2 = \frac{1}{\sqrt{5}} \begin{bmatrix} -1 \\ 2 \end{bmatrix}.$$

Hence $Q = [Q_1 \quad Q_2] = \frac{1}{\sqrt{5}} \begin{bmatrix} 2 & -1 \\ 1 & 2 \end{bmatrix}$ is an orthogonal matrix. We obtain R from equation (*) preceding Theorem 1:

$$L = \begin{bmatrix} \|F_1\| & C_2 \cdot Q_1 \\ 0 & \|F_2\| \end{bmatrix} = \begin{bmatrix} \sqrt{5} & \frac{3}{\sqrt{5}} \\ 0 & \frac{1}{\sqrt{5}} \end{bmatrix} = \frac{1}{\sqrt{5}} \begin{bmatrix} 5 & 3 \\ 0 & 1 \end{bmatrix}.$$

Then $A = QR$.

(d) The columns of A are $C_1 = [1 \quad -1 \quad 0 \quad 1]^T, C_2 = [1 \quad 0 \quad 1 \quad -1]^T$ and $C_3 = [0 \quad 1 \quad 1 \quad 0]^T$. Apply the Gram-Schmidt algorithm

$$F_1 = C_1 = [1 \quad -1 \quad 0 \quad 1]^T$$

$$F_2 = C_2 - \frac{C_2 \cdot F_1}{\|F_1\|^2} F_1 = [1 \quad 0 \quad 1 \quad -1]^T - \tfrac{0}{3} F_1 = [1 \quad 0 \quad 1 \quad -1]^T$$

$$F_3 = C_3 - \frac{C_3 \cdot F_1}{\|F_1\|^2} F_1 - \frac{C_3 \cdot F_2}{\|F_2\|^2} F_2$$

$$= [0 \quad 1 \quad 1 \quad 0]^T - \tfrac{-1}{3} [1 \quad -1 \quad 0 \quad 1]^T - \tfrac{1}{3} [1 \quad 0 \quad 1 \quad -1]^T$$

$$= \tfrac{2}{3} [0 \quad 1 \quad 1 \quad 1].$$

Normalize

$$Q_1 = \frac{1}{\|F_1\|} F_1 = \tfrac{1}{\sqrt{3}} [1 \quad -1 \quad 0 \quad 1]^T$$

$$Q_2 = \frac{1}{\|F_2\|} F_2 = \tfrac{1}{\sqrt{3}} [1 \quad 0 \quad 1 \quad -1]^T$$

$$Q_3 = \frac{1}{\|F_3\|} F_3 = \tfrac{1}{\sqrt{3}} [0 \quad 1 \quad 1 \quad 1]^T.$$

Hence $Q = [Q_1 \quad Q_2 \quad Q_3] = \frac{1}{\sqrt{3}} \begin{bmatrix} 1 & 1 & 0 \\ -1 & 0 & 1 \\ 0 & 1 & 1 \\ 1 & -1 & 1 \end{bmatrix}$ has orthonormal columns. We obtain R

from equation (*) preceding Theorem 1:

$$R = \begin{bmatrix} \|F_1\| & C_2 \cdot Q_1 & C_3 \cdot Q_1 \\ 0 & \|F_2\| & C_3 \cdot Q_2 \\ 0 & 0 & \|F_3\| \end{bmatrix} = \begin{bmatrix} \sqrt{3} & \frac{1}{\sqrt{3}} & \frac{-1}{\sqrt{3}} \\ 0 & \sqrt{3} & \frac{1}{\sqrt{3}} \\ 0 & 0 & \frac{2}{\sqrt{3}} \end{bmatrix} = \frac{1}{\sqrt{3}} \begin{bmatrix} 3 & 0 & -1 \\ 0 & 3 & 1 \\ 0 & 0 & 2 \end{bmatrix}.$$

Then $A = QR$.

Exercises 7.5   Computing Eigenvalues

1(b) $A = \begin{bmatrix} 5 & 2 \\ -3 & -2 \end{bmatrix}$. Then $c_A(x) = \begin{vmatrix} x-5 & -2 \\ 3 & x+2 \end{vmatrix} = (x+1)(x-4)$, so $\lambda_1 = -1$, $\lambda_2 = 4$.

$\lambda_1 = -1 : \begin{bmatrix} -6 & -2 \\ 3 & 1 \end{bmatrix} \rightarrow \begin{bmatrix} 3 & 1 \\ 0 & 0 \end{bmatrix} : X_1 = \begin{bmatrix} -1 \\ 3 \end{bmatrix}$

$\lambda_2 = 4 : \begin{bmatrix} -1 & -2 \\ 3 & 6 \end{bmatrix} \rightarrow \begin{bmatrix} 1 & 2 \\ 0 & 0 \end{bmatrix} ; X_2 = \begin{bmatrix} -2 \\ 1 \end{bmatrix}$.

Starting with $X_0 = \begin{bmatrix} 1 \\ 1 \end{bmatrix}$, the power method gives $X_1 = AX_0$, $X_2 = AX_1, \ldots$:

$$X_1 = \begin{bmatrix} 7 \\ -5 \end{bmatrix}, \quad X_2 = \begin{bmatrix} 25 \\ -11 \end{bmatrix}, \quad X_3 = \begin{bmatrix} 103 \\ -53 \end{bmatrix}, \quad X_4 = \begin{bmatrix} 409 \\ -203 \end{bmatrix}.$$

These are approaching (scalar multiples of) the dominant eigenvector $\begin{bmatrix} 2 \\ -1 \end{bmatrix}$. The Rayleigh

quotients are $r_k = \dfrac{X_k \cdot X_{k+1}}{\|X_k\|^2}$, $k = 0, 1, 2, \ldots$, so $r_0 = 1$, $r_1 = 3.29$, $r_2 = 4.23$, $r_3 = 3.94$.
These are approaching the dominant eigenvalue 4.

(d) $A = \begin{bmatrix} 3 & 1 \\ 1 & 0 \end{bmatrix}$; $c_A(x) = \begin{vmatrix} x-3 & -1 \\ -1 & x \end{vmatrix} = x^2 - 3x - 1$, so the eigenvalues are $\lambda_1 = \frac{1}{2}(3 + \sqrt{13})$,

$\lambda_2 = \frac{1}{2}(3 - \sqrt{13})$. Thus the dominant eigenvalue is $\lambda_1 = \frac{1}{2}(3 + \sqrt{13})$. Since $\lambda_1\lambda_2 = -1$ and $\lambda_1 + \lambda_2 = 3$, we get

$$\begin{bmatrix} \lambda_1 - 3 & -1 \\ -1 & \lambda_1 \end{bmatrix} \rightarrow \begin{bmatrix} 1 & -\lambda_1 \\ 0 & 0 \end{bmatrix}$$

so a dominant eigenvetor is $\begin{bmatrix} \lambda_1 \\ 1 \end{bmatrix}$. We start with $X_0 = \begin{bmatrix} 1 \\ 1 \end{bmatrix}$.

Then $X_{k+1} = AX_k$, $k = 0, 1, \ldots$ gives

$$X_1 = \begin{bmatrix} 4 \\ 1 \end{bmatrix}, \quad X_2 = \begin{bmatrix} 13 \\ 4 \end{bmatrix}, \quad X_3 = \begin{bmatrix} 43 \\ 13 \end{bmatrix}, \quad X_4 = \begin{bmatrix} 142 \\ 43 \end{bmatrix}.$$

These are approaching scalar multiples of the dominant eigenvector $\begin{bmatrix} \lambda_2 \\ 1 \end{bmatrix} = \begin{bmatrix} 3.302776 \\ 1 \end{bmatrix}$.

The Rayleigh quotients are $r_k = \dfrac{X_k \cdot X_{k+1}}{\|X_k\|^2}$:

$$r_0 = 2.5, \quad r_1 = 3.29, \quad r_2 = 3.30270, \quad r_3 = 3.30278.$$

These are rapidly approaching the dominant eigenvalue λ_2.

2(b) $A = \begin{bmatrix} 3 & 1 \\ 1 & 0 \end{bmatrix}$; $c_A(x) = \begin{vmatrix} x-3 & -1 \\ -1 & x \end{vmatrix} = x^2 - 3x - 3$; $\lambda_1 = \frac{1}{2}\left[3 + \sqrt{13}\right] = 3.302776$ and

$\lambda_2 = \frac{1}{2}\left[3 - \sqrt{13}\right] = -0.302776$. The QR-algorithm proceeds as follows:

$$A_1 = \begin{bmatrix} 3 & 1 \\ 1 & 0 \end{bmatrix} = Q_1 R_1 \text{ where } Q_1 = \frac{1}{\sqrt{10}}\begin{bmatrix} 3 & -1 \\ 1 & 3 \end{bmatrix}, \quad R_1 = \frac{1}{\sqrt{10}}\begin{bmatrix} 10 & 3 \\ 0 & -1 \end{bmatrix}.$$

$$A_2 = R_1 Q_1 = \frac{1}{10}\begin{bmatrix} 33 & -1 \\ -1 & -3 \end{bmatrix} = Q_2 R_2 \text{ where } Q_2 = \frac{1}{\sqrt{1090}}\begin{bmatrix} 33 & 1 \\ -1 & 33 \end{bmatrix}, \quad R_2 = \frac{1}{\sqrt{1090}}\begin{bmatrix} 109 & -3 \\ 0 & -10 \end{bmatrix}.$$

$$A_3 = R_2 Q_2 = \frac{1}{109}\begin{bmatrix} 360 & 1 \\ 1 & -33 \end{bmatrix} = \begin{bmatrix} 3.302775 & 0.009174 \\ 0.009174 & -0.302775 \end{bmatrix}.$$

The diagonal entries already approximate λ_1 and λ_2 to 5 decimal places.

4. We prove that $A_k^T = A_k$ for each k by induction in k. If $k = 1$, then $A_1 = A$ is symmetric by hypothesis, so assume $A_k^T = A_k$ for some $k \geq 1$. We have $A_k = Q_k R_k$ so $R_k = Q_k^{-1} A_k = Q_k^T A_k$ because Q_k is orthogonal. Hence

$$A_{k+1} = R_k Q_k = Q_k^T A_k Q_k$$

so

$$A_{k+1}^T = (Q_k^T A_k Q_k)^T = Q_k^T A_k^T Q_k^{TT} = Q_k^T A_k Q_k = A_{k+1}.$$

The eigenvalues of A are all real as A is symmetric, so the QR-algorithm asserts that the A_k converge to an upper triangular matrix T. But T is symmetric (it is the limit of symmetric matrices), so it is diagonal.

Exercises 7.6   Complex Matrices

1(b) $\sqrt{|1-i|^2 + |1+i|^2 + 1^2 + (-1)^2} = \sqrt{(1+1) + (1+1) + 1 + 1} = \sqrt{6}$

(d) $\sqrt{4 + |-i|^2 + |1+i|^2 + |1-i|^2 + |2i|^2} = \sqrt{4 + 1 + (1+1) + (1+1) + 4} = \sqrt{13}$

2(b) Not orthogonal: $\langle (i, -i, 2+i), (i, i, 2-i) \rangle = i(-i) + (-i)(-i) + (2+i)(2+i) = 3 + 4i$

(d) Orthogonal: $\langle 4+4i, 2+i, 2i), (-1+i, 2, 3-2i) \rangle = (4+4i)(-1-i) + (2+i)2 + (2i)(3+2i) = (-8i) + (4+2i) + (-4+6i) = 0$.

3(b) Not a subspace. For example, $i(0,0,1) = (0,0,i)$ is not in U.

(d) If $Z = (v + w, v - 2w, v)$ and $W = (v' + w', v' - 2w', v')$ are in U then

$$Z + W = ((v + v') + w + w'), (v + v') - 2(w + w'), (v + v'))\text{ is in } U$$
$$zZ = (zv + zw, zr - 2zw, zw)\text{ is in } U$$
$$0 = (0 + 0, 0 = 20, 0)\text{ is in } U.$$

Hence U is a subspace.

4(b) $U = \{(iv + w, 0, 2v - w) \mid v, w \in \mathbb{C}\} = \{v(i,0,2) + w(1,0,-1) \mid v, w \in \mathbb{C}\} = \text{span}\{(i,0,2), (1,0,-1)\}$.
If $z(i,0,2) + w(1,0,-1) = (0,0,0)$ $z, w \in \mathbb{C}$ then $iz + w = 0$, $2z - w = 0$. Adding gives $(2+i)z = 0$, so $z = 0$, whence $w = -iz = 0$. Thus $\{(i,0,2), (1,0,-1)\}$ is independent over $\mathbb{C}$, and so is a basis of U. Hence $\dim_{\mathbb{C}} U = 2$.

(d) $U = \{(u, v, w) \mid 2u + (1+i)v - iw = 0; u, v, w \in \mathbb{C}\}$. The condition gives $w = -2iu + (1-i)v$, so

$$U = \{(u, v, -2iu + (1-i)v) \mid u, v \in \mathbb{C}\} = \text{span}\{(1,0,-2i), (0,1,1-i)\}.$$

If $z(1,0,-2i) + w(0,1,i-1)(0,0,0)$ then $z = 0$, $w = 0$, so $\{(1,0,-2i), (0,1,1-i)\}$ is independent over $\mathbb{C}$ and so is a basis osf U. Hence $\dim_{\mathbb{C}} U = 2$.

5(b) $H = \begin{bmatrix} 2 & 3 \\ -3 & 2 \end{bmatrix}$, $H^* = H^T = \begin{bmatrix} 2 & -3 \\ 3 & 2 \end{bmatrix}$, $H^{-1} = \frac{1}{13}\begin{bmatrix} 2 & -3 \\ 3 & 2 \end{bmatrix}$. Hence, H is not Hermitian $(H \neq H^*)$ and not unitary $(H^{-1} \neq H^*)$. Howeer, $HH^* = 13I = H^*H$, so H is normal.

(d) $H = \begin{bmatrix} 1 & -i \\ i & -1 \end{bmatrix}$, $H^* = (\bar{H})^T = \begin{bmatrix} 1 & i \\ -i & -1 \end{bmatrix}^T = \begin{bmatrix} 1 & -i \\ i & -1 \end{bmatrix} = H$. Thus H is Hermitian and so is normal. But, $HH^* = H^2 = 2I$ so H is not unitary.

(f) $H = \begin{bmatrix} 1 & 1+i \\ 1+i & i \end{bmatrix}$. Here $H = H^T$ so $H^* = \bar{H} = \begin{bmatrix} 1 & 1-i \\ 1-i & -i \end{bmatrix} \neq H$ so H is not Hermitian. Next, $HH^* = \begin{bmatrix} 3 & 2-2i \\ 2+i & 3 \end{bmatrix} \neq I$ so H is not unitary. Finally, $H^*H = \begin{bmatrix} 3 & 2+2i \\ 2-2i & 3 \end{bmatrix} \neq HH^*$, so H is not normal.

(h) $H = \frac{1}{\sqrt{2}|z|}\begin{bmatrix} z & z \\ \bar{z} & -\bar{z} \end{bmatrix}$. Here $\bar{H} = \frac{1}{\sqrt{2}|z|}\begin{bmatrix} \bar{z} & \bar{z} \\ z & -z \end{bmatrix}$ so $H^* = \frac{1}{\sqrt{2}|z|}\begin{bmatrix} \bar{z} & z \\ \bar{z} & -z \end{bmatrix}$. Thus $H = H^*$ if and only if $z = \bar{z}$; that is it is Hermitian if and only if z is real. We have $HH^* = \frac{1}{2|z|^2}\begin{bmatrix} 2|z|^2 & 0 \\ 0 & 2|z|^2 \end{bmatrix} = I$, and similarly, $H^*H = I$. Thus it is unitary (and hence normal).

6(b) $Z = \begin{bmatrix} 4 & 3-i \\ 3+i & 1 \end{bmatrix}$, $c_Z(x) = \begin{bmatrix} x-4 & -3+i \\ -3-i & x-1 \end{bmatrix} = x^2 - 5x - 6 = (x+1)(x-6)$.

$\lambda_1 = -1: \begin{bmatrix} -5 & -3+i \\ -3-i & 1 \end{bmatrix} \rightarrow \begin{bmatrix} 3+i & 2 \\ 0 & 0 \end{bmatrix}$; an eigenvector is $X_1 = \begin{bmatrix} -2 \\ 3+i \end{bmatrix}$.

$\lambda_2 = 6: \begin{bmatrix} 2 & -3+i \\ -3-i & 5 \end{bmatrix} \rightarrow \begin{bmatrix} 2 & -3+i \\ 0 & 0 \end{bmatrix}$; an eigenvector is $X_2 = \begin{bmatrix} 3-i \\ 2 \end{bmatrix}$.

As X_1 and X_2 are orthogonal and $X_1 = X_2 = \sqrt{14}$, $U = \frac{1}{\sqrt{14}} \begin{bmatrix} -2 & 3-i \\ 3+i & 2 \end{bmatrix}$ is unitary and

$U^*ZU = \begin{bmatrix} -1 & 0 \\ 0 & 6 \end{bmatrix}$.

(d) $Z = \begin{bmatrix} 2 & 1+i \\ 1-i & 3 \end{bmatrix}$; $c_Z(x) = \begin{vmatrix} x-2 & -1-i \\ -1+i & x-3 \end{vmatrix} = x^2 - 5x + 4 = (x-1)(x-4)$.

$\lambda_1 = 1: \begin{bmatrix} -1 & -1-i \\ -1+i & -2 \end{bmatrix} \rightarrow \begin{bmatrix} 1 & 1+i \\ 0 & 0 \end{bmatrix}$; an eigenvector is $X_1 = \begin{bmatrix} 1+i \\ -1 \end{bmatrix}$.

$\lambda_2 = 4: \begin{bmatrix} 2 & -1-i \\ -1+i & 1 \end{bmatrix} \rightarrow \begin{bmatrix} -1+i & 1 \\ 0 & 0 \end{bmatrix}$; an eigenvector is $X_2 = \begin{bmatrix} 1 \\ 1-i \end{bmatrix}$.

Since X_1 and X_2 are orthogonal and $\|X_1\| = \|X_2\| = \sqrt{3}$, $U = \frac{1}{\sqrt{3}} \begin{bmatrix} 1+i & 1 \\ -1 & 1-i \end{bmatrix}$ is unitary

and $U^*ZU = \begin{bmatrix} 1 & 0 \\ 0 & 4 \end{bmatrix}$.

(f) $Z = \begin{bmatrix} 1 & 0 & 0 \\ 0 & 1 & 1+i \\ 0 & 1-i & 2 \end{bmatrix}$;

$c_Z(x) = \begin{vmatrix} x-1 & 0 & 0 \\ 0 & x-1 & -1-i \\ 0 & -1+i & x-2 \end{vmatrix} = (x-1)(x^2-3x) = (x-1)x(x-3)$.

$\lambda_1 = 1: \begin{bmatrix} 0 & 0 & 0 \\ 0 & 0 & -1-i \\ 0 & -1+i & -1 \end{bmatrix} \rightarrow \begin{bmatrix} 0 & 1 & 0 \\ 0 & 0 & 1 \\ 0 & 0 & 0 \end{bmatrix}$; an eigenvector is $X_1 = \begin{bmatrix} 1 \\ 0 \\ 0 \end{bmatrix}$.

$\lambda_2 = 0: \begin{bmatrix} -1 & 0 & 0 \\ 0 & -1 & -1-i \\ 0 & -1+i & -i \end{bmatrix} \rightarrow \begin{bmatrix} 1 & 0 & 0 \\ 0 & 1 & 1+i \\ 0 & 0 & 0 \end{bmatrix}$; an eigenvector is $X_2 = \begin{bmatrix} 0 \\ 1+i \\ -1 \end{bmatrix}$.

$$\lambda_3 = 3 : \begin{bmatrix} 2 & 0 & 0 \\ 0 & 2 & -1-i \\ 0 & -1+i & 1 \end{bmatrix} \rightarrow \begin{bmatrix} 1 & 0 & 0 \\ 0 & -1+i & 1 \\ 0 & 0 & 0 \end{bmatrix} ; \text{ an eigenvector is } X_3 = \begin{bmatrix} 0 \\ 1 \\ 1-i \end{bmatrix}.$$

Since $\{X_1, X_2, X_3\}$ is orthogonal and $\|X_2\| = \|X_3\| = \sqrt{3}$, $U = \frac{1}{\sqrt{3}} \begin{bmatrix} \sqrt{3} & 0 & 0 \\ 0 & 1+i & 1 \\ 0 & -1 & 1-i \end{bmatrix}$ is

orthogonal and $U^*ZU = \begin{bmatrix} 1 & 0 & 0 \\ 0 & 0 & 0 \\ 0 & 0 & 3 \end{bmatrix}.$

8(b) (1) If $Z = [z_1 \quad z_2 \quad \cdots \quad z_n]$ then $\|Z\|^2 = |z_1|^2 + |z_2|^2 + \cdots + |z_n|^2$. Thus $\|Z\| = 0$ if and only if $|z_1| = \cdots = |z_n| = 0$, if and only if $Z = [0 \quad 0 \quad \cdots \quad 0]$.

(2) By Theorem 1, we have $\langle \lambda Z, W \rangle = \lambda \langle Z, W \rangle$ and $\langle Z, \lambda W \rangle = \bar{\lambda} \langle Z, W \rangle$. Hence

$$\|\lambda Z\|^2 = \langle \lambda Z, \lambda Z \rangle = \lambda \langle Z, \lambda Z \rangle = \lambda \bar{\lambda} \langle Z, Z \rangle = |\lambda|^2 \|Z\|^2.$$

Taking positive square roots gives $\|\lambda Z\| = |\lambda| \|Z\|$.

9(b) If Z is Hermitian then $\bar{Z} = Z^T$. If $Z = [z_{ij}]$, the (k,k)-entry of $\bar{Z}$ is $\bar{z}_{kk}$, and the (k,k)-entry of Z^T is z_{kk}. Thus, $\bar{Z} = Z^T$ implies that $\bar{z}_{kk} = z_{kk}$ for each k; that is z_{kk} is real.

12(b) Let S be skew-Hermitian, that is $S^* = -S$. Then Theorem 3 gives

$$(S^2)^* = (S^*)^2 = (-S)^2 = S^2 \text{ so } S^2 \text{ is Hermitian}$$
$$(iS)^* = (-i)S^* = (-i)(-S) = iS \text{ so } iS \text{ is Hermitian}.$$

(d) If $Z = H + S$ where $H^* = H$ and $S^* = -S$, then $Z^* = H^* + S^* = H - S$, so $Z + Z^* = 2H$ and $Z - Z^* = 2S$. The, $H = \frac{1}{2}(Z + Z^*)$ and $S = \frac{1}{2}(Z + Z^*)$ so the matrices H and S are uniquely determined by the condidtions $Z = H + S$, $H^* = H$, $S^* = -S$, provided such H and S exist. But always,

$$Z = \tfrac{1}{2}(Z + Z^*) + \tfrac{1}{2}(Z - Z^*)$$

and the matices $H = \frac{1}{2}(Z + Z^*)$ and $S = \frac{1}{2}(Z - Z^*)$ are Hermitian and skew-Hermitian respectively:

$$H^* = \tfrac{1}{2}(Z^* + Z^{**}) = \tfrac{1}{2}(Z^* + Z) = H$$
$$S^* = \tfrac{1}{2}(Z^* - Z^{**}) = \tfrac{1}{2}(Z^* - Z) = -S.$$

14(b) If U is unitary, then $U^{-1} = U^*$. We must show that U^{-1} is unitary, that is $(U^{-1})^{-1} = (U^{-1})^*$. But

$$(U^{-1})^{-1} = U = (U^*)^* = (U^{-1})^*.$$

18. If H is Hermitian, the eigenvalues $\lambda_1, \lambda_2, \ldots, \lambda_n$ are all real. These are the roots of $c_H(x)$ so

$$c_H(x) = (x - \lambda_1)(x - \lambda_2) \cdots (x - \lambda_n).$$

It this is multiplied out, each coefficient of $c_H(x)$ is seen to be a sum of products of the λ_i, and so is real.

19(b) Given $A = \begin{bmatrix} 0 & 1 \\ -1 & 0 \end{bmatrix}$, let $U = \begin{bmatrix} a & b \\ c & d \end{bmatrix}$ be invertible and real, and assume ethat $U^{-1}AU = \begin{bmatrix} \lambda & \mu \\ 0 & \nu \end{bmatrix}$. Thus, $AU = U \begin{bmatrix} \lambda & \mu \\ 0 & \nu \end{bmatrix}$ so

$$\begin{bmatrix} c & d \\ -a & -b \end{bmatrix} = \begin{bmatrix} a\lambda & a\mu + b\nu \\ c\lambda & c\mu + d\nu \end{bmatrix}.$$

Equating first column entries gives $c = a\lambda$ and $-a = c\lambda$. Thus, $-a = (a\lambda)\lambda = a\lambda^2$ so $(1 + \lambda^2)a = 0$. Now λ is real (a and c are not both zero so either $\lambda = \frac{c}{a}$ or $\lambda = -\frac{a}{c}$), so $1 + \lambda^2 \neq 0$. Thus $a = 0$ (becuase $(1 + \lambda^2)a = 0$) whence $c = a\lambda = 0$. This contradicts the assumption that A is invertible.

Exercises 7.7   An Application to Quadratic Forms

1(b) $A = \begin{bmatrix} 1 & \frac{1}{2}(1 - 1) \\ \frac{1}{2}(-1 + 1) & 2 \end{bmatrix} = \begin{bmatrix} 1 & 0 \\ 0 & 2 \end{bmatrix}$

(d) $A = \begin{bmatrix} 1 & \frac{1}{2}(2 + 4) & \frac{1}{2}(-1 + 5) \\ \frac{1}{2}(4 + 2) & 1 & \frac{1}{2}(0 - 2) \\ \frac{1}{2}(5 - 1) & \frac{1}{2}(-2 + 0) & 3 \end{bmatrix} = \begin{bmatrix} 1 & 3 & 2 \\ 3 & 1 & -1 \\ 2 & -1 & 3 \end{bmatrix}$

2(b) $q = X^T A X$ where $A = \begin{bmatrix} 1 & 2 \\ 2 & 1 \end{bmatrix}$. $c_A(x) = \begin{vmatrix} x - 1 & -2 \\ -2 & x - 1 \end{vmatrix} = x^2 - 2x - 3 = (x + 1)(x - 3)$

$\lambda_1 = 3 : \begin{bmatrix} 2 & -2 \\ -2 & 2 \end{bmatrix} \rightarrow \begin{bmatrix} 1 & -1 \\ 0 & 0 \end{bmatrix}$; an eigenvector is $X_1 = \begin{bmatrix} 1 \\ 1 \end{bmatrix}$.

$\lambda_2 = -1 : \begin{bmatrix} -2 & -2 \\ -2 & -2 \end{bmatrix} \rightarrow \begin{bmatrix} 1 & 1 \\ 0 & 0 \end{bmatrix}$; an eigenvector is $X_2 = \begin{bmatrix} 1 \\ -1 \end{bmatrix}$.

Hence, $P = \frac{1}{\sqrt{2}} \begin{bmatrix} 1 & 1 \\ 1 & -1 \end{bmatrix}$ is orthogonal and $P^T A P = \begin{bmatrix} 3 & 0 \\ 0 & -1 \end{bmatrix}$. As in Theorem 1, take

$Y = P^T X = \frac{1}{\sqrt{2}} \begin{bmatrix} 1 & 1 \\ 1 & -1 \end{bmatrix} \begin{bmatrix} x_1 \\ x_2 \end{bmatrix} = \frac{1}{\sqrt{2}} \begin{bmatrix} x_1 + x_2 \\ x_1 - x_2 \end{bmatrix}$. Then $y_1 = \frac{1}{\sqrt{2}}(x_1 + x_2)$,

$y_2 = \frac{1}{\sqrt{2}}(x_1 - x_2)$. Finally, $q = 3y_1^2 - y_2^2$, the rank of q is 2 (the number of nonzero eigenvalues) and the index of q is 1 (the number of positive eigenvalues).

(d) $q = X^T A X$ where $A = \begin{bmatrix} 7 & 4 & 4 \\ 4 & 1 & -8 \\ 4 & -8 & 1 \end{bmatrix}$. To find $c_A(x)$, subtract row 2 from row 3 :

$$c_A(x) = \begin{vmatrix} x-7 & -4 & -4 \\ -4 & x-1 & 8 \\ -4 & 8 & x-1 \end{vmatrix} = \begin{vmatrix} x-7 & -4 & -4 \\ -4 & x-1 & 8 \\ 0 & -x+9 & x-9 \end{vmatrix}$$

$$= \begin{vmatrix} x-7 & -8 & -4 \\ -4 & x+7 & 8 \\ 0 & 0 & x-9 \end{vmatrix} = (x-9)^2(x+9)$$

$\lambda_1 = 9 : \begin{bmatrix} 2 & -4 & -4 \\ -4 & 8 & 8 \\ -4 & 8 & 8 \end{bmatrix} \rightarrow \begin{bmatrix} 1 & -2 & -2 \\ 0 & 0 & 0 \\ 0 & 0 & 0 \end{bmatrix}$; two orthogonal eigenvectors are $\begin{bmatrix} 2 \\ 2 \\ -1 \end{bmatrix}$ and

$\begin{bmatrix} 2 \\ -1 \\ 2 \end{bmatrix}$.

$\lambda_2 = -9 : \begin{bmatrix} -16 & -4 & -4 \\ -4 & -10 & 8 \\ -4 & 8 & -10 \end{bmatrix} \rightarrow \begin{bmatrix} 4 & 1 & 1 \\ 0 & -9 & 9 \\ 0 & 9 & -9 \end{bmatrix} \rightarrow \begin{bmatrix} 4 & 0 & 2 \\ 0 & 1 & -1 \\ 0 & 0 & 0 \end{bmatrix}$; an eigenvector is

$\begin{bmatrix} -1 \\ 2 \\ 2 \end{bmatrix}$. These eigenvectors are orthogonal and each has length 3. Hence, $P = \frac{1}{3} \begin{bmatrix} 2 & 2 & -1 \\ 2 & -1 & 2 \\ -1 & 2 & 2 \end{bmatrix}$

is orthogonal and $P^T A P = \begin{bmatrix} 9 & 0 & 0 \\ 0 & 9 & 0 \\ 0 & 0 & -9 \end{bmatrix}$. Thus

$$Y = P^T X = \frac{1}{3} \begin{bmatrix} 2 & 2 & -1 \\ 2 & -1 & 2 \\ -1 & 2 & 2 \end{bmatrix} \begin{bmatrix} x_1 \\ x_2 \\ x_3 \end{bmatrix} = \frac{1}{3} \begin{bmatrix} 2x_1 + 2x_2 - x_3 \\ 2x_1 - x_2 + 2x_3 \\ -x_1 + 2x_2 + 2x_3 \end{bmatrix}$$

so

$$y_1 = \tfrac{1}{3}[2x_1 + 2x_2 - x_3]$$
$$y_2 = \tfrac{1}{3}[2x_1 - x_2 + 2x_3]$$
$$y_3 = \tfrac{1}{3}[-x_1 + 2x_2 + 2x_3]$$

will give $q = 9y_1^2 + 9y_2^2 - 9y_3^2$. The rank of q is 3 and the index of q is 2.

(f) $q = X^T A X$ where $A = \begin{bmatrix} 5 & -2 & -4 \\ -2 & 8 & -2 \\ -4 & -2 & 5 \end{bmatrix}$. To find $c_A(x)$, subtract row 3 from row 1:

$$c_A(x) = \begin{vmatrix} x-5 & 2 & 4 \\ 2 & x-8 & 2 \\ 4 & 2 & x-5 \end{vmatrix} = \begin{vmatrix} x-9 & 0 & -x+9 \\ 2 & x-8 & 2 \\ 4 & 2 & x-5 \end{vmatrix}$$

$$= \begin{vmatrix} x-9 & 0 & 0 \\ 2 & x-8 & 4 \\ 4 & 2 & x-1 \end{vmatrix} = x(x-9)^2.$$

$\lambda_1 = 9: \begin{bmatrix} 4 & 2 & 4 \\ 2 & 1 & 2 \\ 4 & 2 & 4 \end{bmatrix} \to \begin{bmatrix} 2 & 1 & 2 \\ 0 & 0 & 0 \\ 0 & 0 & 0 \end{bmatrix}$; orthogonal eigenvectors are $\begin{bmatrix} -2 \\ 2 \\ 1 \end{bmatrix}$ and $\begin{bmatrix} 1 \\ 2 \\ -2 \end{bmatrix}$.

$\lambda_2 = 0: \begin{bmatrix} -5 & 2 & 4 \\ 2 & -8 & 2 \\ 4 & 2 & -5 \end{bmatrix} \to \begin{bmatrix} 1 & -4 & 1 \\ 0 & -18 & 9 \\ 0 & 18 & -9 \end{bmatrix} \to \begin{bmatrix} 1 & 0 & -1 \\ 0 & 2 & -1 \\ 0 & 0 & 0 \end{bmatrix}$; an eigenvector is $\begin{bmatrix} 2 \\ 1 \\ 2 \end{bmatrix}$.

These eigenvectors are orthogonal and each has length 3. Hence $P = \frac{1}{3}\begin{bmatrix} -2 & 1 & 2 \\ 2 & 2 & 1 \\ 1 & -2 & 2 \end{bmatrix}$ is

orthogonal and $P^T A P = \begin{bmatrix} 9 & 0 & 0 \\ 0 & 9 & 0 \\ 0 & 0 & 0 \end{bmatrix}$. If

$$Y = P^T X = \frac{1}{3}\begin{bmatrix} -2 & 2 & 1 \\ 1 & 2 & -2 \\ 2 & 1 & 2 \end{bmatrix}\begin{bmatrix} x_1 \\ x_2 \\ x_3 \end{bmatrix}$$

then

$$y_1 = \tfrac{1}{3}(-2x_1 + 2x_2 + x_3)$$
$$y_2 = \tfrac{1}{3}(x_1 + 2x_2 - 2x_3)$$
$$y_3 = \tfrac{1}{3}(2x_1 + x_2 + 2x_3)$$

gives $q = 9y_1^2 + 9y_2^2$. The rank and index of q are both 2.

(h) $q = X^T A X$ where $A = \begin{bmatrix} 1 & -1 & 0 \\ -1 & 0 & 1 \\ 0 & 1 & 1 \end{bmatrix}$. To find $c_A(x)$, add row 3 to row 1:

$$c_A(x) = \begin{vmatrix} x-1 & 1 & 0 \\ 1 & x & -1 \\ 0 & -1 & x-1 \end{vmatrix} = \begin{vmatrix} x-1 & 0 & x-1 \\ 1 & x & -1 \\ 0 & -1 & x-1 \end{vmatrix}$$

$$= \begin{vmatrix} x-1 & 0 & 0 \\ 1 & x & -2 \\ 0 & -1 & x-1 \end{vmatrix} = (x-1)(x-2)(x+1)$$

$$\lambda_1 = 2 : \begin{bmatrix} 1 & 1 & 0 \\ 1 & 2 & -1 \\ 0 & -1 & 1 \end{bmatrix} \to \begin{bmatrix} 1 & 1 & 0 \\ 0 & 1 & -1 \\ 0 & -1 & 1 \end{bmatrix} \to \begin{bmatrix} 1 & 0 & 1 \\ 0 & 1 & -1 \\ 0 & 0 & 0 \end{bmatrix} ; \text{ an eigenvector is } \begin{bmatrix} -1 \\ 1 \\ 1 \end{bmatrix}.$$

$$\lambda_2 = 1 : \begin{bmatrix} 0 & 1 & 0 \\ 1 & 1 & -1 \\ 0 & 1 & 0 \end{bmatrix} \to \begin{bmatrix} 1 & 0 & -1 \\ 0 & 1 & 0 \\ 0 & 0 & 0 \end{bmatrix} ; \text{ an eigenvector is } \begin{bmatrix} 1 \\ 0 \\ 1 \end{bmatrix}.$$

$$\lambda_3 = -1 : \begin{bmatrix} -2 & 1 & 0 \\ 1 & -1 & -1 \\ 0 & -1 & -2 \end{bmatrix} \to \begin{bmatrix} 1 & -1 & -1 \\ 0 & -1 & -2 \\ 0 & -1 & -2 \end{bmatrix} \to \begin{bmatrix} 1 & 0 & 1 \\ 0 & 1 & 2 \\ 0 & 0 & 0 \end{bmatrix} ; \text{ an eigenvector } \begin{bmatrix} 1 \\ 2 \\ -1 \end{bmatrix}.$$

Hence,

$$P = \begin{bmatrix} -\frac{1}{\sqrt{3}} & \frac{1}{\sqrt{2}} & \frac{1}{\sqrt{6}} \\ \frac{1}{\sqrt{3}} & 0 & \frac{2}{\sqrt{6}} \\ \frac{1}{\sqrt{3}} & \frac{1}{\sqrt{2}} & -\frac{1}{\sqrt{6}} \end{bmatrix} = \frac{1}{\sqrt{6}} \begin{bmatrix} -\sqrt{2} & \sqrt{3} & 1 \\ \sqrt{2} & 0 & 2 \\ \sqrt{2} & \sqrt{3} & -1 \end{bmatrix}$$

is orthogonal and $P^T A P = \begin{bmatrix} 2 & 0 & 0 \\ 0 & 1 & 0 \\ 0 & 0 & -1 \end{bmatrix}$. If

$$Y = P^T X = \frac{1}{\sqrt{6}} \begin{bmatrix} -\sqrt{2} & \sqrt{2} & \sqrt{2} \\ \sqrt{3} & 0 & \sqrt{3} \\ 1 & 2 & -1 \end{bmatrix} \begin{bmatrix} x_1 \\ x_2 \\ x_3 \end{bmatrix}$$

then

$$y_1 = \tfrac{1}{\sqrt{3}}(-x_1 + x_2 + x_3)$$
$$y_2 = \tfrac{1}{\sqrt{2}}(x_1 + x_3)$$
$$y_3 = \tfrac{1}{\sqrt{6}}(x_1 + 2x_2 + x_3)$$

gives $q = 2y_1^2 + y_2^2 - y_3^2$. Here q has rank 3 and index 2.

3(b) $q = 3x^2 - 4xy = X^T A X$ where $X = \begin{bmatrix} x \\ y \end{bmatrix}$, $A = \begin{bmatrix} 3 & -2 \\ -2 & 0 \end{bmatrix}$. $c_A(t) = \begin{vmatrix} t-3 & 2 \\ 2 & t \end{vmatrix} =$

$(t-4)(t+1)$

$\lambda_1 = 4 : \begin{bmatrix} 1 & 2 \\ 2 & 4 \end{bmatrix} \rightarrow \begin{bmatrix} 1 & 2 \\ 0 & 0 \end{bmatrix}$; an eigenvector is $\begin{bmatrix} 2 \\ -1 \end{bmatrix}$.

$\lambda_2 = -1 : \begin{bmatrix} -4 & 2 \\ 2 & -1 \end{bmatrix} \rightarrow \begin{bmatrix} 2 & -1 \\ 0 & 0 \end{bmatrix}$; an eigenvector is $\begin{bmatrix} 1 \\ 2 \end{bmatrix}$.

Hence, $P = \frac{1}{\sqrt{5}} \begin{bmatrix} 2 & 1 \\ -1 & 2 \end{bmatrix}$ gives $P^T A P = \begin{bmatrix} 4 & 0 \\ 0 & -1 \end{bmatrix}$. If $Y = P^T X = \begin{bmatrix} x_1 \\ y_1 \end{bmatrix}$, then

$x_1 = \frac{1}{\sqrt{5}}(2x - y)$, $y_1 = \frac{1}{\sqrt{5}}(x + 2y)$. The equation $q = 2$ becomes $4x_1^2 - y_1^2 = 2$. This is a hyperbola.

(d) $q = 2x^2 + 4xy + 5y^2 = X^T A X$ where $X = \begin{bmatrix} x \\ y \end{bmatrix}$, $A = \begin{bmatrix} 2 & 2 \\ 2 & 5 \end{bmatrix}$, $c_A(t) = \begin{vmatrix} t-2 & -2 \\ -2 & t-5 \end{vmatrix} =$

$(t-1)(t-6)$.

$\lambda_1 = 6 : \begin{bmatrix} 4 & -2 \\ -2 & 1 \end{bmatrix} \rightarrow \begin{bmatrix} 2 & -1 \\ 0 & 0 \end{bmatrix}$; an eigenvector is $\begin{bmatrix} 1 \\ 2 \end{bmatrix}$.

$\lambda_2 = 1 : \begin{bmatrix} -1 & -2 \\ -2 & -4 \end{bmatrix} \rightarrow \begin{bmatrix} 1 & 2 \\ 0 & 0 \end{bmatrix}$; an eigenvector is $\begin{bmatrix} 2 \\ -1 \end{bmatrix}$.

Hence, $P = \frac{1}{\sqrt{5}} \begin{bmatrix} 1 & 2 \\ 2 & -1 \end{bmatrix}$ gives $P^T A P = \begin{bmatrix} 6 & 0 \\ 0 & 1 \end{bmatrix}$. If $Y = P^T X = \begin{bmatrix} x_1 \\ y_1 \end{bmatrix}$, then $x_1 =$

$\frac{1}{\sqrt{5}}(x + 2y)$, $y_1 = \frac{1}{\sqrt{5}}(2x - y)$ and $q = 1$ becomes $6x_1^2 + y_1^2 = 1$. This is an ellipse.

4. After the rotation, the new variables $X_1 = \begin{bmatrix} x_1 \\ y_1 \end{bmatrix}$ are related to $X = \begin{bmatrix} x \\ y \end{bmatrix}$ by $X = AX_1$

when $A = \begin{bmatrix} \cos\theta & -\sin\theta \\ \sin\theta & \cos\theta \end{bmatrix}$ (this is equation (**) preceding Theorem 2). Thus $x = x_1\cos\theta -$

$y_1\sin\theta$ and $y = x_1\sin\theta + y_1\cos\theta$. If these are substituted in the equation $ax^2 + bxy + cy^2 = d$, the coefficient of $x_1 y_1$ is

$$-2a\sin\theta\cos\theta + b(\cos^2\theta - \sin^2\theta) + 2c\sin\theta\cos\theta = b\cos 2\theta - (a - c)\sin 2\theta.$$

This is zero if θ is chosen so that

$$\cos 2\theta = \frac{a - c}{\sqrt{b^2 + (a - c)^2}} \quad \text{and} \quad \sin 2\theta = \frac{b}{\sqrt{b^2 + (a - c)^2}}.$$

Such an angle 2θ exists because $\left[\frac{a-c}{\sqrt{b^2+(a-c)^2}}\right]^2 + \left[\frac{b}{\sqrt{b^2(a-c)^2}}\right]^2 = 1$.

7(b) The given equation is $X^T A X + B X = 7$ where $X = \begin{bmatrix} x_1 \\ x_2 \\ x_3 \end{bmatrix}$, $A = \begin{bmatrix} 1 & 2 & -2 \\ 2 & 3 & 0 \\ -2 & 0 & 3 \end{bmatrix}$,

$B = \begin{bmatrix} 5 & 0 & -6 \end{bmatrix}$.

$$c_A(x) = \begin{vmatrix} t-1 & -2 & 2 \\ -2 & t-3 & 0 \\ 2 & 0 & t-3 \end{vmatrix} = \begin{vmatrix} t-1 & -2 & 2 \\ -2 & t-3 & 0 \\ 0 & t-3 & t-3 \end{vmatrix} = \begin{vmatrix} t-1 & -4 & 2 \\ -2 & t-3 & 0 \\ 0 & 0 & t-3 \end{vmatrix}$$

$$= (t-3)(t^2 - 4t - 5) = (t-3)(t-5)(t+1).$$

$\lambda_1 = 3 : \begin{bmatrix} 2 & -2 & 2 \\ -2 & 0 & 0 \\ 2 & 0 & 0 \end{bmatrix} \to \begin{bmatrix} 1 & 0 & 0 \\ 0 & 1 & -1 \\ 0 & 0 & 0 \end{bmatrix}$; an eigenvector is $\begin{bmatrix} 0 \\ 1 \\ 1 \end{bmatrix}$.

$\lambda_2 = 5 : \begin{bmatrix} 4 & -2 & 2 \\ -2 & 2 & 0 \\ 2 & 0 & 2 \end{bmatrix} \to \begin{bmatrix} -2 & 2 & 0 \\ 0 & 2 & 2 \\ 0 & 2 & 2 \end{bmatrix} \to \begin{bmatrix} 1 & 0 & 1 \\ 0 & 1 & 1 \\ 0 & 0 & 0 \end{bmatrix}$; an eigenvector is $\begin{bmatrix} 1 \\ 1 \\ -1 \end{bmatrix}$.

$\lambda_3 = -1 : \begin{bmatrix} -2 & -2 & 2 \\ -2 & -4 & 0 \\ 2 & 0 & -4 \end{bmatrix} \to \begin{bmatrix} 1 & 1 & -1 \\ 0 & -2 & -2 \\ 0 & -2 & -2 \end{bmatrix} \to \begin{bmatrix} 1 & 0 & -2 \\ 0 & 1 & 1 \\ 0 & 0 & 0 \end{bmatrix}$: an eigenvector is

$\begin{bmatrix} 2 \\ -1 \\ 1 \end{bmatrix}$.

Hence, $P = \begin{bmatrix} 0 & \frac{1}{\sqrt{3}} & \frac{2}{\sqrt{6}} \\ \frac{1}{\sqrt{2}} & -\frac{1}{\sqrt{3}} & -\frac{1}{\sqrt{6}} \\ \frac{1}{\sqrt{2}} & +\frac{1}{\sqrt{3}} & \frac{1}{\sqrt{6}} \end{bmatrix} = \frac{1}{\sqrt{6}} \begin{bmatrix} 0 & \sqrt{2} & 2 \\ \sqrt{3} & \sqrt{2} & -1 \\ \sqrt{3} & -\sqrt{2} & 1 \end{bmatrix}$ satisfies $P^T A P = \begin{bmatrix} 3 & 0 & 0 \\ 0 & 5 & 0 \\ 0 & 0 & -1 \end{bmatrix}$.

If

$$Y = \begin{bmatrix} y_1 \\ y_2 \\ y_3 \end{bmatrix} = P^T X = \frac{1}{\sqrt{6}} \begin{bmatrix} \sqrt{3}(x_2 + x_3) \\ \sqrt{2}(x_1 + x_2 - x_3) \\ 2x_1 - x_2 + x_3 \end{bmatrix}$$

then

$$y_1 = \frac{1}{\sqrt{2}}(x_2 + x_3)$$
$$y_2 = \frac{1}{\sqrt{3}}(x_1 + x_2 - x_3)$$
$$y_3 = \frac{1}{\sqrt{6}}(2x_1 - x_2 + x_3).$$

As $P^{-1} = P^T$, we have $X = PY$ so substitution in $X^T AX + BX = 7$ gives

$$Y^T(P^T AF)Y + (BF)Y = 7.$$

As $BF = \frac{1}{\sqrt{6}}\begin{bmatrix} -6\sqrt{3} & 11\sqrt{2} & 4 \end{bmatrix} = \begin{bmatrix} -3\sqrt{2} & \frac{11\sqrt{3}}{3} & \frac{2\sqrt{6}}{3} \end{bmatrix}$, this is

$$3y_1^2 + 5y_2^2 - y_3^2 - 3\sqrt{2}y_1 + \tfrac{11}{3}\sqrt{3}y_2 + \tfrac{2}{3}\sqrt{6}y_3 = 7.$$

Exercises 7.8   An Application to Best Approximation and Least Squares

1(b) Here $A = \begin{bmatrix} 3 & 1 & 1 \\ 2 & 3 & -1 \\ 2 & -1 & 1 \\ 3 & -3 & 3 \end{bmatrix}$, $B = \begin{bmatrix} 6 \\ 1 \\ 0 \\ 8 \end{bmatrix}$, $X = \begin{bmatrix} x \\ y \\ z \end{bmatrix}$. Hence, $A^T A = \begin{bmatrix} 26 & -2 & 12 \\ -2 & 20 & -12 \\ 12 & -12 & 12 \end{bmatrix}$.

This is invertible and the inverse is

$$(A^T A)^{-1} = \tfrac{1}{144}\begin{bmatrix} 96 & -120 & -216 \\ -120 & 168 & 288 \\ -216 & 288 & 516 \end{bmatrix} = \tfrac{1}{36}\begin{bmatrix} 24 & -30 & -54 \\ -30 & 42 & 72 \\ -54 & 72 & 129 \end{bmatrix}.$$

Here the (unique) best approximation is

$$Z = (A^T A)^{-1}A^T B = \tfrac{1}{36}\begin{bmatrix} 24 & -30 & -54 \\ -30 & 42 & 72 \\ -54 & 72 & 129 \end{bmatrix}\begin{bmatrix} 44 \\ -15 \\ 29 \end{bmatrix} = \tfrac{1}{36}\begin{bmatrix} -60 \\ 138 \\ 285 \end{bmatrix}.$$

Of course this can be found more efficiently using Gaussian elimination on the normal equations for Z.

2(b) In the notation of Theorem 3: $Y = \begin{bmatrix} 1 \\ 1 \\ 5 \\ 10 \end{bmatrix}$, $M = \begin{bmatrix} 0 & 0^2 & 2^0 \\ 1 & 1^2 & 2^1 \\ 2 & 2^2 & 2^2 \\ 3 & 3^2 & 2^3 \end{bmatrix} = \begin{bmatrix} 0 & 0 & 1 \\ 1 & 1 & 2 \\ 2 & 4 & 4 \\ 3 & 9 & 8 \end{bmatrix}$. Hence,

$M^T M = \begin{bmatrix} 14 & 36 & 34 \\ 36 & 98 & 90 \\ 34 & 90 & 85 \end{bmatrix}$, and $(M^T M)^{-1} = \tfrac{1}{92}\begin{bmatrix} 230 & 0 & -92 \\ 0 & 34 & -36 \\ -92 & -36 & 76 \end{bmatrix}$. Thus, the (unique)

solution to the normal equation is

$$Z = (M^T M)^{-1}M^T Y = \tfrac{1}{92}\begin{bmatrix} 230 & 0 & -92 \\ 0 & 34 & -36 \\ -92 & -36 & 76 \end{bmatrix}\begin{bmatrix} 41 \\ 111 \\ 103 \end{bmatrix} = \tfrac{1}{92}\begin{bmatrix} -46 \\ 66 \\ 60 \end{bmatrix}.$$

The best fitting function is thus $\frac{-46}{92}x + \frac{66}{92}x^2 + \frac{60}{92}2^x$.

3(b) Here $Y = \begin{bmatrix} \frac{1}{2} \\ 1 \\ 5 \\ 9 \end{bmatrix}$, $M = \begin{bmatrix} 1 & (-1)^2 & \sin\left(-\frac{\pi}{2}\right) \\ 1 & 0^2 & \sin(0) \\ 1 & 2^2 & \sin(\pi) \\ 1 & 3^2 & \sin\left(\frac{3\pi}{2}\right) \end{bmatrix} = \begin{bmatrix} 1 & 1 & -1 \\ 1 & 0 & 0 \\ 1 & 4 & 0 \\ 1 & 9 & -1 \end{bmatrix}$. Hence $M^T M = \begin{bmatrix} 4 & 14 & 0 \\ 14 & 98 & -10 \\ 0 & -10 & 2 \end{bmatrix}$

and $(M^T M)^{-1} = \frac{1}{2} \begin{bmatrix} -24 & 7 & 35 \\ 7 & -2 & -10 \\ 35 & -10 & -49 \end{bmatrix}$. Thus, the (unique) solution to the normal equations is

$$Z = (M^T M)^{-1} M^T Y = \frac{1}{2} \begin{bmatrix} -24 & 7 & 35 \\ 7 & -2 & -10 \\ 35 & -10 & -49 \end{bmatrix} \begin{bmatrix} \frac{31}{2} \\ \frac{203}{2} \\ -\frac{19}{2} \end{bmatrix} = \frac{1}{4} \begin{bmatrix} 12 \\ 1 \\ -14 \end{bmatrix}.$$

Hence, the best fitting functions

$$3 + \frac{1}{4}x^2 - \frac{7}{2}\sin\left(\frac{\pi x}{2}\right).$$

4. We want r_0, r_1, r_2, and r_3 to satisfy

$$r_0 + 50r_1 + 18r_2 + 10r_3 = 28$$
$$r_0 + 40r_1 + 20r_2 + 16r_3 = 30$$
$$r_0 + 35r_1 + 14r_2 + 10r_3 = 21$$
$$r_0 + 40r_1 + 12r_2 + 12r_3 = 23$$
$$r_0 + 30r_1 + 16r_2 + 14r_3 = 23.$$

We settle for a best approximation. Here

$$A = \begin{bmatrix} 1 & 50 & 18 & 10 \\ 1 & 40 & 20 & 16 \\ 1 & 35 & 14 & 10 \\ 1 & 40 & 12 & 12 \\ 1 & 30 & 16 & 14 \end{bmatrix} \qquad B = \begin{bmatrix} 28 \\ 30 \\ 21 \\ 23 \\ 23 \end{bmatrix}$$

$$A^T A = \begin{bmatrix} 5 & 195 & 80 & 62 \\ 195 & 7825 & 3150 & 2390 \\ 80 & 3150 & 1320 & 1008 \\ 62 & 2390 & 1008 & 796 \end{bmatrix}.$$

$$(A^T A)^{-1} = \frac{1}{50160} \begin{bmatrix} 1035720 & -16032 & 10080 & -45300 \\ -16032 & 416 & -632 & 800 \\ 10080 & -632 & 2600 & -2180 \\ -45300 & 800 & -2180 & 3950 \end{bmatrix}.$$

So the best approximation

$$Z = (A^T A)^{-1}(A^T B) = \frac{1}{50160}\begin{bmatrix} 1035720 & -16032 & 10080 & -45300 \\ -16032 & 416 & -632 & 800 \\ 10080 & -632 & 2600 & -2180 \\ -45300 & 800 & -2180 & 3950 \end{bmatrix}\begin{bmatrix} 125 \\ 4925 \\ 2042 \\ 1568 \end{bmatrix} = \begin{bmatrix} -5.19 \\ 0.34 \\ 0.51 \\ 0.71 \end{bmatrix}.$$

The best fitting function is

$$y = -5.19 + 0.34x_1 + 0.51x_2 + 0.71x_3.$$

6(b) It suffices to show that the columns of $M = \begin{bmatrix} 1 & e^{x_1} \\ \vdots & \vdots \\ 1 & e^{x_n} \end{bmatrix}$ are independent. If $r_0\begin{bmatrix} 1 \\ \vdots \\ 1 \end{bmatrix} +$

$r_1\begin{bmatrix} e^{x_1} \\ \vdots \\ e^{x_2} \end{bmatrix} = \begin{bmatrix} 0 \\ \vdots \\ 0 \end{bmatrix}$, then $r_0 + r_1 e^{x_i} = 0$ for each i. Thus, $r_1(e^{x_i} - e^{x_j}) = 0$ for all i and j, so

$r_1 = 0$ because two x_i are distinct. Then $r_0 = r_1 e^{x_1} = 0$ too.

Exercises 7.9 An Application to Systems of Differential Equations

1(b) Here $A = \begin{bmatrix} -1 & 5 \\ 1 & 3 \end{bmatrix}$ so $c_A(x) = \begin{vmatrix} x+1 & -5 \\ -1 & x-3 \end{vmatrix} = (x-4)(x+2)$.

$\lambda_1 = 4 : \begin{bmatrix} 5 & -5 \\ -1 & 1 \end{bmatrix} \to \begin{bmatrix} 1 & -1 \\ 0 & 0 \end{bmatrix}$; an eigenvector is $X_1 = \begin{bmatrix} 1 \\ 1 \end{bmatrix}$.

$\lambda_2 = -2 : \begin{bmatrix} -1 & -5 \\ -1 & -5 \end{bmatrix} \to \begin{bmatrix} 1 & 5 \\ 0 & 0 \end{bmatrix}$; an eigenvector is $X_2 = \begin{bmatrix} 5 \\ -1 \end{bmatrix}$.

Thus $P^{-1}AP = \begin{bmatrix} 4 & 0 \\ 0 & -2 \end{bmatrix}$ where $P = \begin{bmatrix} 1 & 5 \\ 1 & -1 \end{bmatrix}$. The general solution is

$$\mathbf{f} = c_1 X_1 e^{\lambda_1 x} + c_2 X_2 e^{\lambda_2 x} = c_1\begin{bmatrix} 1 \\ 1 \end{bmatrix} e^{4x} + c_2\begin{bmatrix} 5 \\ -1 \end{bmatrix} e^{-2x}.$$

Hence, $f_1(x) = c_1 e^{4x} + 5c_2 e^{-2x}$, $f_2(x) = c_1 e^{4x} - c_2 e^{-2x}$. The boundary bondition is $f_1(0) = 1$, $f_2(0) = -1$; that is

$$\begin{bmatrix} 1 \\ -1 \end{bmatrix} = \mathbf{f}(0) = c_1\begin{bmatrix} 1 \\ 1 \end{bmatrix} + c_2\begin{bmatrix} 5 \\ -1 \end{bmatrix}.$$

Thus $c_1 + 5c_2 = 1$, $c_1 - c_2 = -1$; the solution is $c_1 = -\frac{2}{3}$, $c_2 = \frac{1}{3}$, so the specific solution is

$$f_1(x) = \tfrac{1}{3}(5e^{-2x} - 2e^{4x}), \quad f_2(x) = -\tfrac{1}{3}(2e^{4x} + e^{-2x}).$$

(d) Now $A = \begin{bmatrix} 2 & 1 & 2 \\ 2 & 2 & -2 \\ 3 & 1 & 1 \end{bmatrix}$. To evaluate $c_A(x)$, first subtract row 1 from row 3:

$$c_A(x) = \begin{vmatrix} x-2 & -1 & -2 \\ -2 & x-2 & 2 \\ -3 & -1 & x-1 \end{vmatrix} = \begin{vmatrix} x-2 & -1 & -2 \\ -2 & x-2 & 2 \\ -x-1 & 0 & x+1 \end{vmatrix} = \begin{vmatrix} x-4 & -1 & -2 \\ 0 & x-2 & 2 \\ 0 & 0 & x+1 \end{vmatrix}$$

$$= (x+1)(x-2)(x-4).$$

$$\lambda_1 = -1: \begin{bmatrix} -3 & -1 & -2 \\ -2 & -3 & 2 \\ -3 & -1 & -2 \end{bmatrix} \rightarrow \begin{bmatrix} 1 & 5 & -6 \\ 2 & 3 & -2 \\ 0 & 0 & 0 \end{bmatrix} \rightarrow \begin{bmatrix} 1 & 0 & \frac{8}{7} \\ 0 & 1 & -\frac{10}{7} \\ 0 & 0 & 0 \end{bmatrix}; X_1 = \begin{bmatrix} -8 \\ 10 \\ 7 \end{bmatrix}.$$

$$\lambda_2 = 2: \begin{bmatrix} 0 & -1 & -2 \\ -2 & 0 & 2 \\ -3 & -1 & 1 \end{bmatrix} \rightarrow \begin{bmatrix} 1 & 0 & -1 \\ 0 & 1 & 2 \\ 0 & -1 & -2 \end{bmatrix} \rightarrow \begin{bmatrix} 1 & 0 & -1 \\ 0 & 1 & 2 \\ 0 & 0 & 0 \end{bmatrix}; X_2 = \begin{bmatrix} 1 \\ -2 \\ 1 \end{bmatrix}.$$

$$\lambda_3 = 4: \begin{bmatrix} 2 & -1 & -2 \\ -2 & 2 & 2 \\ -3 & -1 & 3 \end{bmatrix} \rightarrow \begin{bmatrix} 2 & -1 & -2 \\ 0 & 2 & 0 \\ -3 & -1 & 3 \end{bmatrix} \rightarrow \begin{bmatrix} 1 & 0 & -1 \\ 0 & 1 & 0 \\ 0 & 0 & 0 \end{bmatrix}; X_3 = \begin{bmatrix} 1 \\ 0 \\ 1 \end{bmatrix}.$$

Thus $P^{-1}AP = \begin{bmatrix} -1 & 0 & 0 \\ 0 & 2 & 0 \\ 0 & 0 & 4 \end{bmatrix}$ where $P = \begin{bmatrix} -8 & 1 & 1 \\ 10 & -2 & 0 \\ 7 & 1 & 1 \end{bmatrix}$. The general solution is

$$\mathbf{f} = c_1 X_1 e^{-x} + c_2 X^2 e^{2x} + c_3 X_3 e^{4x} = c_1 \begin{bmatrix} -8 \\ 10 \\ 7 \end{bmatrix} e^{-x} + c_2 \begin{bmatrix} 1 \\ -2 \\ 1 \end{bmatrix} e^{2x} + c_3 \begin{bmatrix} 1 \\ 0 \\ 1 \end{bmatrix} e^{4x}.$$

That is

$$f_1(x) = -8c_1 e^{-x} + c_2 e^{2x} + c_3 e^{4x}$$
$$f_2(x) = 10c_1 e^{-x} - 2c_2 e^{2x}$$
$$f_3(x) = 7c_1 e^{-x} + c_2 e^{2x} + c_3 e^{4x}.$$

If we insist on the boundary conditions $f_1(0) = f_2(0) = f_3(0) = 1$, we get

$$\begin{array}{rcrcrcl} -8c_1 & + & c_2 & + & c_3 & = & 1 \\ 10c_1 & - & 2c_2 & & & = & 1 \\ 2c_1 & + & c_2 & + & c_3 & = & 1. \end{array}$$

The coefficient matrix is P is invertible, so the solution is unique: $c_1 = 0$, $c_2 = -\frac{1}{2}$, $c_3 = \frac{3}{2}$. Hence

$$f_1(x) = \tfrac{1}{2}(3e^{4x} - e^{2x})$$
$$f_2(x) = e^{2x}$$
$$f_3(x) = \tfrac{1}{2}(3e^{4x} - e^{2x}).$$

Note that $f_1(x) = f_3(x)$ happens to hold.

Chapter 8: Linear Transformations

Exercises 8.1   Examples and Elementary Properties

1(b) $T(X) = XA$ where $X = (x, y, z)$, $A = \begin{bmatrix} 1 \\ 1 \\ -1 \end{bmatrix}$. Hence, matrix algebra gives $T(X + Y) = $

$A(X + Y) = AX + AY = T(X) + T(Y)$ and $T(r\lambda) = A(rX) = rA(X) = rT(X)$.

(d) $T(A+B) = P(A+B)Q = PAQ + PBQ = T(A) + T(B)$; $T(rA) = P(rA)Q = rPAQ = rT(A)$.

(f) Here $T[p(x)] = p(0)$ for all polynomials $p(x)$ in $\mathbf{P}_n$. Thus

$$T[(p + q)(x)] = T[p(x) + q(x)] = p(0) + q(0) = T[p(x)] + T[q(x)]$$
$$T[rp(x)] = rp(0) = r[Tp(x)].$$

(h) Here Z is fixed in $\mathbb{R}^n$ and $T(X) = X \cdot Z$ for all X in $\mathbb{R}^n$. We use Theorem 1, Section 7.1:

$$T(X + Y) = X + Y) \cdot Z = X \cdot Z + Y \cdot Z = T(X) + T(Y)$$
$$T(rX) = (rX) \cdot Z = r(X \cdot Z) = rT(X).$$

(j) If $\mathbf{v}$ is any vector in V, then $\mathbf{v} = r_1\mathbf{e}_1 + \cdots + r_n\mathbf{e}_n$ for unique real numbers $r_1, \ldots, r_n$ and we take $T(\mathbf{v}) = r_1$. Thus $r\mathbf{v} = (rr_1)\mathbf{e}_1 + \cdots + (rr_n)\mathbf{e}_n$ for any r in $\mathbb{R}$, so $T(r\mathbf{v}) = rr_1 = rT(\mathbf{v})$. Similarly, if $\mathbf{w} = s_1\mathbf{e}_1 + \cdots + s_n\mathbf{e}_n$, s_i in $\mathbb{R}$, then $\mathbf{v} + \mathbf{w} = (r_1 + s_1)\mathbf{e}_1 + \cdots + (r_n + s_n)\mathbf{e}_n$. Then

$$T(\mathbf{v} + \mathbf{w}) = r_1 + s_1 = T(\mathbf{v}) + T(\mathbf{w}).$$

2(b) Let $A = \begin{bmatrix} 1 & 0 & 0 & \ldots & 0 \\ 0 & 1 & 0 & \ldots & 0 \\ 0 & 0 & 0 & \ldots & 0 \\ \vdots & \vdots & \vdots & \ddots & \vdots \\ 0 & 0 & 0 & \ldots & 0 \end{bmatrix}$, $B = \begin{bmatrix} 1 & 0 & 0 & \ldots & 0 \\ 0 & -1 & 0 & \ldots & 0 \\ 0 & 0 & 0 & \ldots & 0 \\ \vdots & \vdots & \vdots & \ddots & \vdots \\ 0 & 0 & 0 & \ldots & 0 \end{bmatrix}$, then $A+B = \begin{bmatrix} 2 & 0 & 0 & \ldots & 0 \\ 0 & 0 & 0 & \ldots & 0 \\ 0 & 0 & 0 & \ldots & 0 \\ \vdots & \vdots & \vdots & \ddots & \vdots \\ 0 & 0 & 0 & \ldots & 0 \end{bmatrix}$.

Thus, $T(A) = \text{rank } A = 2$, $T(B) = \text{rank } B = 2$ and $T(A + B) = \text{rank}(A + B) = 1$. Thus $T(A + B) \neq T(A) + T(B)$.

(d) Here $T(\mathbf{v} + \mathbf{w}) = \mathbf{v} + \mathbf{w} + \mathbf{u}$, $T(\mathbf{v}) = \mathbf{v} + \mathbf{u}$, $T(\mathbf{w}) = \mathbf{w} + \mathbf{u}$. Thus if $T(\mathbf{v} + \mathbf{w}) = T(\mathbf{v}) + T(\mathbf{w})$ then $\mathbf{v} + \mathbf{w} + \mathbf{u} = (\mathbf{v} + \mathbf{u}) + (\mathbf{w} + \mathbf{u})$, so $\mathbf{u} = 2\mathbf{u}$, $\mathbf{u} = \mathbf{0}$. This is contrary to assumption. Alternatively, $T(\mathbf{v}) = \mathbf{0} + \mathbf{y} \neq \mathbf{0}$, so T cannot be linear by Theorem 1.

3(b) Because T is linear, $T(3\mathbf{v}_1 + 2\mathbf{v}_2) = 3T(\mathbf{v}_1) + 2T(\mathbf{v}_2) = 3(2) + 2(-3) = 0$.

(d) Since we know the action of T on $\begin{bmatrix} 1 \\ -1 \end{bmatrix}$ and $\begin{bmatrix} 1 \\ 1 \end{bmatrix}$, it suffices to express $\begin{bmatrix} 1 \\ -7 \end{bmatrix}$ as a linear combination of these vectors.

$$\begin{bmatrix} 1 \\ -7 \end{bmatrix} = r \begin{bmatrix} 1 \\ -1 \end{bmatrix} + s \begin{bmatrix} 1 \\ 1 \end{bmatrix}.$$

This is possible: $r = 4$, $s = -3$, so

$$T\begin{bmatrix} 1 \\ -7 \end{bmatrix} = T\left(4\begin{bmatrix} 1 \\ -1 \end{bmatrix} - 3\begin{bmatrix} 1 \\ 1 \end{bmatrix}\right) = 4T\begin{bmatrix} 1 \\ -1 \end{bmatrix} - 3T\begin{bmatrix} 1 \\ 1 \end{bmatrix} = 4\begin{bmatrix} 0 \\ 1 \end{bmatrix} - 3\begin{bmatrix} 1 \\ 1 \end{bmatrix} = \begin{bmatrix} -3 \\ 4 \end{bmatrix}.$$

(f) We now $T(1)$, $T(x+2)$ and $T(x^2+x)$, so we express $2 - x + 3x^2$ as a linear combination of these vectors:

$$2 - x + 3x^2 = r \cdot 1 + s(x+2) + t(x^2+x).$$

This is possible: $r = 10$, $s = -4$ and $t = 3$, so

$$\begin{aligned} T(2 - x + 3x^2) &= T[r \cdot 1 + s(x+2) + t(x^2+x)] \\ &= rT(1) + sT(x+2) + tT(x^2+x) \\ &= 5r + s + 0 \\ &= 46. \end{aligned}$$

4(b) Since $B = \{(2,-1),(1,1)\}$ is a basis of $\mathbb{R}^2$, any vector (x,y) in $\mathbb{R}^2$ is a linear combination $(x,y) = r(2,-1) + s(1,1)$. Indeed: $r = \frac{1}{3}(x-y)$, $s = \frac{1}{3}(x+2y)$. Hence,

$$\begin{aligned} T(x,y) &= T[r(2,-1) + s(1,1)] \\ &= rT(2,-1) + sT(1,1) \\ &= \tfrac{1}{3}(x-y)(1,-1,1) + \tfrac{1}{3}(x+2y)(0,1,0) \\ &= \left(\tfrac{1}{3}(x-y), y, \tfrac{1}{3}(x-y)\right) \\ &= \tfrac{1}{3}(x-y, 3y, x-y). \end{aligned}$$

In particular, $T(\mathbf{v}) = T(-1,2) = \frac{1}{3}(-3,6,-3) = (-1,2,-1)$.

(d) Since $B = \left\{ \begin{bmatrix} 1 & 0 \\ 0 & 0 \end{bmatrix}, \begin{bmatrix} 0 & 1 \\ 1 & 0 \end{bmatrix}, \begin{bmatrix} 1 & 0 \\ 1 & 0 \end{bmatrix}, \begin{bmatrix} 0 & 0 \\ 0 & 1 \end{bmatrix} \right\}$ is a basis of $\mathbf{M}_{22}$, every vector

$\begin{bmatrix} a & b \\ c & d \end{bmatrix}$ is a linear combination $\begin{bmatrix} a & b \\ c & d \end{bmatrix} = r\begin{bmatrix} 1 & 0 \\ 0 & 0 \end{bmatrix} + s\begin{bmatrix} 0 & 1 \\ 1 & 0 \end{bmatrix} + t\begin{bmatrix} 1 & 0 \\ 1 & 0 \end{bmatrix} +$

$u\begin{bmatrix} 0 & 0 \\ 0 & 1 \end{bmatrix}$. Indeed $r = a - c + b$, $s = b$, $t = c - b$, $u = d$. Thus,

$$\begin{aligned} T\begin{bmatrix} a & b \\ c & d \end{bmatrix} &= rT\begin{bmatrix} 1 & 0 \\ 0 & 0 \end{bmatrix} + sT\begin{bmatrix} 0 & 1 \\ 1 & 0 \end{bmatrix} + tT\begin{bmatrix} 1 & 0 \\ 1 & 0 \end{bmatrix} + uT\begin{bmatrix} 0 & 0 \\ 0 & 1 \end{bmatrix} \\ &= (a - c + b) \cdot 3 + b \cdot (-1) + (c - b) \cdot 0 + d \cdot 0 \\ &= 3a + 2b - 3c. \end{aligned}$$

5(b) Since T is linear, the given conditions read

$$T(\mathbf{v}) + 2T(\mathbf{w}) = 3\mathbf{v} - \mathbf{w}$$
$$T(\mathbf{v}) - T(\mathbf{w}) = 2\mathbf{v} - 4\mathbf{w}.$$

Add twice the second equation to the first to get $3T(\mathbf{v}) = 7\mathbf{v} - 9\mathbf{w}$, $T(\mathbf{v}) = \frac{7}{3}\mathbf{v} - 3\mathbf{w}$. Similarly, subtracting the second from the first gives $3T(\mathbf{w}) = \mathbf{v} + 3\mathbf{w}$, $T(\mathbf{w}) = \frac{1}{3}\mathbf{v} + \mathbf{w}$.

8(b) Since $\{\mathbf{v}_1, \ldots, \mathbf{v}_n\}$ is a basis of V, every vector $\mathbf{v}$ in V is a linear combination $\mathbf{v} = r_1\mathbf{v}_1 + \cdots + r_n\mathbf{v}_n$, r_i in $\mathbb{R}$. Hence, as T is linear,

$$T(\mathbf{v}) = r_1 T(\mathbf{v}_1) + \cdots + r_n T(\mathbf{v}_n) = r_1 \mathbf{0} + \cdots + r_n \mathbf{0} + \mathbf{0}.$$

12. $\{1\}$ is a basis of the vector space $\mathbb{R}$. If $T : \mathbb{R} \to V$ is a linear transformation, write $T(1) = \mathbf{v}_0$. Then, for all r in $\mathbb{R}$:

$$T(r) = T(r \cdot 1) = rT(1) = r\mathbf{v}_0.$$

Since $T(r) = r\mathbf{v}$ is linear for each $\mathbf{v}$ in V, this shows that every linear transformation $T : \mathbb{R} \to V$ arises in this way.

15(b) If $\mathbf{v}$ and $\mathbf{v}_1$ are in $T^{-1}(P)$, then $T(\mathbf{v})$ and $T(\mathbf{v}_1)$ are in P. As P is a subspace, $T(\mathbf{v} + \mathbf{v}_1) = T(\mathbf{v}) + T(\mathbf{v}_1)$ and $T(r\mathbf{v}) = rT(\mathbf{v})$ are both in P; that is $\mathbf{v} + \mathbf{v}_1$ and $r\mathbf{v}$ are in $T^{-1}(P)$.

18. Have $T[T(\mathbf{v})] = \mathbf{v}$ for all $\mathbf{v}$ in V. If $\{\mathbf{v}, T(\mathbf{v})\}$ is linearly independent then $T(\mathbf{v}) \neq \mathbf{v}$ (or else $1\mathbf{v} + (-1)T(\mathbf{v}) = 0$) and similarly $T(\mathbf{v}) \neq -\mathbf{v}$. Conversely, assume that $T(\mathbf{v}) \neq \mathbf{v}$ and $T(\mathbf{v}) \neq -\mathbf{v}$. To verify that $\{\mathbf{v}, T(\mathbf{v})\}$ is independent, let $r\mathbf{v} + sT(\mathbf{v}) = 0$. We must show that $r = s = 0$. If $s \neq 0$, then $\frac{r}{s}\mathbf{v} + T(\mathbf{v}) = 0$. Write $a = -\frac{r}{s}$ so that $T(\mathbf{v}) = a\mathbf{v}$. Apply $T : T[T(\mathbf{v})] = aT(\mathbf{v})$, so $\mathbf{v} = aT(\mathbf{v})$ by hypothesis. Hence.

$$\mathbf{v} = aT(\mathbf{v}) = a(\mathbf{v}) = a^2\mathbf{v}$$

so $a^2 = 1$ (as $\mathbf{v} \neq 0$), whence $a = \pm 1$. This means $s = \pm r$, so $r\mathbf{v} + sT(\mathbf{v}) = 0$ gives $\mathbf{v} + T(\mathbf{v}) = \mathbf{0}$ or $\mathbf{v} - T(\mathbf{v}) = \mathbf{0}$, contrary to our assumption. So $s = 0$. Then $r\mathbf{v} + sT(\mathbf{v}) = 0$ gives $r\mathbf{v} = \mathbf{0}$ so $r = 0$, as required.

21(b) Given such a T_1 write $T(x) = a$. We have $T(x^i) = T(x)^i = a^i = E_a(x^i)$ for each i (even $i = 0$). Since $\{1, x, x^2, \ldots, x^i, \ldots, x^n\}$ is a basis of $\mathbb{P}_n$, this gives $T = E_a$ by Theorem 2.

Exercises 8.2   Kernel and Image of a Linear Transformation

1(b) $\ker T_A = \{X \mid AX = 0\}$. We use Gaussian elimination:

$$\begin{bmatrix} 2 & 1 & -1 & 3 & | & 0 \\ 1 & 0 & 3 & 1 & | & 0 \\ 1 & 1 & -4 & 2 & | & 0 \end{bmatrix} \to \begin{bmatrix} 1 & 0 & 3 & 1 & | & 0 \\ 0 & 1 & -7 & 1 & | & 0 \\ 0 & 1 & -7 & 1 & | & 0 \end{bmatrix} \to \begin{bmatrix} 1 & 0 & 3 & 1 & | & 0 \\ 0 & 1 & -7 & 1 & | & 0 \\ 0 & 0 & 0 & 0 & | & 0 \end{bmatrix}.$$

Hence $\ker T_A = \left\{ \begin{bmatrix} -3s - t \\ 7s - t \\ s \\ t \end{bmatrix} \mid s, t \text{ in } \mathbb{R} \right\} = \text{span} \left\{ \begin{bmatrix} -3 \\ 7 \\ 1 \\ 0 \end{bmatrix}, \begin{bmatrix} 1 \\ 1 \\ 0 \\ -1 \end{bmatrix} \right\}$. These vectors are independent so nullity of $T_A = \dim(\ker T_A) = 2$. Next

$$\operatorname{im} T_A = \{AX \mid X \text{ in } \mathbb{R}^4\}$$

$$= \left\{ \begin{bmatrix} 2 & 1 & -1 & 3 \\ 1 & 0 & 3 & 1 \\ 1 & 1 & -4 & 2 \end{bmatrix} \begin{bmatrix} r \\ s \\ t \\ y \end{bmatrix} \mid r, s, t, u \text{ in } \mathbb{R} \right\}$$

$$= \left\{ r \begin{bmatrix} 2 \\ 1 \\ 1 \end{bmatrix} + s \begin{bmatrix} 1 \\ 0 \\ 1 \end{bmatrix} + t \begin{bmatrix} -1 \\ 3 \\ -4 \end{bmatrix} + u \begin{bmatrix} 3 \\ 1 \\ 2 \end{bmatrix} \mid r, s, t, u \text{ in } \mathbb{R} \right\}.$$

Do Gaussian elimination on A^T

$$\begin{bmatrix} 2 & 1 & 1 \\ 1 & 0 & 1 \\ -1 & 3 & -4 \\ 3 & 1 & 2 \end{bmatrix} \rightarrow \begin{bmatrix} 1 & 0 & 1 \\ 0 & 1 & -1 \\ 0 & 3 & -3 \\ 0 & 1 & -1 \end{bmatrix} \rightarrow \begin{bmatrix} 1 & 0 & 1 \\ 0 & 1 & -1 \\ 0 & 0 & 0 \\ 0 & 0 & 0 \end{bmatrix}.$$

Hence, $\operatorname{im} T_A = \operatorname{col} A = \operatorname{span} \left\{ \begin{bmatrix} 1 \\ 0 \\ 1 \end{bmatrix}, \begin{bmatrix} 0 \\ 1 \\ -1 \end{bmatrix} \right\}$. So rank $T_A = \dim(\operatorname{im} T_A) = 2$.

(d) $\ker T_A = \{X \mid AX = 0\}$. We use Gaussian elimination:

$$\begin{bmatrix} 2 & 1 & 0 \\ 1 & -1 & 3 \\ 1 & 2 & -3 \\ 0 & 3 & -6 \end{bmatrix} \rightarrow \begin{bmatrix} 1 & -1 & 3 \\ 0 & 3 & -6 \\ 0 & 3 & -6 \\ 0 & 3 & -6 \end{bmatrix} \rightarrow \begin{bmatrix} 1 & 0 & 1 \\ 0 & 1 & -2 \\ 0 & 0 & 0 \\ 0 & 0 & 0 \end{bmatrix}.$$

Hence, $\ker T_A = \left\{ \begin{bmatrix} -t \\ 2t \\ t \end{bmatrix} \mid t \text{ in } \mathbb{R} \right\} = \operatorname{span} \left\{ \begin{bmatrix} -1 \\ 2 \\ 1 \end{bmatrix} \right\}$. Thus the nullity of T_A is

$\dim(\ker T_A) = 1$. As in (b), $\operatorname{im} T_A = \operatorname{col} A$ and we find a basis by doing Gaussian elimination in A^T:

$$\begin{bmatrix} 2 & 1 & 1 & 0 \\ 1 & -1 & 2 & 3 \\ 0 & 3 & -3 & -6 \end{bmatrix} \rightarrow \begin{bmatrix} 1 & -1 & 2 & 3 \\ 0 & 3 & -3 & -6 \\ 0 & 3 & -3 & -6 \end{bmatrix} \rightarrow \begin{bmatrix} 1 & 0 & 1 & 1 \\ 0 & 1 & -1 & -2 \\ 0 & 0 & 0 & 0 \end{bmatrix}.$$

Hence, $\text{im } T_A = \text{col } A = \text{span}\left\{\begin{bmatrix} 1 \\ 0 \\ 1 \\ 1 \end{bmatrix}, \begin{bmatrix} 0 \\ 1 \\ -1 \\ -2 \end{bmatrix}\right\}$, so rank $T_A = \dim(\text{im } T_A) = 2$.

2(b) Here $T = \mathbf{P}_2 \to \mathbb{R}^2$ given by $T[p(x)] = [p(0) \quad p(1)]$. Hence

$$\ker T = \{p(x) \mid p(0) = p(1) = 0\}.$$

If $p(x) = a + bx + cx^2$ is in $\ker T$, then $0 = p(0) = a$ and $0 = p(1) = a + b + c$. Then $p(x) = bx - bx^2$ and so $\ker T = \text{span}\{x - x^2\}$. Thus $\{x - x^2\}$ is a basis of $\ker T$. Next, im T is a subspace of $\mathbb{R}^2$. We have $[1,0] = T(1 - x)$ and $[0,1] = T[x]$ are both in im T, so im $T = \mathbb{R}^2$. Thus $\{[1,0],[0,1]\}$ is a basis of im T.

(d) Here $T : \mathbb{R}^3 \to \mathbb{R}^4$ given by $T(x,y,z) = (x,x,y,y)$. Thus,

$$\ker T \{(x,y,z) = (0,0,0,0)\} = \{(0,0,z) \mid z \text{ in } \mathbb{R}\} = \text{span } \{(0,0,1)\}.$$

Hence, $\{(0,0,1)\}$ is a basis of $\ker T$. On the other hand,

$$\text{im } T = \{(x,x,y,y) \mid x,y \text{ in } \mathbf{R}\} = \text{span } \{(1,1,0,0),(0,0,1,1)\}.$$

Then $\{(1,1,0,0),(0,0,1,1)\}$ is a basis of im T.

(f) Here $T : \mathbf{M}_{22} \to \mathbb{R}$ is given by $T\begin{bmatrix} a & b \\ c & d \end{bmatrix} = a + d$. Hence

$$\ker T = \left\{\begin{bmatrix} a & b \\ c & d \end{bmatrix} \mid a + d = 0\right\} = \left\{\begin{bmatrix} a & b \\ c & -a \end{bmatrix} \mid a,b,c \text{ in } \mathbb{R}\right\}$$

$$= \text{span}\left\{\begin{bmatrix} 1 & 0 \\ 0 & -1 \end{bmatrix}, \begin{bmatrix} 0 & 1 \\ 0 & 0 \end{bmatrix}, \begin{bmatrix} 0 & 0 \\ 1 & 0 \end{bmatrix}\right\}.$$

Hence, $\left\{\begin{bmatrix} 1 & 0 \\ 0 & -1 \end{bmatrix}, \begin{bmatrix} 0 & 1 \\ 0 & 0 \end{bmatrix}, \begin{bmatrix} 0 & 0 \\ 1 & 0 \end{bmatrix}\right\}$ is a basis of $\ker T$ (being independent). On the other hand,

$$\text{im } T = \left\{a + d \mid \begin{bmatrix} a & b \\ c & d \end{bmatrix} \text{ in } \mathbf{M}_{22}\right\} = \mathbb{R}.$$

So $\{1\}$ is a basis of im T.

(h) $T : \mathbb{R}^n \to \mathbb{R}$, $T(r_1, r_2, \ldots, r_n) = r_1 + r_2 + \cdots + r_n$. Hence,

$$\ker T = \{(r_1, r_2, \ldots, r_n) \mid r_1 + r_2 + \cdots + r_n = 0\}$$
$$= \{(r_1, r_2, \ldots, r_{n-1}, -r_1 - \cdots - r_{n-1} \mid r_i \text{ in } \mathbb{R}\}$$
$$= \text{span } \{(1,0,0,\ldots,-1),(0,1,0,\ldots,-1),\ldots,(0,0,1,\ldots,-1)\}.$$

This is a basis of ker T. On the other hand,

$$\text{im } T = \{r_1 + \cdots + r_n \mid (r_1, r_2, \ldots, r_n) \text{ is in } \mathbb{R}^n\} = \mathbb{R}.$$

Thus $\{1\}$ is a basis of im T.

(j) $T : \mathbf{M}_{22} \to \mathbf{M}_{22}$ is given by $T(X) = XA$ where $A = \begin{bmatrix} 1 & 1 \\ 0 & 0 \end{bmatrix}$. Writing $X = \begin{bmatrix} x & y \\ z & w \end{bmatrix}$:

$$\ker T = \{X \mid XA = 0\} = \left\{ \begin{bmatrix} x & y \\ z & w \end{bmatrix} \middle| \begin{bmatrix} x & x \\ z & x \end{bmatrix} = \begin{bmatrix} 0 & 0 \\ 0 & 0 \end{bmatrix} \right\} = \left\{ \begin{bmatrix} 0 & y \\ 0 & z \end{bmatrix} \middle| y, z \text{ in } \mathbb{R} \right\}$$

$$= \text{span} \left\{ \begin{bmatrix} 0 & 1 \\ 0 & 0 \end{bmatrix}, \begin{bmatrix} 0 & 0 \\ 0 & 1 \end{bmatrix} \right\}.$$

Thus, $\left\{ \begin{bmatrix} 0 & 1 \\ 0 & 0 \end{bmatrix}, \begin{bmatrix} 0 & 0 \\ 0 & 1 \end{bmatrix} \right\}$ is a basis of ker T (being independent). On the other hand,

$$\text{im } T = \{XA \mid X \text{ in } \mathbf{M}_{22}\} = \left\{ \begin{bmatrix} x & x \\ z & z \end{bmatrix} \middle| x, z \text{ in } \mathbb{R} \right\} = \text{span} \left\{ \begin{bmatrix} 1 & 1 \\ 0 & 0 \end{bmatrix}, \begin{bmatrix} 0 & 0 \\ 1 & 1 \end{bmatrix} \right\}.$$

Thus, $\left\{ \begin{bmatrix} 1 & 1 \\ 0 & 0 \end{bmatrix}, \begin{bmatrix} 0 & 0 \\ 1 & 1 \end{bmatrix} \right\}$ is a basis of im T.

3(b) Have $T : V \to \mathbb{R}^2$ given by $T(\mathbf{v}) = (P(\mathbf{v}), Q(\mathbf{v}))$ where $P : V \to \mathbb{R}$ and $Q : V \to \mathbb{R}$ are linear transformations. T is linear by (a). Now

$$\begin{aligned} \ker T &= \{\mathbf{v} \mid T(\mathbf{v}) = (0, 0)\} \\ &= \{\mathbf{v} \mid P(\mathbf{v}) = 0 \text{ and } Q(\mathbf{v}) = 0\} \\ &= \{\mathbf{v} \mid P(\mathbf{v}) = 0\} \cap \{\mathbf{v} \mid Q(\mathbf{v}) = 0\} \\ &= \ker P \cap \ker Q. \end{aligned}$$

4(b) $\ker T = \{(x, y, z) \mid x + y + z = 0, 2x - y + 3z = 0, z - 3y = 0, 3x + 4z = 0\}$. Solving:

$$\begin{bmatrix} 1 & 1 & 1 & 0 \\ 2 & -1 & 3 & 0 \\ 0 & -3 & 1 & 0 \\ 3 & 0 & 4 & 0 \end{bmatrix} \to \begin{bmatrix} 1 & 1 & 1 & 0 \\ 0 & -3 & 1 & 0 \\ 0 & -3 & 1 & 0 \\ 0 & -3 & 1 & 0 \end{bmatrix} \to \begin{bmatrix} 1 & 0 & \frac{4}{3} & 0 \\ 0 & 1 & -\frac{1}{3} & 0 \\ 0 & 0 & 0 & 0 \\ 0 & 0 & 0 & 0 \end{bmatrix}.$$

Hence, $\ker T = \{(-4t, t, 3t) \mid t \text{ in } \mathbb{R}\} = \text{span}\{(-4, 1, 3)\}$. Hence, $\{(1, 0, 0), (0, 1, 0), (-4, 1, 3)\}$ is one basis of $\mathbb{R}^3$ containing a basis of ker T. Thus

$$\{T(1, 0, 0), T(0, 1, 0)\} = \{(1, 2, 0, 3), (1, -1, -3, 0)\}$$

is a basis of im T by Theorem 5.

6(b) Yes. $\dim(\operatorname{im} T) = \dim V - \dim(\ker T) = 5 - 2 = 3$. As $\dim W = 3$ and $\operatorname{im} T$ is a 3-dimensional subspace, $\operatorname{im} T = W$. Thus, T is onto.

(d) No. If $\ker T = V$ then $T(\mathbf{v}) = \mathbf{0}$ for all $\mathbf{v}$ in V, so $T = 0$ is the zero transformation. But W need not be the zero space. For example, $T : \mathbb{R}^2 \to \mathbb{R}^2$ defined by $T(x,y) = (0,0)$ for all (x,y) in $\mathbb{R}^2$.

(f) No. $\operatorname{im} T \subseteq \ker T$ means $T(\mathbf{v})$ is in $\ker T$ for all $\mathbf{v}$ in V; that is $T[T(\mathbf{v})] = 0$ for all $\mathbf{v}$ in V.

However, T need not be the zero transformation. The matrix $A = \begin{bmatrix} 0 & 1 \\ 0 & 0 \end{bmatrix}$ satisfies $A^2 = 0$

so $T_A : \mathbb{R}^2 \to \mathbb{R}^2$ given by $T_A(X) = AX$ satisfies $T_A[T_A(X)] = A[AX] = A^2 X = 0X = 0$ for

all X. But $T_A \begin{bmatrix} 0 \\ 1 \end{bmatrix} = \begin{bmatrix} 0 & 1 \\ 0 & 0 \end{bmatrix} \begin{bmatrix} 0 \\ 1 \end{bmatrix} = \begin{bmatrix} 1 \\ 0 \end{bmatrix} \neq \begin{bmatrix} 0 \\ 0 \end{bmatrix}$.

(h) Yes. We always have $\dim(\operatorname{im} T) \leq \dim W$ (because $\operatorname{im} T$ is a subspace of W). Since $\dim(\ker T) \leq \dim W$ also holds in this case:

$$\dim V = \dim(\ker T) + \dim(\operatorname{im} T) \leq \dim W + W = 2\dim W.$$

Hence $\dim W \geq \frac{1}{2}\dim V$.

(j) No. $T : \mathbb{R}^2 \to \mathbb{R}^2$ given by $T(x,y) = (x,0)$ is not one-to-one (because $\ker T = \{(0,y) \mid y \text{ in } \mathbb{R}\}$ is not 0).

(l) No. $T : \mathbb{R}^2 \to \mathbb{R}^2$ given by $T(x,y) = (x,0)$ is not onto.

7(b) Given $\mathbf{w}$ in W, we must show that it is a linear combination of $T(\mathbf{v}_1),\ldots,T(\mathbf{v}_n)$. As T is onto, $\mathbf{w} = T(\mathbf{v})$ for some $\mathbf{v}$ in V. Since $V = \operatorname{span}\{\mathbf{v}_1,\ldots,\mathbf{v}_n\}$ we can write $\mathbf{v} = r_1\mathbf{v}_1 + \cdots + r_n\mathbf{v}_n$, r_i in $\mathbb{R}$, so

$$\mathbf{w} = T(\mathbf{v}) = T(r_1\mathbf{v}_1 + \cdots + r_n\mathbf{v}_n) = r_1 T(\mathbf{v}_1) + \cdots + r_n T(\mathbf{v}_n).$$

8(b) If T is onto, let $\mathbf{v}$ be any vector in V. Then $\mathbf{v} = T(r_1,\ldots,r_n)$ for some $(r_1,\ldots,r_n)$ in $\mathbb{R}^n$; that is $\mathbf{v} = r_1\mathbf{v}_1 + \cdots + r_n\mathbf{v}_n$ is in $\operatorname{span}\{\mathbf{v}_1,\ldots,\mathbf{v}_n\}$. Thus $V = \operatorname{span}\{\mathbf{v}_1,\ldots,\mathbf{v}_n\}$. Conversely, if $V = \operatorname{span}\{\mathbf{v}_1,\ldots,\mathbf{v}_n\}$, let $\mathbf{v}$ be any vector in V. Then $\mathbf{v}$ is in $\operatorname{span}\{\mathbf{v}_1,\ldots,\mathbf{v}_n\}$ so $r_1,\ldots,r_n$ exist in $\mathbb{R}$ such that

$$\mathbf{v} = r_1\mathbf{v}_1 + + r_n\mathbf{v}_n = T(r_1,\ldots,r_n).$$

Thus T is onto.

10. The trace map $T : \mathbf{M}_{22} \to \mathbb{R}$ is linear (Example 3, Section 8.1) and it is onto (for example, $r = \operatorname{tr}[\operatorname{diag}(r,0,\ldots,0)] = T[\operatorname{diag}(r,0,\ldots,0)]$ for any r in $\mathbb{R}$). Hence, $\dim(\ker T) = \dim M_{nn} - \dim(\operatorname{im} T) = n^2 - \dim(\mathbb{R}) = n^2 - 1$.

15(b) Write $B = \{x - 1, x^2 - 1, \ldots, x^n - 1\}$. Then $T(x^k - 1) = 1 - 1 = 0$ for all k, so each polynomial in B is in $\ker T$. Moreover, the polynomials in B are independent (they have distinct degrees) so, as $\dim(\ker T) = n$ by (a), B is a basis of $\ker T$.

22. Given a column $Y \neq 0$ in $\mathbb{R}^n$, define $T : \mathbf{M}_{mn} \to \mathbb{R}^m$ by $T(A) = AY$ for all A in $\mathbf{M}_{mn}$. This is linear and $\ker T = U$, so the dimension theorem gives

$$mn = \dim(\mathbf{M}_{mn}) = \dim(\ker T) + \dim(\operatorname{im} T) = \dim U + \dim(\operatorname{im} T).$$

Hence, it suffices to show that $\dim(\operatorname{im} T) = m$, equivalently (since $\operatorname{im} T \subseteq \mathbb{R}^m$) that T is onto. So let X be a column in $\mathbb{R}^m$, we must find a matrix A in $\mathbf{M}_{mn}$ such that $AY = X$. Write A in terms of its columns as $A = [C_1 \quad C_2 \quad \ldots \quad C_n]$ and write $Y = [y_1 \quad y_2 \quad \ldots \quad y_n]^T$. Then the requirement that $AY = X$ becomes

$$X = [C_1 \quad C_2 \quad \ldots \quad C_n] \begin{bmatrix} y_1 \\ y_2 \\ \vdots \\ y_n \end{bmatrix} = y_1 C_1 + y_2 C_2 + \cdots + y_n C_n. \tag{$*$}$$

Since $Y \neq 0$, let $y_k \neq 0$. Then $(*)$ is satisfied if we choose $C_k = y_k^{-1} X$ and $C_j = 0$ if $j \neq k$. Hence T is onto as required.

28(b) By Theorem 4, Section 6.4, let $\{\mathbf{u}_1, \ldots, \mathbf{u}_m, \ldots, \mathbf{u}_n\}$ be a basis of V where $\{\mathbf{u}_1, \ldots, \mathbf{u}_m\}$ is a basis of U. By Theorem 3, Section 8.1, there is a linear transformation $S : V \to V$ such that

$$S(\mathbf{u}_i) = \mathbf{u}_i \quad \text{if } 1 \leq i \leq m$$
$$S(\mathbf{u}_i) = \mathbf{0} \quad \text{if } i > m.$$

Hence, $\mathbf{u}_i$ is in $\operatorname{im} S$ for $1 \leq i \leq m$, whence $U \subseteq \operatorname{im} S$. On the other hand, if $\mathbf{w}$ is in $\operatorname{im} S$, write $\mathbf{w} = S(\mathbf{v})$, $\mathbf{v}$ in V. Then r_i exist in $\mathbb{R}$ such that

$$\mathbf{v} = r_1 \mathbf{u}_1 + \cdots + r_m \mathbf{u}_m + \cdots + r_n \mathbf{u}_n$$

so

$$\mathbf{w} = r_1 S(\mathbf{u}_1) + \cdots + r_m S(\mathbf{u}_m) + \cdots + r_n S(\mathbf{u}_n)$$
$$= r_1 \mathbf{u}_1 + \cdots + r_m \mathbf{u}_m + \mathbf{0}.$$

It follows that $\mathbf{w}$ is in U, so $\operatorname{im} S \subseteq U$. Then $u = \operatorname{im} S$ as required.

Exercises 8.3   Isomorphisms and Composition

1(b) T is one-to-one because $T(x, y, z) = (0, 0, 0)$ means $x = x + y = x + y + z = 0$, whence $x = y = z = 0$. Now T is onto by Theorem 3.

Alternatively: $\{T(1, 0, 0), T(0, 1, 0), T(0, 01)\} = \{(1, 1, 1), (0, 1, 1), (0, 0, 1)\}$ is independent, so T is an isomorphism by Theorem 1.

(d) T is one-to-one because $T(X) = 0$ implies $UXV = 0$, whence $X = 0$ as U and V are invertible. Now Theorem 3 implies that T is onto and so is an isomorphism.

(f) T is one-to-one because $T(\mathbf{v}) = \mathbf{0}$ implies $k\mathbf{v} = \mathbf{0}$, so $\mathbf{v} = \mathbf{0}$ because $k \neq 0$. Hence, T is onto if $\dim V$ is finite (by Theorem 3) and so is an isomorphism. In general, T is onto because $T(k^{-1}\mathbf{v}) = k(k^{-1}\mathbf{v}) = \mathbf{v}$ holds for all $\mathbf{v}$ in V. Thus T is an isomorphism in any case.

(h) T is onto because $T(A^T) = (A^T)^T = A$ for every $n \times m$ matrix A (note that A^T is in $\mathbf{M}_{mn}$ so $T(A^T)$ makes sense). Hence, T is one-to-one by Theorem 3, and so is an isomorphism. (A direct proof that T is one-to-one: $T(A) = 0$ implies $A^T = 0$, whence $A = 0$.)

4(b) $ST(x, y, z) = S(x + y, 0, y + z) = (x + y, 0, y + z)$; $TS(x, y, z) = T(x, 0, z) = (x, 0, z)$. These are not equal (if $y \neq 0$) so $ST \neq TS$.

(d) $ST \begin{bmatrix} a & b \\ c & d \end{bmatrix} = S \begin{bmatrix} c & a \\ d & b \end{bmatrix} = \begin{bmatrix} c & 0 \\ 0 & b \end{bmatrix}$; $TS \begin{bmatrix} a & b \\ c & d \end{bmatrix} = T \begin{bmatrix} a & 0 \\ 0 & d \end{bmatrix} = \begin{bmatrix} 0 & a \\ d & 0 \end{bmatrix}$. These are not equal for some values of a, b, c and d (nearly all) so $ST \neq TS$.

5(b) $T^2(x, y) = T[T(x, y)] = T(x + y, 0) = [x + y + 0, 0] = (x + y, 0) = T(x, y)$. This holds for all (x, y), whence $T^2 = T$.

(d) $T^2 \begin{bmatrix} a & b \\ c & d \end{bmatrix} = T\left(T \begin{bmatrix} a & b \\ c & d \end{bmatrix}\right) = T \begin{bmatrix} \frac{1}{2} \begin{bmatrix} a + c & b + d \\ a + c & b + d \end{bmatrix} \end{bmatrix} = \frac{1}{2}T \begin{bmatrix} a + c & b + d \\ a + c & b + d \end{bmatrix}$

$= \frac{1}{4} \begin{bmatrix} (a + c) + (a + c) & (b + d) + (b + d) \\ (a + c) + (a + c) & (b + d) + (b + d) \end{bmatrix} = \frac{1}{2} \begin{bmatrix} a + c & b + d \\ a + c & b + d \end{bmatrix} = T \begin{bmatrix} a & b \\ c & d \end{bmatrix}$. This holds

for all $\begin{bmatrix} a & b \\ c & d \end{bmatrix}$, so $T^2 = T$.

6(b) No inverse. For example $T(1, -1, 1, -1) = (0, 0, 0, 0)$ so $(1, -1, 1, -1)$ is a nonzero vector in $\ker T$. Hence T is not one-to-one, and so has no inverse.

(d) T is one-to-one because $T \begin{bmatrix} a & b \\ c & d \end{bmatrix} = \begin{bmatrix} 0 & 0 \\ 0 & 0 \end{bmatrix}$ implies $a + 2c + 0 = 3c - a$ and $b + 2d = 0 = 3d - b$, whence $a = b = c = d = 0$. Thus T is an isomorphism by Theorem 3. If $T^{-1} \begin{bmatrix} a & b \\ c & d \end{bmatrix} = \begin{bmatrix} x & y \\ z & w \end{bmatrix}$, then $\begin{bmatrix} a & b \\ c & d \end{bmatrix} = T \begin{bmatrix} x & y \\ z & w \end{bmatrix} = \begin{bmatrix} x + 2z & y + 2w \\ 3z - x & 3w - y \end{bmatrix}$. Thus

$$x + 2z = a$$
$$-x + 3z = c$$
$$y + 2w = b$$
$$-y + 3w = d.$$

The solution is $x = \frac{1}{5}(3a - 2c)$, $z = \frac{1}{5}(a + c)$, $y = \frac{1}{5}(3b - 2d)$, $w = \frac{1}{5}(b + d)$. Hence

$$T^{-1} \begin{bmatrix} a & b \\ c & d \end{bmatrix} = \frac{1}{5} \begin{bmatrix} 3a - 2c & 3b - 2d \\ a + c & b + d \end{bmatrix}.$$

(f) T is one-to-one because, if $p(x)$ in $\mathbf{P}_2$ satisfies $T(p) = 0$, then $p(0) = p(1) = p(-1) = 0$. If $p(x) = a + bx + cx^2$, this means $a = 0$, $a + b + c = 0$ and $a - b + c = 0$, whence $a = b = c = 0$, and $p(x) = 0$. Hence, T^{-1} exists by Theorem 3. If $T^{-1}(a, b, c) = r + sx + tx^2$, then

$$(a, b, c) = T(r + sx + tx^2) = (r, r + s + t, r - s + t).$$

Then $r = a$, $r + s + t = b$, $r - s + t = c$, whence $r = a$, $s = \frac{1}{2}(b - c)$, $t = \frac{1}{2}(b + c - 2a)$. Finally

$$T^{-1}(a, b, c) = a + \tfrac{1}{2}(b - c)x + \tfrac{1}{2}(b + c - 2a)x^2.$$

7(b) $T^2(x, y) = T[T(x, y)] = T(ky - x, y) = [ky - (ky - x), y] = (x, y) = 1_{\mathbb{R}^2}(x, y)$. Hence, $T^2 = 1_{\mathbb{R}^2}$, and so $T^{-1} = T$.

8(b) $T^2(x, y, z, w) = T[T[x, y, z, w]] = T(-y, x - y, z, -w) = (-(x - y), -y - (x - y), z, -(-w)) = (y - x, -x, z, w)$.
$T^3(x, y, z, w) = T[T^2(x, y, z, w)] = T(y - x, -x, z, w) = (x, y, z, -w)$.
$T^6(x, y, z, w) = T^3[T^3(x, y, z, w)] = T^3[x, y, z, -w] = (x, y, z, w) = 1_{\mathbb{R}^4}(x, y, z, w)$. Hence, $T^6 = 1_{\mathbb{R}^4}$ so $T^{-1} = T^5$. Explicitly:

$$T^{-1}(x, y, z, w) = T^2[T^3(x, y, z, w)] = T^2(x, y, z, -w) = (y - x, -x, z, -w).$$

9(b) Define $S : \mathbf{M}_{nn} \to \mathbf{M}_{nn}$ by $S(A) = U^{-1}A$. Then

$$ST(A) = S(T(A)) = U^{-1}(UA) = A = 1_{\mathbf{M}nn}(A) \qquad \text{so } ST = 1_{\mathbf{M}_{nn}}$$
$$TS(A) = T(S(A)) = U(U^{-1}A) = A = 1_{\mathbf{M}nn}(A) \qquad \text{so } TS = 1_{\mathbf{M}_{nn}}.$$

Hence, T is invertible and $T^{-1} = S$.

10(b) Given $V \xrightarrow{T} W \xrightarrow{S} U$ with T and S both onto, we are to show that $ST : V \to U$ is onto. Given $\mathbf{u}$ in U, we have $\mathbf{u} = S(\mathbf{w})$ for some $\mathbf{w}$ in W because S is onto; then $\mathbf{w} = T(\mathbf{v})$ for some $\mathbf{v}$ in V because T is onto. Hence,

$$ST(\mathbf{v}) = S[T(\mathbf{v})] = S[\mathbf{w}] = \mathbf{u}.$$

This shows that ST is onto.

13(b) Given $V \xrightarrow{T} U \xrightarrow{S} W$ with ST onto, let $\mathbf{w}$ be a vector in W. Then $\mathbf{w} = ST(\mathbf{v})$ for some $\mathbf{v}$ in V because ST is onto, whence $\mathbf{w} = S[T(\mathbf{v})]$, so S is onto. Now the dimension theorem applied to S gives

$$\dim U = \dim(\ker S) + \dim(\operatorname{im} S) = \dim(\ker S) + \dim W$$

because $\operatorname{im} S = W$ (S is onto). As $\dim(\ker S) \geq 0$, this gives $\dim U \geq \dim W$.

14. If $T^2 = 1_V$ then $TT = 1_V$ so T is invertible and $T^{-1} = T$ by the definition of the inverse of a transformation. Conversely, if $T^{-1} = T$ then $T^2 = TT^{-1} = 1_V$.

16. Theorem 5, Section 7.2 shows that $\{T(\mathbf{e}_1), T(\mathbf{e}_2), \dots, Y(\mathbf{e}_r)\}$ is a basis of $\operatorname{im} T$. Write $U = \operatorname{span}\{\mathbf{e}_1, \dots, \mathbf{e}_r\}$. Then $B = \{\mathbf{e}_1, \dots, \mathbf{e}_r\}$ is a basis of U and $T : U \to \operatorname{im} T$ carries B to the basis $\{T(\mathbf{e}_1), \dots, T(\mathbf{e}_r)\}$. Thus $T : U \to \operatorname{im} T$ is itself an isomorphism. Note that $T : V \to W$ may not be an isomorphism, but restricting T to the subspace U of V does result in an isomorphism in this case.

19(b) We have $V = \{(x,y) \mid x, y \text{ in } \mathbb{R}\}$ with a new addition and scalar multiplication:

$$(x,y) \oplus (x_1, y_1) = (x + x_1, y + y_1 + 1)$$
$$a \odot (x,y) = (ax, ay + a - 1).$$

We use the notation $\oplus$ and $\odot$ for clarity. Define

$$T : V \to \mathbb{R}^2$$

by $T(x,y) = (x, y + 1)$. Then T is a linear transformation because:

$$\begin{aligned}
T[(x,y) \oplus (x_1, y_1)] &= T(x + x_1, y + y_1 + 1) \\
&= (x + x_1, (y + y_1 + 1) + 1) \\
&= (x, y + 1) + (x_1, y_1 + 1) \\
&= T(x,y) + T(x_1, y_1)
\end{aligned}$$

$$\begin{aligned}
T(a \odot (x,y)) &= T(ax, ay + a - 1) \\
&= (ax, ay + a) \\
&= a(x, y + 1) \\
&= aT(x,y).
\end{aligned}$$

Moreover T is one-to-one because $T(x,y) = (0,0)$ means $x = 0 = y + 1$, so $(x,y) = (0,1)$, the zero vector of V. (Alternatively, $T(x,y) = T(x_1, y_1)$ implies $(x, y + 1) = (x_1, y_1 + 1)$, whence $x = x_1, y = y_1$.) As T is clearly onto $\mathbb{R}^2$, it is an isomorphism.

26(b) If $p(x)$ is in $\ker T$, then $p(x) = -xp'(x)$. If we write $p(x) = a_0 + a_1 x + \cdots + a_n x^n$, this becomes

$$a_0 + a_1 x + \cdots + a_{n-1} x^{n-1} + a_n x^n = -a_1 x - 2a_2 x^2 - \cdots - na_n x^n.$$

Equating coefficients gives $a_0 = 0$, $a_1 = -a_1$, $a_2 = -2a_2$, $\ldots$, $a_n = -na_n$. Hence, $a_0 = a_1 = \cdots = a_n = 0$, so $p(x) = 0$. Thus $\ker T = 0$, so T is one-to-one. As $T : \mathbf{P}_n \to \mathbf{P}_n$ and $\dim \mathbf{P}_n$ is finite, this implies that T is also onto, and so is an isomorphism.

27(b) If $TS = 1_W$ then, given $\mathbf{w}$ in W, $T[S(\mathbf{w})] = \mathbf{w}$, so T is onto. Conversely, if T is onto, choose a basis $\{\mathbf{e}_1, \ldots, \mathbf{e}_r, \ldots, \mathbf{e}_n\}$ of V such that $\{\mathbf{e}_{r+1}, \ldots, \mathbf{e}_n\}$ is a basis of $\ker T$. By Theorem 5, §8.2, $\{T(\mathbf{e}_1), \ldots, T(\mathbf{e}_n)\}$ is a basis of $\text{im } T = W$ (as T is onto). Hence, a linear transformation $S : W \to V$ exists such that $S[T(\mathbf{e}_i)] = \mathbf{e}_i$ for $i = 1, 2, \ldots, r$. We claim that $TS = 1_W$, and we show this by verifying that these transformations agree on the basis $\{T(\mathbf{e}_1), \ldots, T(\mathbf{e}_r)\}$ of W. Indeed

$$TS[T(\mathbf{e}_i)] = T\{S[T(\mathbf{e}_i)]\} = T(\mathbf{e}_i) = 1_W[T(\mathbf{e}_i)]$$

for $i = 1, 2, \ldots, n$.

28(b) If $T = SR$, then every vector $T(\mathbf{v})$ in $\text{im } T$ has the form $T(\mathbf{v}) = S[R(\mathbf{v})]$, whence $\text{im } T \subseteq \text{im } S$. Since R is invertible, $S = TR^{-1}$ implies $\text{im } S \subseteq \text{im } T$, so $\text{im } S = \text{im } T$.

   Conversely, assume that $\text{im } S = \text{im } T$. The dimension theorem gives $\dim(\ker S) = n - \dim(\text{im } S) = n - \dim(\text{im } T) = \dim(\ker T)$. Hence, let $\{\mathbf{e}_1, \ldots, \mathbf{e}_r, \ldots, \mathbf{e}_n\}$ and $\{\mathbf{f}_1, \ldots, \mathbf{f}_r, \ldots, \mathbf{f}_n\}$

be bases of V such that $\{\mathbf{e}_{r+1}, \ldots, \mathbf{e}_n\}$ and $\{\mathbf{f}_{r+1}, \ldots, \mathbf{f}_n\}$ are bases of $\ker S$ and $\ker T$, respectively. By Theorem 5, §8.2, $\{S(\mathbf{e}_1), \ldots, S(\mathbf{e}_r)\}$ and $\{T(\mathbf{f}_1), \ldots, T(\mathbf{f}_r)\}$ are both bases of $\operatorname{im} S = \operatorname{im} T$. So let $\mathbf{g}_1, \ldots, \mathbf{g}_r$ in V be such that

$$S(\mathbf{e}_i) = T(\mathbf{g}_i)$$

for each $i = 1, 2, \ldots, r$.

Claim: $B = \{\mathbf{g}_1, \ldots, \mathbf{g}_r, \mathbf{f}_{r+1}, \ldots, \mathbf{f}_n\}$ is a basis of V.

Proof: It suffices (by Theorem 3, §6.4) to show that B is independent. If $a_1\mathbf{g}_1 + \cdots + a_r\mathbf{g}_r + b_{r+1}\mathbf{f}_{r+1} + \cdots + b_n\mathbf{f}_n = \mathbf{0}$, apply T to get

$$\begin{aligned}\mathbf{0} &= a_1 T(\mathbf{g}_1) + \cdots + a_r T(\mathbf{g}_r) + b_{r+1} T(\mathbf{f}_{r+1}) + \cdots + b_n T(\mathbf{f}_n) \\ &= a_1 T(\mathbf{g}_1) + \cdots + a_r T(\mathbf{g}_r) + \mathbf{0}\end{aligned}$$

because $T(\mathbf{f}_j) = 0$ if $j > r$. Hence $a_1 = \cdots = a_r = 0$; whence $\mathbf{0} = b_{r+1}\mathbf{f}_{r+1} + \cdots + b_n\mathbf{f}_n$. This gives $b_{r+1} = \cdots = b_n = 0$ and so proves the claim.

By the claim, we can define $R : V \to V$ by

$$\begin{aligned}R(\mathbf{g}_i) &= \mathbf{e}_i \quad \text{for } i = 1, 2, \ldots, r \\ R(\mathbf{f}_j) &= \mathbf{e}_j \quad \text{for } j = r+1, \ldots, n.\end{aligned}$$

Then R is an isomorphism by Theorem 1, §8.3, and we claim that $SR = T$. We show this by verifying that SR and T have the same effort on the basis B in the claim. The definition of R gives

$$\begin{aligned}SR(\mathbf{g}_i) &= S[R(\mathbf{g}_i)] = S(\mathbf{e}_i) = T(\mathbf{g}_i) \quad \text{for } i = 1, 2, \ldots, r \\ SR(\mathbf{f}_j) &= S[\mathbf{e}_j] = \mathbf{0} = T(\mathbf{f}_j) \quad\quad\quad \text{for } j = r+1, \ldots, n.\end{aligned}$$

Hence $SR = T$.

29. As in the hint, let $\{\mathbf{e}_1, \mathbf{e}_2, \ldots, \mathbf{e}_r, \ldots, \mathbf{e}_n\}$ be a basis of V where $\{\mathbf{e}_{r+1}, \ldots, \mathbf{e}_n\}$ is a basis of $\ker T$. Then $\{T(\mathbf{e}_1), \ldots, T(\mathbf{e}_r)\}$ is linearly independent by Theorem 5, §8.2, so extend it to a basis $\{T(\mathbf{e}_1), \ldots, T(\mathbf{e}_r), \mathbf{w}_{r+1}, \ldots, \mathbf{w}_n\}$ of V. Then define $S : V \to V$ by

$$\begin{aligned}S[T(\mathbf{e}_i)] &= \mathbf{e}_i \quad \text{for } 1 \le i \le r \\ S(\mathbf{w}_j) &= \mathbf{e}_j \quad\ \text{for } r+1 \le j \le n.\end{aligned}$$

Then, S is an isomorphism (by Theorem 1) and we claim that $TST = T$. We verify this by showing that TST and T agree on the basis $\{\mathbf{e}_1, \ldots, \mathbf{e}_r, \ldots, \mathbf{e}_n\}$ of V (and invoking Theorem 2, §8.1).

$$\begin{aligned}\text{If } 1 \le i \le r\text{:} \quad &TST(\mathbf{e}_i) = T\{S[T(\mathbf{e}_i)]\} = T(\mathbf{e}_i) \\ \text{If } r+1 \le j \le n\text{:} \quad &TST(\mathbf{e}_j) = TS[T(\mathbf{e}_j)] = TS[\mathbf{0}] = \mathbf{0} = T(\mathbf{e}_j)\end{aligned}$$

where, at the end, we use the fact that $\mathbf{e}_j$ is in $\ker T$ for $r+1 \le j \le n$.

Exercises 8.4   The Matrix of a Linear Transformation

1(b) $C_B(\mathbf{v}) = \begin{bmatrix} a \\ 2b - c \\ c - b \end{bmatrix}$ because $\mathbf{v} = ax^2 + bx + c = ax^2 + (2b - c)(x + 1) + (c - b)(x + 2)$.

(d) $C_B(\mathbf{v}) = \frac{1}{2} \begin{bmatrix} a - b \\ a + b \\ -a + 3b + 2c \end{bmatrix}$ because

$\mathbf{v} = (a, b, c) = \frac{1}{2} [(a - b)(1, -1, 2) + (a + b)(1, 1, -1) + (-a + 3b + 2c)(0, 0, 1))]$

2(b) $M_{DB}(T) = \begin{bmatrix} C_D[T(1)] & C_D[T(x)] & C_D[T(x^2)] \end{bmatrix} = \begin{bmatrix} 2 & 1 & 3 \\ -1 & 0 & -2 \end{bmatrix}$. Comparing columns gives

$$C_D[T(1)] = \begin{bmatrix} 2 \\ -1 \end{bmatrix} \qquad C_D[T(x)] = \begin{bmatrix} 1 \\ 0 \end{bmatrix} \qquad C_D[T(x^2)] = \begin{bmatrix} 3 \\ -2 \end{bmatrix}.$$

Hence

$$T(1) = 2(1, 1) - (0, 1) = (2, 1)$$
$$T(x) = 1(1, 1) + 0(0, 1) = (1, 1)$$
$$T(x^2) = 3(1, 1) - 2(0, 1) = (3, 1).$$

Thus

$$T(a + bx + cx^2) = aT(1) + bT(x) + cT(x^2)$$
$$= a(2, 1) + b(1, 1) + c(3, 1)$$
$$= (2a + b + 3c, a + b + c).$$

3(b) $M_{DB}(T) = \begin{bmatrix} C_D \left\{ T \begin{bmatrix} 1 & 0 \\ 0 & 0 \end{bmatrix} \right\} & C_D \left\{ T \begin{bmatrix} 0 & 1 \\ 0 & 0 \end{bmatrix} \right\} & C_D \left\{ T \begin{bmatrix} 0 & 0 \\ 1 & 0 \end{bmatrix} \right\} & C_D \left\{ T \begin{bmatrix} 0 & 0 \\ 0 & 1 \end{bmatrix} \right\} \end{bmatrix}$

$= \begin{bmatrix} C_D \left\{ \begin{bmatrix} 1 & 0 \\ 0 & 0 \end{bmatrix} \right\} & C \left\{ \begin{bmatrix} 0 & 0 \\ 1 & 0 \end{bmatrix} \right\} & C \left\{ \begin{bmatrix} 0 & 1 \\ 0 & 0 \end{bmatrix} \right\} & C_D \left\{ \begin{bmatrix} 0 & 0 \\ 0 & 1 \end{bmatrix} \right\} \end{bmatrix}$

$= \begin{bmatrix} 1 & 0 & 0 & 0 \\ 0 & 0 & 1 & 0 \\ 0 & 1 & 0 & 0 \\ 0 & 0 & 0 & 1 \end{bmatrix}.$

(d) $M_{DB}(T) = \begin{bmatrix} C_D[T(1)] & C_D[T(x)] & C_D[T(x^2)] \end{bmatrix} = \begin{bmatrix} C_D(1) & C_D(x + 1) & C_D(x^2 + 2x + 1) \end{bmatrix}$

$= \begin{bmatrix} 1 & 1 & 1 \\ 0 & 1 & 2 \\ 0 & 0 & 1 \end{bmatrix}.$

4(b) $M_{DB}(T) = [C_D[T(1,1)] \quad C_D[T(1,0)]] = [C_D(1,5,4,1) \quad C_D(2,3,0,1)] = \begin{bmatrix} 1 & 2 \\ .5 & 3 \\ 4 & 0 \\ 1 & 1 \end{bmatrix}.$

We have $\mathbf{v} = (a,b) = b(1,1) + (a-b)(1,0)$ so $C_B(\mathbf{v}) = \begin{bmatrix} b \\ a-b \end{bmatrix}$. Hence,

$$C_D[T(\mathbf{v})] = M_{DB}(T)C_B(\mathbf{v}) = \begin{bmatrix} 1 & 2 \\ 5 & 3 \\ 4 & 0 \\ 1 & 1 \end{bmatrix} \begin{bmatrix} b \\ a-b \end{bmatrix} = \begin{bmatrix} 2a - b \\ 3a + 2b \\ 4b \\ a \end{bmatrix}.$$

Finally, we recover the action of T:

$$T(\mathbf{v}) = (2a - b)(1,0,0,0) + (3a + 2b)(0,1,0,0) + 4b(0,0,1,0) + a(0,0,0,1)$$
$$= (2a - b, 3a + 2b, 4b, a).$$

(d) $\begin{aligned} M_{DB}(T) &= [C_D[T(1)] \quad C_D[T(x)] \quad C_D[T(x^2)]] \\ &= [C_D(1,0) \quad C_D(1,0) \quad C_D(0,1)] \\ &= \begin{bmatrix} \frac{1}{2} & \frac{1}{2} & -\frac{1}{2} \\ \frac{1}{2} & \frac{1}{2} & \frac{1}{2} \end{bmatrix} \\ &= \frac{1}{2} \begin{bmatrix} 1 & 1 & -1 \\ 1 & 1 & 1 \end{bmatrix}. \end{aligned}$

We have $\mathbf{v} = a + bx + cx^2$ so $C_B(\mathbf{v}) = \begin{bmatrix} a \\ b \\ c \end{bmatrix}$. Hence

$$C_D[T(\mathbf{v})] = M_{DB}(T)C_B(\mathbf{v}) = \frac{1}{2}\begin{bmatrix} 1 & 1 & -1 \\ 1 & 1 & 1 \end{bmatrix}\begin{bmatrix} a \\ b \\ c \end{bmatrix} = \frac{1}{2}\begin{bmatrix} a + b - c \\ a + b + c \end{bmatrix}.$$

Finally, we recover the action of T:

$$T(\mathbf{v}) = \tfrac{1}{2}(a + b - c)(1,-1) + \tfrac{1}{2}(a + b + c)(1,1) = (a + b, c).$$

(f) $M_{DB}(T) = \left[C_D\left\{ T\begin{bmatrix} 1 & 0 \\ 0 & 0 \end{bmatrix} \right\} \quad C_D\left\{ T\begin{bmatrix} 0 & 1 \\ 0 & 0 \end{bmatrix} \right\} \quad C_D\left\{ T\begin{bmatrix} 0 & 0 \\ 1 & 0 \end{bmatrix} \right\} \quad C_D\left\{ T\begin{bmatrix} 0 & 0 \\ 0 & 1 \end{bmatrix} \right\} \right]$

$= \left[C_D\begin{bmatrix} 1 & 0 \\ 0 & 0 \end{bmatrix} \quad C_D\begin{bmatrix} 0 & 1 \\ 1 & 0 \end{bmatrix} \quad C_D\begin{bmatrix} 0 & 1 \\ 1 & 0 \end{bmatrix} \quad C_D\begin{bmatrix} 0 & 0 \\ 0 & 1 \end{bmatrix} \right]$

$= \begin{bmatrix} 1 & 0 & 0 & 0 \\ 0 & 1 & 1 & 0 \\ 0 & 1 & 1 & 0 \\ 0 & 0 & 0 & 1 \end{bmatrix}.$

We have $\mathbf{v} = \begin{bmatrix} a & b \\ c & d \end{bmatrix} = a\begin{bmatrix} 1 & 0 \\ 0 & 0 \end{bmatrix} + b\begin{bmatrix} 0 & 1 \\ 0 & 0 \end{bmatrix} + c\begin{bmatrix} 0 & 0 \\ 1 & 0 \end{bmatrix} + d\begin{bmatrix} 0 & 0 \\ 0 & 1 \end{bmatrix},$

so $C_B(\mathbf{v}) = \begin{bmatrix} a \\ b \\ c \\ d \end{bmatrix}$. Hence

$$C_D[T(\mathbf{v})] = M_{DB}(T)C_B(\mathbf{v}) = \begin{bmatrix} 1 & 0 & 0 & 0 \\ 0 & 1 & 1 & 0 \\ 0 & 1 & 1 & 0 \\ 0 & 0 & 0 & 1 \end{bmatrix}\begin{bmatrix} a \\ b \\ c \\ d \end{bmatrix} = \begin{bmatrix} a \\ b+c \\ b+c \\ d \end{bmatrix}.$$

Finally, we recover the action of T:

$$T(\mathbf{v}) = a\begin{bmatrix} 1 & 0 \\ 0 & 0 \end{bmatrix} + (b+c)\begin{bmatrix} 0 & 1 \\ 0 & 0 \end{bmatrix} + (b+c)\begin{bmatrix} 0 & 0 \\ 1 & 0 \end{bmatrix} + d\begin{bmatrix} 0 & 0 \\ 0 & 1 \end{bmatrix} = \begin{bmatrix} a & b+c \\ b+c & d \end{bmatrix}.$$

5(b) Have $\mathbb{R}^3 \xrightarrow{T} \mathbb{R}^4 \xrightarrow{S} \mathbb{R}^2$. Let B, D, E be the standard bases. Then

$$M_{ED}(S) = [C_E[S(1,0,0,0)] \quad C_E[S(0,1,0,0)] \quad C_E[S(0,0,1,0)] \quad C_E S(0,0,0,1)]]$$
$$= [C_E(1,0) \quad C_E(1,0) \quad C_E(0,1) \quad C_E(0,-1)]$$
$$= \begin{bmatrix} 1 & 1 & 0 & 0 \\ 0 & 0 & 1 & -1 \end{bmatrix}$$

$$M_{DB}(T) = [C_D[T(1,0,0)] \quad C_D[T(0,1,0)] \quad C_D[T(0,0,1)]]$$
$$= [C_D(1,0,1,-1) \quad C_D(1,1,0,1) \quad C_D(0,1,1,0)]$$
$$= \begin{bmatrix} 1 & 1 & 0 \\ 0 & 1 & 1 \\ 1 & 0 & 1 \\ -1 & 1 & 0 \end{bmatrix}.$$

We have $ST(a,b,c) = S(a+b, c+b, a+c, b-a) = (a+2b+c, 2a-b+c)$. Hence

$$M_{EB}(ST) = [C_E[ST(1,0,0)] \quad C_E[ST(0,1,0)] \quad C_E[ST(0,0,1)]]$$
$$= [C_E(1,2)] \quad C_E(2,-1) \quad C_E(1,1)$$
$$= \begin{bmatrix} 1 & 2 & 1 \\ 2 & -1 & 1 \end{bmatrix}.$$

With this we confirm Theorem 3 as follows:

$$M_{ED}(S)M_{DB}(T) = \begin{bmatrix} 1 & 1 & 0 & 0 \\ 0 & 0 & 1 & -1 \end{bmatrix} \begin{bmatrix} 1 & 1 & 0 \\ 0 & 1 & 1 \\ 1 & 0 & 1 \\ -1 & 1 & 0 \end{bmatrix} = \begin{bmatrix} 1 & 2 & 1 \\ 2 & -1 & 1 \end{bmatrix} = M_{EB}(ST).$$

(d) Have $\mathbb{R}^3 \xrightarrow{T} \mathbf{P}_2 \xrightarrow{S} \mathbb{R}^2$ with bases $B = \{(1,0,0),(0,1,0),(0,0,1)\}$, $D = \{1, x, x^2\}$, $E = \{(1,0),(0,1)\}$.

$$M_{ED}(S) = [C_E[S(1)] \quad C_E[S(x)] \quad C_E[S(x^2)]]$$
$$= [C_E(1,0) \quad C_E(-1,0) \quad C_E(0,1)]$$
$$= \begin{bmatrix} 1 & -1 & 0 \\ 0 & 0 & 1 \end{bmatrix}$$

$$M_{DB}(T) = [C_D[T(1,0,0)] \quad C_D[T(0,1,0)] \quad C_D[T(0,0,1)]]$$
$$= [C_D(1-x) \quad C_D(-1+x^2) \quad C_D(x)]$$
$$= \begin{bmatrix} 1 & -1 & 0 \\ -1 & 0 & 1 \\ 0 & 1 & 0 \end{bmatrix}.$$

The action of ST is $ST(a, b, c) = S\left[(a - b) + (c - a)x + bx^2\right] = (2a - b - c, b)$. Hence,

$$\begin{aligned} M_{EB}(ST) &= [C_E[ST(1,0,0)] \quad C_E[ST(0,1,0)] \quad C_E[ST(0,0,1)]] \\ &= [C_E(2,0) \quad C_E(-1,1) \quad C_E(-1,0)] \\ &= \begin{bmatrix} 2 & -1 & -1 \\ 0 & 1 & 0 \end{bmatrix}. \end{aligned}$$

Hence, we verify Theorem 3 as follows:

$$M_{ED}(S)M_{DB}(T) = \begin{bmatrix} 1 & -1 & 0 \\ 0 & 0 & 1 \end{bmatrix} \begin{bmatrix} 1 & -1 & 0 \\ -1 & 0 & 1 \\ 0 & 1 & 0 \end{bmatrix} = \begin{bmatrix} 2 & -1 & -1 \\ 0 & 1 & 0 \end{bmatrix} = M_{EB}(ST).$$

7(b)
$$\begin{aligned} M_{DB}(T) &= [C_D[T(1,0,0)] \quad C_D[T(0,1,0)] \quad C_D[T(0,0,1)]] \\ &= [C_D(0,1,1) \quad C_D(1,0,1) \quad C(1,1,0)] \\ &= \begin{bmatrix} 0 & 1 & 1 \\ 1 & 0 & 1 \\ 1 & 1 & 0 \end{bmatrix}. \end{aligned}$$

If $T^{-1}(a, b, c) = (x, y, z)$ then $(a, b, c) = T(x, y, z) = (y + z, x + z, x + y)$. Hence, $y + z = a$, $x + z = b$, $x + y = c$. The solution is

$$T^{-1}(a, b, c) = (x, y, z) = \tfrac{1}{2}(b + c - a, a + c - b, a + b - c).$$

Hence,

$$\begin{aligned} M_{BD}(T^{-1}) &= \left[C_B\left[T^{-1}(1,0,0)\right] \quad C_B\left[T^{-1}(0,1,0)\right] \quad C_B\left[T^{-1}(0,0,1)\right]\right] \\ &= \left[C_B\left(-\tfrac{1}{2}, \tfrac{1}{2}, \tfrac{1}{2}\right) \quad C_B\left(\tfrac{1}{2}, -\tfrac{1}{2}, \tfrac{1}{2}\right) \quad C_B\left(\tfrac{1}{2}, \tfrac{1}{2}, -\tfrac{1}{2}\right)\right] \\ &= \tfrac{1}{2}\begin{bmatrix} -1 & 1 & 1 \\ 1 & -1 & 1 \\ 1 & 1 & -1 \end{bmatrix}. \end{aligned}$$

This matrix is $M_{DB}(T)^{-1}$ as Theorem 4 asserts.

(d)
$$\begin{aligned} M_{DB}(T) &= [C_D[T(1)] \quad C_D[T(x)] \quad C[T(x^2)]] \\ &= [C_D(1,0,0) \quad C_D(1,1,0) \quad C_D(1,1,1)] \\ &= \begin{bmatrix} 1 & 1 & 1 \\ 0 & 1 & 1 \\ 0 & 0 & 1 \end{bmatrix}. \end{aligned}$$

If $T^{-1}(a, b, c) = r + sx + tx^2$, then $(a, b, c) = T(r + sx + tx^2) = (r + s + t, s + t, t)$. Hence, $r + s + t = a$, $s + t = b$, $t = c$; the solution is $t = c$, $s = b - c$, $r = a - b$. Thus,

$$T^{-1}(a, b, c) = r + sx + tx^2 = (a - b) + (b - c)x + cx^2.$$

Hence,

$$\begin{aligned} M_{BD}(T^{-1}) &= \left[C_B \left[T^{-1}(1,0,0) \right] \quad C_B \left[T^{-1}(0,1,0) \right] \quad C_B \left[T^{-1}(0,0,1) \right] \right] \\ &= \left[C_B(1) \quad C_B(-1+x) \quad C_B(-x+x^2) \right] \\ &= \begin{bmatrix} 1 & -1 & 0 \\ 0 & 1 & -1 \\ 0 & 0 & 1 \end{bmatrix}. \end{aligned}$$

This matrix is $M_{DB}(T)^{-1}$ as Theorem 4 asserts.

8(b) $\quad M_{DB}(T) = \left[C_D \left\{ T \begin{bmatrix} 1 & 0 \\ 0 & 0 \end{bmatrix} \right\} \quad C_D \left\{ T \begin{bmatrix} 0 & 1 \\ 0 & 0 \end{bmatrix} \right\} \quad C_D \left\{ T \begin{bmatrix} 0 & 0 \\ 1 & 0 \end{bmatrix} \right\} \quad C_D \left\{ T \begin{bmatrix} 0 & 0 \\ 0 & 1 \end{bmatrix} \right\} \right]$

$$\begin{aligned} &= \left[C_D(1,0,0,0) \quad C_D(1,1,0,0) \quad C_D(1,1,1,0) \quad C_D(0,0,0,1) \right] \\ &= = \begin{bmatrix} 1 & 1 & 1 & 0 \\ 0 & 1 & 1 & 0 \\ 0 & 0 & 1 & 0 \\ 0 & 0 & 0 & 1 \end{bmatrix}. \end{aligned}$$

This is invertible and the matrix inversion algorithm (and Theorem 4) gives

$$M_{DB}(T^{-1}) = [M_{DB}(T)]^{-1} = \begin{bmatrix} 1 & -1 & 0 & 0 \\ 0 & 1 & -1 & 0 \\ 0 & 0 & 1 & 0 \\ 0 & 0 & 0 & 1 \end{bmatrix}.$$

If $\mathbf{v} = (a,b,c,d)$ then

$$C_B \left[T^{-1}(\mathbf{v}) \right] = M_{DB}(T^{-1}) C_D(\mathbf{v}) = \begin{bmatrix} 1 & -1 & 0 & 0 \\ 0 & 1 & -1 & 0 \\ 0 & 0 & 1 & 0 \\ 0 & 0 & 0 & 1 \end{bmatrix} \begin{bmatrix} a \\ b \\ c \\ d \end{bmatrix} = \begin{bmatrix} a-b \\ b-c \\ c \\ d \end{bmatrix}.$$

Hence, we get a formula for the action of T^{-1}:

$$\begin{aligned} T^{-1}(a,b,c,d) = T^{-1}(\mathbf{v}) &= (a-b) \begin{bmatrix} 1 & 0 \\ 0 & 0 \end{bmatrix} + (b-c) \begin{bmatrix} 0 & 1 \\ 0 & 0 \end{bmatrix} + c \begin{bmatrix} 0 & 0 \\ 1 & 0 \end{bmatrix} + d \begin{bmatrix} 0 & 0 \\ 0 & 1 \end{bmatrix} \\ &= \begin{bmatrix} a-b & b-c \\ c & d \end{bmatrix}. \end{aligned}$$

12. Since $D = \{T(\mathbf{e}_1), \dots, T(\mathbf{e}_n)\}$, we have $C_D \left[T(\mathbf{e}_j) \right] = C_j = $ column j of I_n. Hence,

$$\begin{aligned} M_{DB}(T) &= \left[C_D \left[T(\mathbf{e}_1) \right] \quad C_D \left[T(\mathbf{e}_2) \right] \quad \dots \quad C_D \left[T(\mathbf{e}_n) \right] \right] \\ &= \left[C_1 \quad C_2 \quad \dots \quad C_n \right] = I_n. \end{aligned}$$

16(b) Define $T : \mathbf{P}_n \to \mathbb{R}^{n+1}$ by $T[p(x)] = (p(a_0), p(a_1), \ldots, p(a_n))$, where $a_0, \ldots, a_n$ are fixed distinct real numbers. If $B = \{1, x, \ldots, x^n\}$ and $D \subseteq \mathbb{R}^{n+1}$ is the standard basis,

$$M_{DB}(T)(= \begin{bmatrix} C_D[T(1)] & C_D[T(x)] & C_D[T(x^2)] & \ldots & C_D[T(x^2)] \end{bmatrix}$$
$$= \begin{bmatrix} C_D(1, 1, \ldots, 1) & C_D(a_0, a_1, \ldots, a_n) & C_D(a_0^2, a_1^2, \ldots, a_n^2) & \ldots & C_D(a_0^n, a_1^n, \ldots, a_n^n) \end{bmatrix}$$
$$= \begin{bmatrix} 1 & a_0 & a_0^2 & \ldots & a_0^n \\ 1 & a_1 & a_1^2 & \ldots & a_1^n \\ \vdots & \vdots & \vdots & \ddots & \vdots \\ 1 & a_n & a_n^2 & \ldots & a_n^n \end{bmatrix}$$

Since the a_i are distinct, this matrix has nonzero determinant by Theorem 2, §3.4. Hence, T is an isomorphism by Theorem 4.

20(d) Assume that $V \xrightarrow{R} W \xrightarrow{S,T} U$. Recall that $S + T : W \to U$ is defined by $(S + T)(\mathbf{w}) = S(\mathbf{w}) + T(\mathbf{w})$ for all $\mathbf{w}$ in W. Hence, for $\mathbf{v}$ in V:

$$\begin{aligned} [(S + T) R](\mathbf{v}) &= (S + T)[R(\mathbf{v})] \\ &= S[R(\mathbf{v})] + T[R(\mathbf{v})] \\ &= (SR)(\mathbf{v}) + (TR)(\mathbf{v}) \\ &= (SR + TR)(\mathbf{v}). \end{aligned}$$

Hence, $(S + T)R = SR + TR$.

21(b) If P and Q are subspaces of a vector space W, recall that $P + Q = \{p + q \mid p \text{ in } P, q \text{ in } Q\}$ is a subspace of W (Exercise 22, §6.4). Now let $\mathbf{w}$ be any vector in $\operatorname{im}(S + T)$. Then $\mathbf{w} = (S + T)(\mathbf{v}) = S(\mathbf{v}) + T(\mathbf{v})$ for some $\mathbf{v}$ in V, whence $\mathbf{w}$ is in $\operatorname{im} S + \operatorname{im} T$. Thus, $\operatorname{im}(S + T) \subseteq \operatorname{im} S + \operatorname{im} T$.

22(b) If T is in X_1^0, then $T(\mathbf{v}) = \mathbf{0}$ for all $\mathbf{v}$ in X_1. As $X \subseteq X_1$, this implies that $T(\mathbf{v}) = \mathbf{0}$ for all $\mathbf{v}$ in X; that is T is in X^0. Hence, $X_1^0 \subseteq X^0$.

24(b) We have $R : V \to \mathbf{L}(\mathbb{R}, V)$ defined by $R(\mathbf{v}) = S_\mathbf{v}$. Here $S_\mathbf{v} : \mathbb{R} \to V$ is defined by $S_\mathbf{v}(r) = r\mathbf{v}$.

$\underline{R \text{ is a linear transformation}}$: The requirements that $R(\mathbf{v} + \mathbf{w}) = R(\mathbf{v}) + R(\mathbf{w})$ and $R(a\mathbf{v}) = aR(\mathbf{v})$ translate to $S_{\mathbf{v}+\mathbf{w}} = S_\mathbf{v} + S_\mathbf{w}$ and $S_{a\mathbf{v}} = aS_\mathbf{v}$. If r is arbitrary in $\mathbb{R}$:

$$S_{\mathbf{v}+\mathbf{w}}(r) = r(\mathbf{v} + \mathbf{w}) = r\mathbf{v} + r\mathbf{w} = S_\mathbf{v}(r) + S_\mathbf{w}(r) = (S_\mathbf{v} + S_\mathbf{w})(r)$$
$$S_{a\mathbf{v}}(r) = r(a\mathbf{v}) = a(r\mathbf{v}) = a[S_\mathbf{v}(r)] = (aS_\mathbf{v})(r).$$

Hence, $S_{\mathbf{v}+\mathbf{w}} = S_\mathbf{v} + S_\mathbf{w}$ and $S_{a\mathbf{v}} = aS_\mathbf{v}$ so R is linear.

$\underline{R \text{ is one-to-one}}$: If $R(\mathbf{v})$ then $S_\mathbf{v} = 0$ is the zero transformation $\mathbb{R} \to V$. Hence, $0 = S_\mathbf{v}(r) = r\mathbf{v}$ for all r; taking $r = 1$ gives $\mathbf{v} = \mathbf{0}$. Thus $\ker R = 0$.

$\underline{R \text{ is onto}}$: Given T in $\mathbf{L}(\mathbb{R}, V)$, we must find $\mathbf{v}$ in V such that $T = R(\mathbf{v})$; that is $T = S_\mathbf{v}$. Now $T : \mathbb{R} \to V$ is a linear transformation and we take $\mathbf{v} = T(1)$. Then, for r in $\mathbb{R}$:

$$S_\mathbf{v}(r) = r\mathbf{v} = rT(1) = T(r \cdot 1) = T(r).$$

Hence, $S_\mathbf{v} = T$ as required.

25(b) Given the linear transformation $T : \mathbb{R} \to V$ and an ordered basis $B = \{e_1, e_2, \ldots, e_n\}$ of V, write $T(1) = a_1 e_1 + a_2 e_2 + \cdots + a_n e_n$ where the a_i are in $\mathbb{R}$. We must show that $T = a_1 S_1 + a_2 S_2 + + a_n S_n$ where $S_i(r) = r e_i$ for all r in $\mathbb{R}$. We have

$$(a_1 S_1 + a_2 S_2 + \cdots + a_n S_n)(r) = a_1 S_1(r) + a_2 S_2(r) + \cdots + a_n S_n(r)$$
$$= a_1(r e_1) + a_2(r e_2) + \cdots + a_n(r e_n)$$
$$= r T(1)$$
$$= T(r)$$

for all r in $\mathbb{R}$. Hence $a_1 S_1 + a_2 S_2 + \cdots + a_n S_n = T$.

27(b) Given $\mathbf{v}$ in V, write $\mathbf{v} = r_1 e_1 + r_2 e_2 + \cdots + r_n e_n$, r_i in $\mathbb{R}$. We must show that $r_j = E_j(\mathbf{v})$ for each j. To see this, apply the linear transformation E_j:

$$E_j(\mathbf{v}) = E_j(r_1 e_1 + r_2 e_2 + + r_j e_j + \cdots + r_n e_n)$$
$$= r_1 E_j(e_1) + r_2 E_j(e_2) + \cdots + r_j E_j(e_j) + \cdots + r_n E_j(e_n)$$
$$= r_1 \cdot 0 + r_2 \cdot 0 + \cdots + r_j \cdot 1 + \cdots + r_n \cdot 0$$
$$= r_j$$

using the definition of E_j.

Exercises 8.5 Change of Basis

1(b) $P_{D \leftarrow B} = \begin{bmatrix} C_D(1) & C_D(1+x) & C_D(x^2) \end{bmatrix} = \begin{bmatrix} -\frac{3}{2} & -1 & \frac{1}{2} \\ 1 & 1 & 0 \\ 0 & 0 & 1 \end{bmatrix} = \frac{1}{2} \begin{bmatrix} -3 & -2 & 1 \\ 2 & 2 & 0 \\ 0 & 0 & 2 \end{bmatrix}$ because

$$x = -\frac{3}{2} \cdot 2 + 1(x + 3) + 0(x^2 - 1)$$
$$1 + x = (-1) \cdot 2 + 1(x + 3) + 0(x^2 - 1)$$
$$x^2 = \frac{1}{2} \cdot 2 + 0(x + 3) + 1(x^2 - 1).$$

Given $\mathbf{v} = 1 + x + x^2$, we have

$$C_B(\mathbf{v}) = \begin{bmatrix} 0 \\ 1 \\ 1 \end{bmatrix} \quad \text{and} \quad C_D(\mathbf{v}) = \begin{bmatrix} -\frac{1}{2} \\ 1 \\ 1 \end{bmatrix}$$

because $\mathbf{v} = 0 \cdot x + 1(1 + x) + 1 \cdot x^2$ and $\mathbf{v} = -\frac{1}{2} \cdot 2 + 1 \cdot (x + 3) + 1(x^2 - 1)$. Hence

$$P_{D \leftarrow B} C_B(\mathbf{v}) = \frac{1}{2} \begin{bmatrix} -3 & -2 & 1 \\ 2 & 2 & 0 \\ 0 & 0 & 2 \end{bmatrix} \begin{bmatrix} 0 \\ 1 \\ 1 \end{bmatrix} = \frac{1}{2} \begin{bmatrix} -1 \\ 2 \\ 2 \end{bmatrix} = C_D(\mathbf{v})$$

as expected.

4(b) $P_{B \leftarrow D} = \begin{bmatrix} C_B(1 + x + x^2) & C_B(1 - x) & C_B(-1 + x^2) \end{bmatrix} = \begin{bmatrix} 1 & 1 & -1 \\ 1 & -1 & 0 \\ 1 & 0 & 1 \end{bmatrix}$

$P_{D \leftarrow B} = \begin{bmatrix} C_D(1) & C_D(x) & C_D(x^2) \end{bmatrix} = \frac{1}{3} \begin{bmatrix} 1 & 1 & 1 \\ 1 & -2 & 1 \\ -1 & -1 & 2 \end{bmatrix}$ because

$$1 = \tfrac{1}{3}\left[(1 + x + x^2) + (1 - x) - (-1 + x^2) \right]$$
$$x = \tfrac{1}{3}\left[(1 + x + x^2) - 2(1 - x) - (-1 + x^2) \right]$$
$$x^2 = \tfrac{1}{3}\left[(1 + x + x^2) + (1 - x) + 2(-1 + x^2) \right].$$

The fact that $P_{D \leftarrow B} = (P_{B \leftarrow D})^{-1}$ is verified by multiplying these matrices. Next:

$$P_{E \leftarrow D} = \begin{bmatrix} C_E(1 + x + x^2) & C_E(1 - x) & C_E(-1 + x^2) \end{bmatrix} = \begin{bmatrix} 1 & 0 & 1 \\ 1 & -1 & 0 \\ 0 & 1 & -1 \end{bmatrix}$$

$$P_{E \leftarrow B} = \begin{bmatrix} C_E(1) & C_E(x) & C_E(x^2) \end{bmatrix} = \begin{bmatrix} 0 & 0 & 1 \\ 0 & 1 & 0 \\ 1 & 0 & 0 \end{bmatrix}$$

where we note the order of the vectors in $E = \{x^2, x, 1\}$. Finally, matrix multiplication verifies that $P_{E \leftarrow D} P_{D \leftarrow B} = P_{E \leftarrow B}$.

5(b) Let $B = \{(1, 2, -1), (2, 3, 0), (1, 0, 2)\}$ be the basis formed by the transposes of the columns of A. Since D is the standard basis:

$$P_{D \leftarrow B} = \begin{bmatrix} C_D(1, 2, -1) & C_D(2, 3, 0) & C_D(1, 0, 2) \end{bmatrix} = \begin{bmatrix} 1 & 2 & 1 \\ 2 & 3 & 0 \\ -1 & 0 & 2 \end{bmatrix} = A.$$

Hence Theorem 2 gives

$$A^{-1} = (P_{D \leftarrow B})^{-1} = P_{B \leftarrow D} = \begin{bmatrix} C_B(1, 0, 0) & C_B(0, 1, 0) & C_B(0, 0, 1) \end{bmatrix} = \begin{bmatrix} 6 & -4 & -3 \\ -4 & 3 & 2 \\ 3 & -2 & -1 \end{bmatrix}$$

because

$$(1, 0, 0) = 6(1, 2, -1) - 4(2, 3, 0) + 3(1, 0, 2)$$
$$(0, 1, 0) = -4(1, 2, -1) + 3(2, 3, 0) - 2(1, 0, 2)$$
$$(0, 0, 1) = -3(1, 2, -1) + 2(2, 3, 0) - 1(1, 0, 2).$$

7(b) Since $B_0 = \{1, x, x^2\}$, we have

$$P = P_{B_0 \leftarrow B} = \begin{bmatrix} C_{B_0}(1-x^2) & C_{B_0}(1+x) & C_{B_0}(2x+x^2) \end{bmatrix} = \begin{bmatrix} 1 & 1 & 0 \\ 0 & 1 & 2 \\ -1 & 0 & 1 \end{bmatrix}$$

$$\begin{aligned}
M_{B_0}(T) &= \begin{bmatrix} C_{B_0}[T(1)] & C_{B_0}[T(x)] & C_{B_0}[T(x^2)] \end{bmatrix} \\
&= \begin{bmatrix} C_{B_0}(1+x^2) & C_{B_0}(1+x) & C_{B_0}(x+x^2) \end{bmatrix} \\
&= \begin{bmatrix} 1 & 1 & 0 \\ 0 & 1 & 1 \\ 1 & 0 & 1 \end{bmatrix}.
\end{aligned}$$

Finally

$$\begin{aligned}
M_B(T) &= \begin{bmatrix} C_B[T(1-x^2)] & C_B[T(1+x)] & C_B[T(2x+x^2)] \end{bmatrix} \\
&= \begin{bmatrix} C_B(1-x) & C_B(2+x+x^2) & C_B(2+3x+x^2) \end{bmatrix} \\
&= \begin{bmatrix} -2 & -3 & -1 \\ 3 & 5 & 3 \\ -2 & -2 & 0 \end{bmatrix}
\end{aligned}$$

because

$$\begin{aligned}
1 - x &= -2(1-x^2) + 3(1+x) - 2(2x+x^2) \\
2 + x + x^2 &= -3(1-x^2) + 5(1+x) - 2(2x+x^2) \\
2 + 3x + x^2 &= -1(1-x^2) + 3(1+x) + 0(2x+x^2).
\end{aligned}$$

The verification that $P^{-1}M_{B_0}(T)P = M_B(T)$ is equvalent to checking that $M_{B_0}(T)P = PM_B(T)$, and so can be seen by matrix multiplication.

8(b) $P^{-1}AF = \begin{bmatrix} 5 & -2 \\ -7 & 3 \end{bmatrix}\begin{bmatrix} 29 & -12 \\ 70 & -29 \end{bmatrix}\begin{bmatrix} 3 & 2 \\ 7 & 5 \end{bmatrix} = \begin{bmatrix} 5 & -2 \\ 7 & -3 \end{bmatrix}\begin{bmatrix} 3 & 2 \\ 7 & 5 \end{bmatrix} = \begin{bmatrix} 1 & 0 \\ 0 & -1 \end{bmatrix} =$

D. Let $B = \left\{ \begin{bmatrix} 3 \\ 7 \end{bmatrix}, \begin{bmatrix} 2 \\ 5 \end{bmatrix} \right\}$ consist of the columns of P. These are eigenvectors of A corresponding to the eigenvalues $1, -1$ respectively. Hence,

$$M_B(T_A) = \begin{bmatrix} C_B\left(T_A\begin{bmatrix} 3 \\ 7 \end{bmatrix}\right) & C_B\left(T_A\begin{bmatrix} 2 \\ 5 \end{bmatrix}\right) \end{bmatrix} = \begin{bmatrix} C_B\begin{bmatrix} 3 \\ 7 \end{bmatrix} & C_B\begin{bmatrix} -2 \\ -5 \end{bmatrix} \end{bmatrix} = \begin{bmatrix} 1 & 0 \\ 0 & -1 \end{bmatrix}.$$

9(b) Choose a basis of $\mathbb{R}^2$, say $B = \{(1,0), (0,1)\}$, and compute

$$M_B(T) = \begin{bmatrix} C_B[T(1,0)] & C_B[T(0,1)] \end{bmatrix} = \begin{bmatrix} C_B(3,2) & C_B(5,3) \end{bmatrix} = \begin{bmatrix} 3 & 5 \\ 2 & 3 \end{bmatrix}.$$

Hence, $c_T(x) = c_{M_B(T)}(x) = \begin{vmatrix} x-3 & -5 \\ -2 & x-3 \end{vmatrix} = x^2 - 6x - 1$. Note that the calculation is easy because B is the standard basis, but any basis could be used.

(d) Use the basis $B = \{1, x, x^2\}$ of $\mathbf{P}_2$ and compute

$$M_B(T) = [C_B[T(1)] \quad C_B[T(x)] \quad C_B[T(x^2)]]$$
$$= [C_B(1 + x - 2x^2) \quad C_B(1 - 2x + x^2) \quad C_B(-2 + x)]$$
$$= \begin{bmatrix} 1 & 1 & -2 \\ 1 & -2 & 1 \\ -2 & 1 & 0 \end{bmatrix}.$$

Hence,

$$c_T(x) = c_{M_B(T)}(x) = \begin{vmatrix} x-1 & -1 & 2 \\ -1 & x+2 & -1 \\ 2 & -1 & x \end{vmatrix} = \begin{vmatrix} x-1 & -1 & 2 \\ -1 & x+2 & -1 \\ -x+3 & 0 & x-2 \end{vmatrix}$$
$$= x^3 + x^2 - 8x - 3.$$

(f) Use $B = \left\{ \begin{bmatrix} 1 & 0 \\ 0 & 0 \end{bmatrix}, \begin{bmatrix} 0 & 1 \\ 0 & 0 \end{bmatrix}, \begin{bmatrix} 0 & 0 \\ 1 & 0 \end{bmatrix}, \begin{bmatrix} 0 & 0 \\ 0 & 1 \end{bmatrix} \right\}$ and compute

$$M_B(T) = \left[C_B \left\{ T \begin{bmatrix} 1 & 0 \\ 0 & 0 \end{bmatrix} \right\} \quad C_B \left\{ T \begin{bmatrix} 0 & 1 \\ 0 & 0 \end{bmatrix} \right\} \quad C_B \left\{ T \begin{bmatrix} 0 & 0 \\ 1 & 0 \end{bmatrix} \right\} \quad C_B \left\{ T \begin{bmatrix} 0 & 0 \\ 0 & 1 \end{bmatrix} \right\} \right]$$

$$= \left[C_B \begin{bmatrix} 1 & 0 \\ 1 & 0 \end{bmatrix} \quad C_B \begin{bmatrix} 0 & 1 \\ 0 & 1 \end{bmatrix} \quad C_B \begin{bmatrix} -1 & 0 \\ -1 & 0 \end{bmatrix} \quad C_B \begin{bmatrix} 0 & -1 \\ 0 & -1 \end{bmatrix} \right]$$

$$= \begin{bmatrix} 1 & 0 & -1 & 0 \\ 0 & 1 & 0 & -1 \\ 1 & 0 & -1 & 0 \\ 0 & 1 & 0 & -1 \end{bmatrix}.$$

Hence,

$$c_T(x) = c_{M_B(T)}(x) = \begin{vmatrix} x-1 & 0 & 1 & 0 \\ 0 & x-1 & 0 & 1 \\ -1 & 0 & x+1 & 0 \\ 0 & -1 & 0 & x+1 \end{vmatrix}$$

$$= (x-1) \begin{vmatrix} x-1 & 0 & 1 \\ 0 & x+1 & 0 \\ -1 & 0 & x+1 \end{vmatrix} + \begin{vmatrix} 0 & x-1 & 1 \\ -1 & 0 & 0 \\ 0 & -1 & x+1 \end{vmatrix} = x^4.$$

16. As in the discussion preceding Theorem 1, write each $\mathbf{b}_j$ as a linear combination of the $\mathbf{d}_i$

$$\mathbf{b}_j = p_{1j}\mathbf{d}_1 + p_{2j}\mathbf{d}_2 + \cdots + p_{nj}\mathbf{d}_n$$

where each p_{ij} is in $\mathbb{R}$. Then write $P = [p_{ij}]$. By the definition of these p_{ij}

$$C_D(\mathbf{b}_j) = \begin{bmatrix} p_{1j} \\ p_{2j} \\ \vdots \\ p_{nj} \end{bmatrix} = \text{column } j \text{ of } P.$$

Thus,

$$P = [C_D(\mathbf{p}_1) \quad C_D(\mathbf{p}_2) \quad \cdots \quad C_D(\mathbf{p}_n)]$$

in block form. Given $\mathbf{v}$ in V, write it as

$$\mathbf{v} = v_1\mathbf{b}_1 + v_2\mathbf{b}_2 + \cdots + v_n\mathbf{b}_n,$$

so $C_B(\mathbf{v}) = [v_j]$. Thus

$$\mathbf{v} = \sum_{j=1}^{n} v_j\mathbf{b}_j = \sum_{j=1}^{n} v_j \left(\sum_{i=1}^{n} p_{ij}\mathbf{d}_i \right) = \sum_{i=1}^{n} \left(\sum_{j=1}^{n} p_{ij}v_j \right) \mathbf{d}_i,$$

The coefficients of $\mathbf{d}_i$ are the entries of the column $P[v_j] = PC_B(\mathbf{v})$. They are also the entries of $C_D(\mathbf{v})$, so

$$C_D(\mathbf{v}) = PC_B(\mathbf{v}).$$

Exercises 8.6   Invariant Subspaces and Direct Sums

3(b) Given $\mathbf{v}$ in $S(U)$, we must show that $T(\mathbf{v})$ is also in $S(U)$. We have $\mathbf{v} = S(\mathbf{u})$ for some u in U. As $ST = TS$:

$$T(\mathbf{v}) = T[S(\mathbf{u})] = (TS)(\mathbf{u}) = (ST)(\mathbf{u}) = S[T(\mathbf{u})].$$

As $T(\mathbf{u})$ is in U (because U is T-invariant), this shows that $T(\mathbf{v}) = S[T(\mathbf{u})]$ is in $S(U)$.

6. Suppose that a subspace U of V is T-invariant for every linear transformation $T : V \to V$. If $U \neq 0$, we must show that $U = V$. Choose $\mathbf{u} \neq 0$ in U, and (by Theorem 2, §6.4) extend $\{\mathbf{u}\}$ to a basis $\{\mathbf{u}, \mathbf{e}_2, \ldots, \mathbf{e}_n\}$ of V. Now let $\mathbf{v}$ be any vector in V. Then (by Theorem 3, §8.1) there is a linear transformation $T : V \to V$ such that $T(\mathbf{u}) = \mathbf{v}$ and $T(\mathbf{e}_i) = \mathbf{0}$ for each i. Then $\mathbf{v} = T(\mathbf{u})$ lies in U because U is T-invariant. As $\mathbf{v}$ was arbitrary, it follows that $V = U$. (Remark: The only place we used the hypothesis that V is finite dimensional is in extending $\{\mathbf{u}\}$ to a basis of V. In fact, this is true for any vector space.)

8(b) We have $U = \text{span}\{1 - 2x^2, x + x^2\}$. To show that U is T-invariant, it suffices (by Example 3) to show that $T(1 - 2x^2)$ and $T(x + x^2)$ both lie in U. We have

$$\left.\begin{array}{l} T(1 - 2x^2) = 3 + 3x - 3x^2 = 3(1 - 2x^2) + 3(x + x^2) \\[2mm] T(x + x^2) = -1 + 2x^2 = -(1 - 2x^2) \end{array}\right\} \qquad (*)$$

So both as in U. Hence, U is T-invariant. To get a block triangular matrix for T extend the basis $\{1 - 2x^2, x + x^2\}$ of U to a basis B of V in any way at all, say

$$B = \{1 - 2x^2, x + x^2, x^2\}.$$

Then, using (*), we have

$$M_B(T) = \begin{bmatrix} C_B\left[T(1 - 2x^2)\right] & C_B\left[T(x + x^2)\right] & C_B\left[T(x^2)\right] \end{bmatrix} = \begin{bmatrix} 3 & -1 & 1 \\ 3 & 0 & 1 \\ 0 & 0 & 3 \end{bmatrix}$$

where the last column is because $T(x^2) = 1 + x + 2x^2 = (1 - 2x^2) + (x + x^2) + 3(x^2)$. Finally,

$$c_T(x) = \begin{vmatrix} x - 3 & 1 & -1 \\ -3 & x & -1 \\ 0 & 0 & x - 3 \end{vmatrix} = (x - 3) \begin{vmatrix} x - 3 & 1 \\ -3 & x \end{vmatrix} = (x - 3)(x^2 - 3x + 3).$$

9(b) If U if T-invariant and $U \neq 0$, $U \neq \mathbb{R}^2$, then $\dim U = 1$. Thus $U = \mathbb{R}\mathbf{u}$ where $\mathbf{u} \neq \mathbf{0}$. Thus $T_A(\mathbf{u})$ is in $\mathbb{R}\mathbf{u}$ (because U is T-invariant), say $T_A(\mathbf{u}) = r\mathbf{u}$, that is $A\mathbf{u} = r\mathbf{u}$, whence $(rI - A)\mathbf{u} = 0$. But $\det(rI - A) = \begin{vmatrix} r - \theta & -\sin\theta \\ \sin\theta & r - \cos\theta \end{vmatrix} = (r - \cos\theta)^2 + \sin^2\theta \neq 0$ because $\sin\theta \neq 0$ $(0 < \theta < \pi)$. Hence, $(rI - A)\mathbf{u} = 0$ implies $\mathbf{u} = \mathbf{0}$, a contradiction. So $U = 0$ or $U = \mathbb{R}^2$.

10(b) If $\mathbf{v}$ is in $U \cap W$, then $\mathbf{v} = (a, a, b, b) = (c, d, c, -d)$ for some a, b, c, d. Hence $a = b = c = d = 0$, so $U \cap W = 0$. To see that $\mathbb{R}^4 = U + W$, we have (after solving systems of equations)

$$(1, 0, 0, 0) = \tfrac{1}{2}(1, 1, -1, -1) + \tfrac{1}{2}(1, -1, 1, 1) \text{ is in } U + W$$
$$(0, 1, 0, 0) = \tfrac{1}{2}(1, 1, 1, 1) + \tfrac{1}{2}(-1, 1, -1, -1) \text{ is in } U + W$$
$$(0, 0, 1, 0) = \tfrac{1}{2}(-1, -1, 1, 1) + \tfrac{1}{2}(1, 1, 1, -1) \text{ is in } U + W$$
$$(0, 0, 0, 1) = \tfrac{1}{2}(1, 1, 1, 1) + \tfrac{1}{2}(-1, -1, -1, 1) \text{ is in } U + W.$$

Hence, $\mathbb{R}^4 = U + W$. A simpler argument is as follows. As $\dim U = 2 = \dim W$, the subspace $U \oplus W$ has dimension $2 + 2 = 4$ by Theorem 7. Hence $U \oplus W = \mathbb{R}^4$ because $\dim \mathbb{R}^4 = 4$.

(d) If A is in $U \cap W$, then $A = \begin{bmatrix} a & a \\ b & b \end{bmatrix} = \begin{bmatrix} c & d \\ -c & d \end{bmatrix}$ for some a, b, c, d, whence $a = b = c = d = 0$. Thus, $U \cap W = 0$. Thus, by Theorem 7

$$\dim(U \oplus W) = \dim U + \dim W = 2 + 2 = 4.$$

Since $\dim \mathbf{M}_{22} = 4$, we have $U \oplus W = \mathbf{M}_{22}$. Again, as in (b), we could show directly that each of $\begin{bmatrix} 1 & 0 \\ 0 & 0 \end{bmatrix}, \begin{bmatrix} 0 & 1 \\ 0 & 0 \end{bmatrix}, \begin{bmatrix} 0 & 0 \\ 1 & 0 \end{bmatrix}, \begin{bmatrix} 0 & 0 \\ 0 & 1 \end{bmatrix}$ is in $U + W$.

14. First U is a subspace $AE = A$ and $A_1 E = A_1$ implies that $(A + A_1)E = AE + A_1 E = A + A_1$ and $(rA)E = r(AE) = rA$ for all $r \in \mathbb{R}$. Similarly, W is a subspace because $BE = 0 = B_1 E$ implies that $(B + B_1)E = BE + B_1 E = 0 + 0 = 0$ and $(rB)E = r(BE) = r0 = 0$ for all $r \in \mathbb{R}$. These calculations hold for *any* matrix E; if $E^2 = E$ we get $\mathbf{M}_{22} = U \oplus W$. First $U \cap W = 0$ because X in $U \cap W$ implies $X = XE = 0$. To prove that $U + W = \mathbf{M}_{22}$ let X be any matrix in $\mathbf{M}_{22}$. Then:

$$XE \text{ is in } U \quad \text{because} \quad (XE)E = XE^2 = XE$$

$$X - XE \text{ is in } W \quad \text{because} \quad (X - XE)E = XE - XE^2 = XE - XE = 0.$$

As $X = XE + (X - XE)$, we have X in $U + W$. Thus $\mathbf{M}_{22} = U + W$.

18(b) If $\begin{bmatrix} x \\ y \end{bmatrix}$ is in $\ker T_A$, then $A \begin{bmatrix} x \\ y \end{bmatrix} = 0$, that is $\begin{bmatrix} y \\ 0 \end{bmatrix} = \begin{bmatrix} 0 \\ 0 \end{bmatrix}$,

so $\begin{bmatrix} x \\ y \end{bmatrix} = \begin{bmatrix} x \\ 0 \end{bmatrix} \in \mathbb{R} \begin{bmatrix} 1 \\ 0 \end{bmatrix}$. Thus $\ker T_A \subseteq \mathbb{R} \begin{bmatrix} 1 \\ 0 \end{bmatrix}$. The other inclusion follows because

$A \begin{bmatrix} 1 \\ 0 \end{bmatrix} = 0$. Hence, $\ker T_A = \mathbb{R} \begin{bmatrix} 1 \\ 0 \end{bmatrix}$, which shows that $\mathbb{R} \begin{bmatrix} 1 \\ 0 \end{bmatrix}$ is T_A-invariant, by exercise 2. Now suppose that U is any T_A-invariant subspace, $U \neq 0$, $U \neq \mathbb{R}^2$. Then $\dim U = 1$, say $U = \mathbb{R}\mathbf{p}$, $\mathbf{p} \neq \mathbf{0}$. Then, $\mathbf{p}$ is in U so $A^2 \mathbf{p} = \lambda A\mathbf{p}$, $0\mathbf{p} = \lambda(\lambda\mathbf{p})$, $\mathbf{0} = \lambda^2 \mathbf{p}$. Thus $\lambda^2 = 0$, whence $\lambda = 0$ and $A\mathbf{p} = \lambda\mathbf{p} = \mathbf{0}$. Hence $\mathbf{p}$ is in $\ker T_A$, whence $U \subseteq \ker T_A$. But $\dim U = 1 = \dim(\ker T_A)$, so $U = \ker T_A$.

20. Let B_1 be a basis of U and extend it to a basis S of V. Then $M_B(T) = \begin{bmatrix} M_{B_1}(T) & Y \\ 0 & Z \end{bmatrix}$ by Theorem 1. Since we are writing T_1 for the restriction of T to U, $M_{B_1}(T) = M_{B_1}(T_1)$. Hence,

$$c_T(x) = \det[xI - M_B(T)] = \det \begin{bmatrix} xI - M_{B_1}(T) & -x \\ 0 & xI - Z \end{bmatrix}$$

$$= \det[xI - M_{B_1}(T_1)] \det[xI - Z] = c_{T_1}(x) \cdot q(x)$$

where $q(x) = \det[xI - Z]$.

22(b) $T^2[p(x)] = T\{T[p(x)]\} = T[p(-x)] = p(-(-x)) = p(x) = 1_{\mathbf{P}_3}(p(x))$. Hence, $T^2 = 1_{\mathbf{P}_3}$. As in Example 10, let

$$U_1 = \{p(x) \mid T[p(x)] = p(x)\} = \{p(x) \mid p(-x) = p(x)\}$$
$$U_2 = \{p(x) \mid T[p(x)] = -p(x)\} = \{p(x) \mid p(-x) = -p(x)\}.$$

These are the subspaces of even and odd polynomials in $\mathbf{P}_3$, respectively, and have bases $B_1 = \{1, x^2\}$ and $B_2 = \{x, x^3\}$. Hence, use the ordered basis $B = \{1, x^2; x, x^3\}$ of $\mathbf{P}_3$. Then

$$M_B(T) = \begin{bmatrix} M_{B_1}(T) & 0 \\ 0 & M_{B_2}(T) \end{bmatrix} = \begin{bmatrix} I_2 & 0 \\ 0 & -I_2 \end{bmatrix}$$

as in Example 10. More explicitly,

$$M_B(T) = \begin{bmatrix} C_B[t(1)] & C_B[T(x^2)] & C_B[T(x)] & C_B[T(x^3)] \end{bmatrix}$$
$$= \begin{bmatrix} C_B(1) & C_B(x^2) & C_B(-x) & C_B(-x^3) \end{bmatrix}$$

$$= \begin{bmatrix} 1 & 0 & 0 & 0 \\ 0 & 1 & 0 & 0 \\ 0 & 0 & -1 & 0 \\ 0 & 0 & 0 & -1 \end{bmatrix}$$

$$= \begin{bmatrix} I_2 & 0 \\ 0 & -I_2 \end{bmatrix}.$$

29(b) We have $T_{f,z}^2[\mathbf{v}] = T_{f,z}\{T_{f,z}(\mathbf{v})\} = T_{f,z}[f(\mathbf{v})z] = f[f(\mathbf{v})z]z = f(\mathbf{v})f(z)z$. This equals $T_{f,z}(\mathbf{v}) = f(\mathbf{v})z$ for all $\mathbf{v}$ if and only if

$$f(\mathbf{v})(z - f(z)z) = 0$$

for all $\mathbf{v}$. Since $f \neq 0$, $f(\mathbf{v}) \neq 0$ for some $\mathbf{v}$, so this holds if and only if

$$z = f(z)z.$$

As $z \neq 0$, this holds if and only if $f(z) = 1$.

30(b) Let λ be an eigenvalue of T. If A is in $E_\lambda(T)$ then $T(A) = \lambda A$; that is $UA = \lambda A$. If we write $A = [\mathbf{p}_1 \quad \mathbf{p}_2 \quad \cdots \quad \mathbf{p}_n]$ in terms of its columns $\mathbf{p}_1, \mathbf{p}_2, \dots, \mathbf{p}_n$, then $UA = \lambda A$ becomes

$$U[\mathbf{p}_1 \quad \mathbf{p}_2 \quad \cdots \quad \mathbf{p}_n] = \lambda[\mathbf{p}_1 \quad \mathbf{p}_2 \cdots \quad \mathbf{p}_n]$$
$$[U\mathbf{p}_1 \quad U\mathbf{p}_2 \quad \cdots \quad U\mathbf{p}_n] = [\lambda\mathbf{p}_1 \quad \lambda\mathbf{p}_2 \quad \cdots \quad \lambda\mathbf{p}_n].$$

Comparing columns gives $U\mathbf{p}_i = \lambda\mathbf{p}_i$ for each i; that is $\mathbf{p}_i$ is in $E_\lambda(U)$ for each i. Conversely, if $\mathbf{p}_1, \mathbf{p}_2, \dots, \mathbf{p}_n$ are all in $E_\lambda(U)$ then $U\mathbf{p}_i = \lambda\mathbf{p}_i$ for each i, so $T(A) = UA = \lambda A$ as above. Thus A is in $E_\lambda(T)$.

Exercises 8.7   Block Triangular Form

1(b) $c_A(x) = \begin{vmatrix} x+5 & -3 & -1 \\ 4 & x-2 & -1 \\ 4 & -3 & x \end{vmatrix} = \begin{vmatrix} x+1 & -x-1 & 0 \\ 4 & x-2 & -1 \\ 4 & -3 & x \end{vmatrix} = \begin{vmatrix} x+1 & 0 & 0 \\ 4 & x+2 & -1 \\ 4 & 1 & x \end{vmatrix} = (x+1)^3.$

Hence, $\lambda_1 = -1$ and we are in case $k = 1$ of the triangulation algorithm.

$$-I - A = \begin{bmatrix} 4 & -3 & -1 \\ 4 & -3 & -1 \\ 4 & -3 & -1 \end{bmatrix} \rightarrow \begin{bmatrix} 4 & -3 & -1 \\ 0 & 0 & 0 \\ 0 & 0 & 0 \end{bmatrix}; \quad P_{11} = \begin{bmatrix} 1 \\ 1 \\ 1 \end{bmatrix}, \quad F_{12} = \begin{bmatrix} 0 \\ 1 \\ -3 \end{bmatrix}.$$

Hence, $\{P_{11}, F_{12}\}$ is a basis of $\text{null}(-I-A)$. We now expand this to a basis of $\text{null}\big[(-I-A)^2\big]$. However, $(-I-A)^2 = 0$ so $\text{null}\big[(-I-A)^2\big] = \mathbb{R}^3$. Hence, in this case, we expand $\{P_{11}, F_{12}\}$ to any basis $\{P_{11}, F_{12}, F_{13}\}$ of $\mathbb{R}^3$, say by taking $P_{13} = \begin{bmatrix} 0 \\ 0 \\ 1 \end{bmatrix}$. Hence

$$P = [P_{11} \quad P_{12} \quad P_{13}] = \begin{bmatrix} 1 & 0 & 0 \\ 1 & 1 & 0 \\ 1 & -3 & 1 \end{bmatrix} \text{ satisfies } P^{-1}AF = \begin{bmatrix} -1 & 0 & 1 \\ 0 & -1 & 0 \\ 0 & 0 & -1 \end{bmatrix}$$

as may be verified.

(d) $c_A(x) = \begin{vmatrix} x+3 & 1 & 0 \\ -4 & x+1 & -3 \\ -4 & 2 & x-4 \end{vmatrix} = \begin{vmatrix} x+3 & 1 & 0 \\ -4 & x+1 & -3 \\ 0 & -x+1 & x-1 \end{vmatrix}$

$= \begin{vmatrix} x+3 & 1 & 0 \\ -4 & x-2 & -3 \\ 0 & 0 & x-1 \end{vmatrix} = (x-1)^2(x+2).$

Hence $\lambda_1 = 1$, $\lambda_3 = -2$, and we are in case $k = 2$ of the triangulation algorithm.

$$I - A = \begin{bmatrix} 4 & 1 & 0 \\ -4 & 2 & -3 \\ -4 & 2 & -3 \end{bmatrix} \rightarrow \begin{bmatrix} 4 & 1 & 0 \\ 0 & 3 & -3 \\ 0 & 0 & 0 \end{bmatrix}; \quad P_{11} = \begin{bmatrix} -1 \\ 4 \\ 4 \end{bmatrix}.$$

Thus, $\text{null}(I-A) = \text{span}\{P_{11}\}$. We enlarge $\{P_{11}\}$ to a basis of $\text{null}\big[(I-A)^2\big]$

$$(I-A)^{12} = \begin{bmatrix} 12 & 6 & -3 \\ -12 & -6 & 3 \\ -12 & -6 & 3 \end{bmatrix} \rightarrow \begin{bmatrix} 4 & 2 & -1 \\ 0 & 0 & 0 \\ 0 & 0 & 0 \end{bmatrix}; \quad P_{11} = \begin{bmatrix} -1 \\ 4 \\ 4 \end{bmatrix}, \quad F_{12} = \begin{bmatrix} 0 \\ 1 \\ 2 \end{bmatrix}.$$

Thus, $\text{null}\big[(I-A)^2\big] = \text{span}\{P_{11}, F_{12}\}$. As $\dim[G_{\lambda_1}(A)] = 2$ in this case (by Lemma 1), we have $G_{\lambda_1}(A) = \text{span}\{P_{11}, F_{12}\}$. However, it is instructive to continue the process:

$$(I-A)^2 = 3\begin{bmatrix} 4 & 2 & -1 \\ -4 & -2 & 1 \\ -4 & -2 & 1 \end{bmatrix}$$

whence

$$(I-A)^3 = 9\begin{bmatrix} 4 & 2 & -1 \\ -4 & -2 & 1 \\ -4 & -2 & 1 \end{bmatrix} = 3(I-A)^2.$$

This continutes to give $(I - A)^4 = 3^2(I - A)^2, \ldots$, and in general $(I - A)^k = 3^{k-2}(I - A)^2$ for $k \geq 2$. Thus $\text{null}\left[(I - A)^k\right] = \text{null}\left[(I - A)^2\right]$ for all $k \geq 2$, so

$$G_{\lambda_1}(A) = \text{null}\left[(I - A)^2\right] = \text{span}\{P_{11}, F_{12}\}.$$

as we expected. Turning to $\lambda_2 = -2$:

$$-2I - A = \begin{bmatrix} 1 & 1 & 0 \\ -4 & -1 & -3 \\ -4 & 2 & -6 \end{bmatrix} \rightarrow \begin{bmatrix} 1 & 1 & 0 \\ 0 & 3 & -3 \\ 0 & 6 & -6 \end{bmatrix} \rightarrow \begin{bmatrix} 1 & 1 & 0 \\ 0 & 1 & -1 \\ 0 & 0 & 0 \end{bmatrix} ; \quad P_{21} = \begin{bmatrix} -1 \\ 1 \\ 1 \end{bmatrix}.$$

Hence, $\text{null}[-2I - A] = \text{span}\{P_{21}\}$. We need go no further with this as $\{P_{11}, F_{12}, P_{21}\}$ is a basis of $\mathbb{R}^3$. Hence

$$P = [P_{11} \quad P_{12} \quad P_{21}] = \begin{bmatrix} -1 & 0 & -1 \\ 4 & 1 & 1 \\ 4 & 2 & 1 \end{bmatrix} \text{ satisfies } P^{-1}AF = \begin{bmatrix} 1 & 1 & 0 \\ 0 & 1 & 0 \\ 0 & 0 & -2 \end{bmatrix}$$

as may be verified.

(f) To evaluate $c_A(x)$, we begin by adding column 4 to column 1:

$$c_A(x) = \begin{vmatrix} x+3 & -6 & -3 & -2 \\ 2 & x-3 & -2 & -2 \\ 1 & -3 & x & -1 \\ 1 & -1 & -2 & x \end{vmatrix} = \begin{vmatrix} x+1 & -6 & -3 & -2 \\ 0 & x-3 & -2 & -2 \\ 0 & -3 & x & -1 \\ x+1 & -1 & -2 & x \end{vmatrix} = \begin{vmatrix} x+1 & -6 & -3 & -2 \\ 0 & x-3 & -2 & -2 \\ 0 & -3 & x & -1 \\ 0 & 5 & 1 & x+2 \end{vmatrix}$$

$$= (x+1) \begin{vmatrix} x-3 & -2 & -2 \\ -3 & x & -1 \\ 5 & 1 & x+2 \end{vmatrix} = (x+1) \begin{vmatrix} x-3 & -2 & 0 \\ -3 & x & -x-1 \\ 5 & 1 & x+1 \end{vmatrix} = (x+1) \begin{vmatrix} x-3 & -2 & 0 \\ 2 & x+1 & 0 \\ 5 & 1 & x+ \end{vmatrix}$$

$$= (x+1)^1 \begin{vmatrix} x-3 & -2 \\ 2 & x+1 \end{vmatrix} = (x+1)^2(x-1)^2.$$

Hence, $\lambda_1 = -1$, $\lambda_2 = 1$ and we are in case $k = 2$ of the triangulation algorithm. We omit the details of the row reductions:

$$-I - A = \begin{bmatrix} 2 & -6 & -3 & -2 \\ 2 & -4 & -2 & -2 \\ 1 & -3 & -1 & -1 \\ 1 & -1 & -2 & -1 \end{bmatrix} \rightarrow \begin{bmatrix} 1 & 0 & 0 & -1 \\ 0 & 1 & 0 & 0 \\ 0 & 0 & 1 & 0 \\ 0 & 0 & 0 & 0 \end{bmatrix} ; \quad P_{11} = \begin{bmatrix} 1 \\ 0 \\ 0 \\ 1 \end{bmatrix}$$

$$(-I - A)^2 = \begin{bmatrix} -13 & 23 & 13 & 13 \\ -8 & 12 & 8 & 8 \\ -6 & 10 & 6 & 6 \\ -3 & 5 & 3 & 3 \end{bmatrix} \rightarrow \begin{bmatrix} 1 & 0 & -1 & -1 \\ 0 & 1 & 0 & 0 \\ 0 & 0 & 0 & 0 \\ 0 & 0 & 0 & 0 \end{bmatrix} ; \quad P_{11} = \begin{bmatrix} 1 \\ 0 \\ 0 \\ 1 \end{bmatrix}, \quad F_{12} = \begin{bmatrix} 1 \\ 0 \\ 1 \\ 0 \end{bmatrix}.$$

We have dim $[G\lambda_1(A)] = 2$ as $\lambda_1 = -1$ has multiplicity 2 in $c_A(x)$, so $G_{\lambda_1}(A) = \text{span}\{P_{11}, F_{12}\}$. Turning to $\lambda_2 = 1$:

$$I - A = \begin{bmatrix} 4 & -6 & -3 & -2 \\ 2 & -2 & -2 & -2 \\ 1 & -3 & 1 & -1 \\ 1 & -1 & -2 & 1 \end{bmatrix} \rightarrow \begin{bmatrix} 1 & 0 & 0 & -5 \\ 0 & 1 & 0 & -2 \\ 0 & 0 & 1 & -2 \\ 0 & 0 & 0 & 0 \end{bmatrix}; \quad P_{21} = \begin{bmatrix} 5 \\ 2 \\ 2 \\ 1 \end{bmatrix}$$

$$(I - A)^2 = \begin{bmatrix} -1 & -1 & 1 & 5 \\ 0 & 0 & 0 & 0 \\ -2 & -2 & 2 & -6 \\ 1 & 1 & -5 & 3 \end{bmatrix} \rightarrow \begin{bmatrix} 1 & 1 & 0 & 0 \\ 0 & 0 & 1 & 0 \\ 0 & 0 & 0 & 1 \\ 0 & 0 & 0 & 0 \end{bmatrix}; \quad P_{21} = \begin{bmatrix} 5 \\ 2 \\ 2 \\ 1 \end{bmatrix}, \quad F_{22} = \begin{bmatrix} 1 \\ -1 \\ 0 \\ 0 \end{bmatrix}.$$

Hence, $G_{\lambda_2}(A) = \text{span}\{P_{21}, F_{22}\}$ using Lemma 1. Finally, then

$$P = [P_{11}, F_{12}, P_{21}, F_{22}] = \begin{bmatrix} 1 & 1 & 5 & 1 \\ 0 & 0 & 2 & -1 \\ 0 & 1 & 2 & 0 \\ 1 & 0 & 1 & 0 \end{bmatrix} \text{ gives } P^{-1}AF = \begin{bmatrix} -1 & 1 & 0 & 0 \\ 0 & -1 & 0 & 0 \\ 0 & 0 & 1 & -2 \\ 0 & 0 & 0 & 1 \end{bmatrix}$$

as may be verified.

4. Let B be any basis of V and write $A = M_B(T)$. Then $c_T(x) = c_A(x)$ and this is a polynomial: $c_T(x) = a_0 + a_1 x + \cdots + a_n x^n$ for some a_i in $\mathbb{R}$. Since $M_B : \mathbf{L}(V, V) \to M_{nn}$ is an isomorphism of vector spaces (Exercise 26, §8.4) and $M_B(T^k) = M_B(T)^k$ for $k \geq 1$, we get

$$\begin{aligned} M_B[c_T(T)] &= M_B[a_0 1_V + a_1 T + \cdots + a_n T^n] \\ &= a_0 M_B(1_V) + a_1 M_B(T) + \cdots + a_n M_B(T)^n \\ &= a_0 I + a_1 A + \cdots + a_n A^n \\ &= c_A(A) \\ &= 0 \end{aligned}$$

by the Cayley-Hamilton Theorem. Hence $c_T(T) = 0$ because M_B is one-to-one.

6(b) We have $B = \{T^m(\mathbf{v}), T^{m-1}(\mathbf{v}), \ldots, T(\mathbf{v}), \mathbf{v}\}$. Observe that

$$T[T^i(\mathbf{v})] = T^{i+1}(\mathbf{v})$$

so T carries each vector in B to the preceding vector in B, (and $T(T^m(\mathbf{v})) = \mathbf{0}$ because $T^{m+1} = 0$). Since B is a basis of $W = \text{span } B$, this shows that W is T-invariant and that the matrix of $T : W \to W$ with respect to B is

$$
M_B(T) =
\begin{bmatrix}
0 & 1 & 0 & \cdots & 0 \\
0 & 0 & 1 & \cdots & 0 \\
0 & 0 & 0 & \cdots & 0 \\
\vdots & \vdots & \vdots & \ddots & \vdots \\
0 & 0 & 0 & \cdots & 1 \\
0 & 0 & 0 & \cdots & 0
\end{bmatrix}.
$$

This is a Jordan block with $\lambda = 0$.

Exercises 8.8   More on Linear Recurrences

1(b) The associated polynomial is

$$p(x) = x^3 - 7x + 6 = (x-1)(x-2)(x+3).$$

Hence, $\{[1), [2^n), [(-3)^n)\}$ is a basis of the space of all solutions to the recurrence. The general solution is thus,

$$[x_n) = a[1) + b[2^n) + c[(-3)^n)$$

where a, b and c are constants. The requirement that $x_0 = 1$, $x_1 = 2$, $x_2 = 1$ determines a, b, and c. We have

$$x_n = a + b2^n + c(-3)^n$$

for all $n \geq 0$. So taking $n = 0, 1, 2$ gives

$$
\begin{aligned}
a + b + c &= x_0 = 1 \\
a + 2b - 3c &= x_1 = 2 \\
a + 4b + 9c &= x_2 = 1.
\end{aligned}
$$

The solution is $a = \frac{15}{20}$, $b = \frac{8}{20}$, $c = -\frac{3}{20}$, so

$$x_n = \tfrac{1}{20}(15 + 2^{n+3} + (-3)^{n+1}) \quad n \geq 0.$$

2(b) The associated polynomial is

$$p(x) = x^3 - 3x + 2 = (x-1)^2(x+2).$$

As 1 is a double root of $p(x)$, $[1^n) = [1)$ and $[n1^n) = [n)$ are solutions to the recurrence by Theorem 3. Similarly, $[(-2)^n)$ is a solution, so $\{[1), [n), [(-2)^n)\}$ is a basis for the space of solutions by Theorem 4. The required sequence has the form

$$[x_n) = a[1) + b[n) + c[(-2)^n)$$

for constants a, b, c. Thus, $x_n = a + bn + c(-2)^n$ for $n \geq 0$, so taking $n = 0, 1, 2$, we get

$$
\begin{array}{ccccccc}
a & + & & & c & = & x_0 & = & 1 \\
a & + & b & - & 2c & = & x_1 & = & -1 \\
a & + & 2b & + & 4c & = & x_2 & = & 1.
\end{array}
$$

The solution is $a = \frac{5}{9}$, $b = -\frac{6}{9}$, $c = \frac{4}{9}$, so

$$x_n = \frac{1}{9}\left[5 - 6n + (-2)^{n+2}\right] \quad n \geq 0.$$

(d) The associated polynomial is

$$p(x) = x^3 - 3x^2 + 3x - 1 = (x-1)^3.$$

Hence, $[1^n) = [1)$, $[n1^n) = [n)$ and $[n^2 1^n) = [n^2)$ are solutions and so $\{[1), [n), [n^2)\}$ is a basis for the space of solutions. Thus

$$x_n = a \cdot 1 + bn + cn^2,$$

a, b, c constants. As $x_0 = 1$, $x_1 = -1$, $x_2 = 1$, we obtain

$$
\begin{array}{ccccccc}
a & & & & & = & x_0 & = & 1 \\
a & + & b & + & c & = & x_1 & = & -1 \\
z & + & 2b & + & 4c & = & x_2 & = & 1.
\end{array}
$$

The solution is $a = 1$, $b = -4$, $c = 2$, so

$$x_n = 1 - 4n + 2n^2 \quad n \geq 0.$$

This can be written

$$x_n = 2(n-1)^2 - 1.$$

3(b) The associated polynomial is

$$p(x) = x^2 - (a+b)x + ab = (x-a)(x-b).$$

Hence, as $a \neq b$, $\{[a^n), [b^n)\}$ is a basis for the space of solutions.

4(b) The recurrence $x_{n+4} = -x_{n+2} + 2x_{n+3}$ has $r_0 = 0$ as there is no term x_n. If we write $y_n = x_{n+2}$, the recurrence becomes

$$y_{n+2} = -y_n + 2y_{n+1}.$$

Now the associated polynomial is $x^2 - 2x + 1 = (x-1)^2$ so basis sequences for the solution space for y_n are

$$[1^n) = [1, 1, 1, 1, \ldots)$$
$$[n1^n) = [0, 1, 2, 3, \ldots).$$

As $y_n = x_{n+2}$, corresponding basis sequences for x_n are

$$[0,0,1,1,1,1,\dots)$$
$$[0,0,0,1,2,3,\dots).$$

Also,

$$[1,0,0,0,0,0,\dots)$$
$$[0,1,0,0,0,0,\dots)$$

are solutions for x_n, so these four sequences form a basis for the solution space for x_n.

9. The sequence has length 2 and associated polynomial $x^2 + 1$. The roots are nonreal: $\lambda_1 = i$ and $\lambda_2 = -i$. Hence, by Remark 2,

$$[i^n + (-i)^n) = [2,0,-2,0,2,0,-2,0,\dots)$$
$$[i(i^n - (-i)^n)) = [0,-2,0,2,0,-2,0,2,\dots)$$

are solutions. They are independent as is easily verified, so they are a basis for the space of solutions.

Chapter 9: Inner Product Spaces

Exercises 9.1　　Inner Products and Norms

1(b) P5 fails: $\langle (0,1,0),(0,1,0) \rangle = -1$.

The other axioms hold. Write $\mathbf{x} = (x_1,x_2,x_3)$, $\mathbf{y} = (y_1,y_2,y_3)$ and $\mathbf{z} = (z_1,z_2,z_3)$.

P1 holds: $\langle \mathbf{x}, \mathbf{y} \rangle = x_1 y_1 - x_2 y_2 + x_3 y_3$ is real for all x,y in $\mathbb{R}^n$.

P2 holds: $\langle \mathbf{x}, \mathbf{y} \rangle = x_1 y_1 - x_2 y_2 + x_3 y_3 = y_1 x_1 - y_2 x_2 + y_3 x_3 = \langle \mathbf{y}, \mathbf{x} \rangle$.

P3 holds: $\langle \mathbf{x} + \mathbf{y}, \mathbf{z} \rangle = (x_1 + y_1)z_1 - (x_2 + y_2)z_2 + (x_3 + y_3)z_3$

$$= (x_1 z_1 - x_2 z_2 + x_3 z_3) + (y_1 z_1 - y_2 z_2 + y_3 z_3) = \langle \mathbf{x}, \mathbf{z} \rangle + \langle \mathbf{y}, \mathbf{z} \rangle.$$

P4 holds: $\langle r\mathbf{x}, \mathbf{y} \rangle = (rx_1)y_1 - (rx_2)y_2 + (rx_3)y_3 = r(x_1 y_1 - x_2 y_2 + x_3 y_3) = r\langle \mathbf{x}, \mathbf{y} \rangle$.

(d) P5 fails: $\langle x - 2, x \rangle = (1 - 2)(1) = -1$

P1 holds: $\langle p(x), q(x) \rangle = p(1)q(1)$ is real

P2 holds: $\langle p(x), q(x) \rangle = p(1)q(1) = q(1)p(1) = \langle q(x), p(x) \rangle$

P3 holds: $\langle p(x) + r(x), q(x) \rangle = (p(1) + r(1))q(1) = p(1)q(1) + r(1)q(1)$

$$= \langle p(x), q(x) \rangle + \langle r(x), q(x) \rangle$$

P4 holds: $\langle rp(x), q(x) \rangle = [rp(1)]q(1) = r[p(1)q(1)] = r\langle p(x), q(x) \rangle$

(f) P5 fails: If $f(x) = x$, $g(x) = x - 1$ then $\langle f, g \rangle = 1(-1) + 0 \cdot 0 = -1$

P1 holds: $\langle f, g \rangle = f(1)g(0) + f(0)g(1)$ is real

P2 holds: $\langle f, g \rangle = f(1)g(0) + f(0)g(1) = g(1)f(0) + g(0)f(1) = \langle g, f \rangle$

P3 holds: $\langle f + h, g \rangle = (f + h)(1)g(0) + (f + h)(0)g(1) = [f(1) + h(1)]g(0) + [f(0) + h(0)]g(1)$

$$= [f(1)g(0) + f(0)g(1)] + [h(1)g(0) + h(0)g(1)] = \langle f, g \rangle + \langle h, g \rangle$$

P4 holds: $\langle rf, h \rangle = (rf)(1)g(0) + (rf)(0)g(1) = [r \cdot f(1)]g(0) + [rf(0)]g(1)$

$$= r[f(1)g(0) + f(0)g(1)] = r\langle f, g \rangle$$

3(b) $\|f\|^2 = \int_{-\pi}^{\pi} \cos^2 x \, dx = \int_{-\pi}^{\pi} \frac{1}{2}[1 + \cos(2x)] \, dx = \frac{1}{2}\left[x + \frac{1}{2}\sin(2x)\right]_{-\pi}^{\pi} = \pi$. Hence $\hat{f} = \frac{1}{\sqrt{\pi}}f$ is a unit vector.

(d) $\|\mathbf{v}\|^2 = \langle \mathbf{v}, \mathbf{v} \rangle = \mathbf{v}^T \begin{bmatrix} 1 & -1 \\ -1 & 2 \end{bmatrix} \mathbf{v} = \begin{bmatrix} 3 & -1 \end{bmatrix} \begin{bmatrix} 1 & -1 \\ -1 & 2 \end{bmatrix} \begin{bmatrix} 3 \\ -1 \end{bmatrix} = 17$. Hence $\hat{\mathbf{v}} =$

$\frac{1}{\|\mathbf{v}\|}\mathbf{v} = \frac{1}{\sqrt{17}} \begin{bmatrix} 3 \\ -1 \end{bmatrix}$ is a unit vector in this space.

4(c) $d(\mathbf{u}, \mathbf{v}) = \|\mathbf{u} - \mathbf{v}\| = \|(1,2,-1,2) - (2,1,-1,3)\| = \|(-1,1,0,-1)\| = \sqrt{3}$.

(d) $\|f - g\|^2 = \int_{-\pi}^{\pi} (1 - \cos x)^2 \, dx = \int_{-\pi}^{\pi} \left[\frac{3}{2} - 2\cos x + \frac{1}{2}\cos(2x)\right] dx = 3\pi$. Hence $d(f,g) = \sqrt{3\pi}$.

8. The space D_n uses pointwise addition and scalar multiplication:

$$(f + g)(k) = f(k) + g(k) \quad \text{and} \quad (rf)(k) = rf(k)$$

for all $k = 1, 2, \dots, n$.

P1. $\langle f, g \rangle = f(1)g(1) + f(2)g(2) + \cdots + f(n)g(n)$ is real

P2. $\langle f,g \rangle = f(1)g(1) + f(2)g(2) + \cdots + f(n)g(n)) = g(1)f(1) + g(2)f(2) + \cdots + g(n)f(n) = \langle g,f \rangle$.

P3. $\begin{aligned} \langle f+h,g \rangle &= (f+h)(1)g(1) + (f+h)(2)g(2) + \cdots + (f+h)(n)g(n) \\ &= [f(1)+h(1)]g(1) + [f(2)+h(2)]g(2) + \cdots + [f(n)+h(n)]g(n) \\ &= [f(1)g(1) + f(2)g(2) + \cdots + f(n)g(n)] + [h(1)g(1) + h(2)g(2) + \cdots + h(n)g(n)] \\ &= \langle f,g \rangle + \langle h,g \rangle \end{aligned}$

P4. $\begin{aligned} \langle rf,g \rangle &= (rf)(1)g(1) + (rf)(2)g(2) + \cdots + (rf)(n)g(n) \\ &= [rf(1)]g(1) + [rf(2)]g(2) + \cdots + [rf(n)]g(n) \\ &= r[f(1)g(1) + f(2)g(2) + \cdots + f(n)g(n)] = r\langle f,g \rangle \end{aligned}$

P5. $\langle f,f \rangle = f(1)^2 + f(2)^2 + \cdots + f(n)^2 \geq 0$ for all f. If $\langle f,f \rangle = 0$ then

$$f(1) = f(2) = \cdots = f(n) = 0 \text{ (as the } f(k) \text{ are real numbers) so } f = 0.$$

12(b) We need only verify P5. [P1-P4 hold for any symmetric matrix A by (the discussion preceding) Theorem 2.] If $\mathbf{v} = \begin{bmatrix} v_1 \\ v_2 \end{bmatrix}$:

$$\begin{aligned} \langle \mathbf{v}, \mathbf{v} \rangle = \mathbf{v}^T A \mathbf{v} &= [v_1, v_2] \begin{bmatrix} 5 & -3 \\ -3 & 2 \end{bmatrix} \begin{bmatrix} v_1 \\ v_2 \end{bmatrix} \\ &= 5v_1^2 - 6v_1v_2 + 2v_2^2 \\ &= 5\left[v_1^2 - \tfrac{6}{5}v_1v_2 + \tfrac{9}{25}v_2^2 \right] - \tfrac{9}{5}v_2^2 + 2v_2^2 \\ &= 5\left(v_1 - \tfrac{3}{5}v_2 \right)^2 + \tfrac{1}{5}v_2^2 \\ &= \tfrac{1}{5}\left[(5v_1 - 3v_2)^2 + v_2^2 \right]. \end{aligned}$$

Thus, $\langle \mathbf{v}, \mathbf{v} \rangle \geq 0$ for all $\mathbf{v}$; and $\langle \mathbf{v}, \mathbf{v} \rangle = 0$ if and only if $5v_1 - 3v_2 = 0 = v_2$; that is if and only if $v_1 = v_2 = 0$ ($\mathbf{v} = \mathbf{0}$). So P5 holds.

(d) As in (b), consider $\mathbf{v} = \begin{bmatrix} v_1 \\ v_2 \end{bmatrix}$.

$$\begin{aligned} \langle \mathbf{v}, \mathbf{v} \rangle &= [v_1 \quad v_2] \begin{bmatrix} 3 & 4 \\ 4 & 6 \end{bmatrix} \begin{bmatrix} v_1 \\ v_2 \end{bmatrix} \\ &= 3v_1^2 + 8v_1v_2 + 6v_2^2 \\ &= \left(3v_1^2 + \tfrac{8}{3}v_1v_2 + \tfrac{16}{9}v_2^2 \right) - \tfrac{16}{3}v_2^2 + 6v_2^2 \\ &= 3\left(v_1 + \tfrac{4}{5}v_2 \right)^2 + \tfrac{2}{3}v_2^2 \\ &= \tfrac{1}{3}\left[(3v_1 + 4v_2)^2 + 2v_2^2 \right]. \end{aligned}$$

Thus, $\langle \mathbf{v}, \mathbf{v} \rangle \geq 0$ for all $\mathbf{v}$; and $(\mathbf{v}, \mathbf{v}) = 0$ if and only if $3v_1 + 4v_2 = 0 = v_2$; that is if and only if $\mathbf{v} = \mathbf{0}$. Hence P5 holds. The other axioms hold because A is symmetric.

13(b) If $A = \begin{bmatrix} a_{11} & a_{12} \\ a_{21} & a_{22} \end{bmatrix}$, then a_{ij} is the coefficient if $v_i w_j$ in $\langle \mathbf{v}, \mathbf{w} \rangle$. Here $a_{11} = 1$, $a_{12} = -1 = a_{21}$,

and $a_{22} = 2$. Thus, $A = \begin{bmatrix} 1 & -1 \\ -1 & 2 \end{bmatrix}$. Note that $a_{12} = a_{21}$, so A is symmetric.

(d) As in (b): $A = \begin{bmatrix} 1 & 0 & -2 \\ 0 & 2 & 0 \\ -2 & 0 & 5 \end{bmatrix}$.

16(b)
$$\begin{aligned}
\langle \mathbf{u} - 2\mathbf{v} - \mathbf{w}, 3\mathbf{w} - \mathbf{v} \rangle &= 3\langle \mathbf{u}, \mathbf{w} \rangle - 6\langle \mathbf{v}, \mathbf{w} \rangle - 3\langle \mathbf{w}, \mathbf{w} \rangle - \langle \mathbf{u}, \mathbf{v} \rangle + 2\langle \mathbf{v}, \mathbf{v} \rangle + \langle \mathbf{w}, \mathbf{v} \rangle \\
&= 3\langle \mathbf{u}, \mathbf{w} \rangle - 5\langle \mathbf{v}, \mathbf{w} \rangle - 3\|\mathbf{w}\|^2 - \langle \mathbf{u}, \mathbf{v} \rangle + 2\|\mathbf{v}\|^2 \\
&= 3 \cdot 0 - 5 \cdot 3 - 3 \cdot 3 - (-1) + 2 \cdot 4 \\
&= -15
\end{aligned}$$

20. (1) $\langle \mathbf{u}, \mathbf{v} + \mathbf{w} \rangle \overset{P2}{=} \langle \mathbf{v} + \mathbf{w}, \mathbf{u} \rangle \overset{P3}{=} \langle \mathbf{v}, \mathbf{u} \rangle + \langle \mathbf{w}, \mathbf{u} \rangle \overset{P2}{=} \langle \mathbf{u}, \mathbf{v} \rangle + \langle \mathbf{u}, \mathbf{w} \rangle$

(2) $\langle \mathbf{v}, r\mathbf{w} \rangle \overset{P2}{=} \langle r\mathbf{w}, \mathbf{v} \rangle \overset{P4}{=} r\langle \mathbf{w}, \mathbf{v} \rangle \overset{P2}{=} r\langle \mathbf{v}, \mathbf{w} \rangle$

(3) By (1): $\langle \mathbf{v}, \mathbf{0} \rangle = \langle \mathbf{v}, \mathbf{0} + \mathbf{0} \rangle \overset{(1)}{=} \langle \mathbf{v}, \mathbf{0} \rangle + \langle \mathbf{v}, \mathbf{0} \rangle$. Hence $\langle \mathbf{v}, \mathbf{0} \rangle = 0$. Now $\langle \mathbf{0}, \mathbf{v} \rangle = 0$ by P2.

(4) If $\mathbf{v} = \mathbf{0}$ then $\langle \mathbf{v}, \mathbf{v} \rangle = \langle \mathbf{0}, \mathbf{0} \rangle = 0$ by (3). If $\langle \mathbf{v}, \mathbf{v} \rangle = 0$ then it is impossible by P5 that $\mathbf{v} \neq \mathbf{0}$, so $\mathbf{v} = \mathbf{0}$.

22(b)
$$\begin{aligned}
\langle 3\mathbf{u} - 4\mathbf{v}, 5\mathbf{u} + \mathbf{v} \rangle &= 15\langle \mathbf{u}, \mathbf{u} \rangle + 3\langle \mathbf{u}, \mathbf{v} \rangle - 20\langle \mathbf{v}, \mathbf{u} \rangle - 4\langle \mathbf{v}, \mathbf{v} \rangle \\
&= 15\|\mathbf{u}\|^2 - 17\langle \mathbf{u}, \mathbf{v} \rangle - 4\|\mathbf{v}\|^2.
\end{aligned}$$

26(b) Here

$$\begin{aligned}
W &= \left\{ \mathbf{w} \mid \mathbf{w} \text{ in } \mathbb{R}^3 \text{ and } \mathbf{v} \cdot \mathbf{w} = 0 \right\} \\
&= \{(x, y, z) \mid x - y + 2z = 0\} \\
&= \{(s, s + 2t, t) \mid s, t \text{ in } \mathbb{R}\} \\
&= \text{span } B
\end{aligned}$$

where $B = \{(1, 1, 0), (0, 2, 1)\}$. Then B is the desired basis because B is independent $[s(1, 1, 0) + t(0, 2, 1) = (s, s + 2t, t) = (0, 0, 0)]$ implies $s = t = 0]$.

28. Write $\mathbf{u} = \mathbf{v} - \mathbf{w}$; we show that $\mathbf{u} = \mathbf{0}$. We know

$$\langle \mathbf{u}, \mathbf{v}_i \rangle = \langle \mathbf{v} - \mathbf{w}, \mathbf{v}_i \rangle = \langle \mathbf{v}, \mathbf{v}_i \rangle - \langle \mathbf{w}, \mathbf{v}_i \rangle = 0$$

for each i. As $V = \text{span}\{\mathbf{v}_1, \dots, \mathbf{v}_n\}$, write $\mathbf{u} = r\mathbf{v}_1 + \cdots + r_n\mathbf{v}_n$, r_i in $\mathbb{R}$. Then

$$\begin{aligned}
\|\mathbf{u}\|^2 = \langle \mathbf{u}, \mathbf{u} \rangle &= \langle \mathbf{u}, r_1\mathbf{v}_1 + \cdots + r_n\mathbf{v}_n \rangle \\
&= r_1\langle \mathbf{u}, \mathbf{v}_1 \rangle + \cdots + r_n\langle \mathbf{u}, \mathbf{v}_n \rangle \\
&= r_1 \cdot 0 + \cdots + r_n \cdot 0 \\
&= 0.
\end{aligned}$$

Thus, $\|\mathbf{u}\| = 0$, so $\mathbf{u} = \mathbf{0}$.

32(b) $\cos \dfrac{\mathbf{v} \cdot \mathbf{w}}{\|\mathbf{v}\| \, \|\mathbf{w}\|} = \dfrac{2+2+0+2+0}{\sqrt{16}\sqrt{9}} = \dfrac{6}{4\cdot 3} = \dfrac{1}{2}$. As $0 \le \theta \le \pi$, this implies that $\theta = \dfrac{\pi}{3}$ or $60°$.

Exercises 9.2   Orthogonal Sets of Vectors

1(b) B is an orthogonal set because (writing $\mathbf{e}_1 = \begin{bmatrix} 1 \\ 1 \\ 1 \end{bmatrix}$, $\mathbf{e}_2 = \begin{bmatrix} -1 \\ 0 \\ 1 \end{bmatrix}$ and $\mathbf{e}_3 = \begin{bmatrix} 1 \\ -6 \\ 1 \end{bmatrix}$)

$$\langle \mathbf{e}_1, \mathbf{e}_2 \rangle = \begin{bmatrix} 1 & 1 & 1 \end{bmatrix} \begin{bmatrix} 2 & 0 & 1 \\ 0 & 1 & 0 \\ 1 & 0 & 2 \end{bmatrix} \begin{bmatrix} -1 \\ 0 \\ 1 \end{bmatrix} = \begin{bmatrix} 3 & 1 & 3 \end{bmatrix} \begin{bmatrix} -1 \\ 0 \\ 1 \end{bmatrix} = 0$$

$$\langle \mathbf{e}_1, \mathbf{e}_3 \rangle = \begin{bmatrix} 1 & 1 & 1 \end{bmatrix} \begin{bmatrix} 2 & 0 & 1 \\ 0 & 1 & 0 \\ 1 & 0 & 2 \end{bmatrix} \begin{bmatrix} 1 \\ -6 \\ 1 \end{bmatrix} = \begin{bmatrix} 3 & 1 & 3 \end{bmatrix} \begin{bmatrix} 1 \\ -6 \\ 1 \end{bmatrix} = 0$$

$$\langle \mathbf{e}_2, \mathbf{e}_3 \rangle = \begin{bmatrix} -1 & 0 & 1 \end{bmatrix} \begin{bmatrix} 2 & 0 & 1 \\ 0 & 1 & 0 \\ 1 & 0 & 2 \end{bmatrix} \begin{bmatrix} 1 \\ -6 \\ 1 \end{bmatrix} = \begin{bmatrix} -1 & 0 & 1 \end{bmatrix} \begin{bmatrix} 1 \\ -6 \\ 1 \end{bmatrix} = 0.$$

Thus, B is an orthogonal basis of V and the expansion theorem gives

$$\mathbf{v} = \frac{\langle \mathbf{v}, \mathbf{e}_1 \rangle}{\|\mathbf{e}_1\|^2} \mathbf{e}_1 + \frac{\langle \mathbf{v}, \mathbf{e}_2 \rangle}{\|\mathbf{e}_2\|^2} \mathbf{e}_2 + \frac{\langle \mathbf{v}, \mathbf{e}_3 \rangle}{\|\mathbf{e}_3\|^2} \mathbf{e}_3$$

$$= \frac{3a+b+3c}{7} \mathbf{e}_1 + \frac{c-a}{2} \mathbf{e}_2 + \frac{3a-6b+3c}{42} \mathbf{e}_3$$

$$= \tfrac{1}{14}\{(6a+2b+6c)\mathbf{e}_1 + (7c-7a)\mathbf{e}_2 + (a-2b+c)\mathbf{e}_3].$$

(d) Observe first that $\left\langle \begin{bmatrix} a & b \\ c & d \end{bmatrix}, \begin{bmatrix} a' & b' \\ c' & d' \end{bmatrix} \right\rangle = aa'+bb'+cc'+dd'$. Now write $B = \{\mathbf{e}_1, \mathbf{e}_2, \mathbf{e}_3, \mathbf{e}_4\}$

where $\mathbf{e}_1 = \begin{bmatrix} 1 & 0 \\ 0 & 1 \end{bmatrix}$, $\mathbf{e}_2 = \begin{bmatrix} 1 & 0 \\ 0 & -1 \end{bmatrix}$, $\mathbf{e}_3 = \begin{bmatrix} 0 & 1 \\ 1 & 0 \end{bmatrix}$, $\mathbf{e}_4 = \begin{bmatrix} 0 & 1 \\ -1 & 0 \end{bmatrix}$. Then B is orthogonal because

$$\langle \mathbf{e}_1, \mathbf{e}_2 \rangle = 1+0+0-1 = 0 \qquad \langle \mathbf{e}_2, \mathbf{e}_3 \rangle = 0+0+0+0 = 0$$
$$\langle \mathbf{e}_1, \mathbf{e}_3 \rangle = 0+0+0+0 = 0 \qquad \langle \mathbf{e}_2, \mathbf{e}_4 \rangle = 0+0+0+0 = 0$$
$$\langle \mathbf{e}_1, \mathbf{e}_4 \rangle = 0+0+0+0 = 0 \qquad \langle \mathbf{e}_3, \mathbf{e}_4 \rangle = 0+1-1+0 = 0.$$

The expansion theorem gives

$$\mathbf{v} = \frac{\langle \mathbf{v}, \mathbf{e}_1 \rangle}{\|\mathbf{e}_1\|^2} \mathbf{e}_1 + \frac{\langle \mathbf{v}, \mathbf{e}_2 \rangle}{\|\mathbf{e}_2\|^2} \mathbf{e}_2 + \frac{\langle \mathbf{v}, \mathbf{e}_3 \rangle}{\|\mathbf{e}_3\|^2} \mathbf{e}_3 + \frac{\langle \mathbf{v}, \mathbf{e}_4 \rangle}{\|\mathbf{e}_4\|^2} \mathbf{e}_4$$

$$= \left(\frac{a+d}{2}\right) \mathbf{e}_1 + \left(\frac{a-d}{2}\right) \mathbf{e}_2 + \left(\frac{b+c}{2}\right) \mathbf{e}_3 + \left(\frac{b-c}{2}\right) \mathbf{e}_4.$$

2(b) Write $\mathbf{b}_1 = (1,1,1)$, $\mathbf{b}_2 = (1,-1,1)$, $\mathbf{b}_3 = (1,1,0)$. Then

$$
\begin{aligned}
\mathbf{e}_1 &= \mathbf{b}_1 = (1,1,1) \\
\mathbf{e}_2 &= \mathbf{b}_2 - \frac{\langle \mathbf{b}_2, \mathbf{e}_1 \rangle}{\|\mathbf{e}_1\|^2} \mathbf{e}_1 \\
&= (1,-1,1) - \tfrac{4}{6}(1,1,1) \\
&= \tfrac{1}{3}(1,-5,1) \\
\mathbf{e}_3 &= \mathbf{b}_3 - \frac{\langle \mathbf{b}_3, \mathbf{e}_1 \rangle}{\|\mathbf{e}_1\|^2} \mathbf{e}_1 - \frac{\langle \mathbf{b}_3, \mathbf{e}_2 \rangle}{\|\mathbf{e}_2\|^2} \mathbf{e}_2 \\
&= (1,1,0) - \tfrac{3}{6}(1,1,1) - \frac{\tfrac{1}{3}(-3)}{\tfrac{1}{9}(30)} \cdot \tfrac{1}{3}(1,-5,1) \\
&= \tfrac{1}{10}[(10,10,9) - (5,5,5) + (1,-5,1)] \\
&= \tfrac{1}{5}(3,0,-2).
\end{aligned}
$$

The Gram-Schmidt algorithm gives $\{\mathbf{e}_1, \mathbf{e}_2, \mathbf{e}_3\}$; it may be convenient to avoid fractions and use $\{(1,1,1), (1,-5,1), (3,0,-2)\}$.

3(b) Note that $\left\langle \begin{bmatrix} a & b \\ c & d \end{bmatrix}, \begin{bmatrix} a' & b' \\ c' & d' \end{bmatrix} \right\rangle = aa' + bb' + cc' + dd'$. For convenience write $\mathbf{b}_1 = \begin{bmatrix} 1 & 1 \\ 0 & 1 \end{bmatrix}$, $\mathbf{b}_2 = \begin{bmatrix} 1 & 0 \\ 1 & 1 \end{bmatrix}$, $\mathbf{b}_3 = \begin{bmatrix} 1 & 0 \\ 0 & 1 \end{bmatrix}$, $\mathbf{b}_4 = \begin{bmatrix} 1 & 0 \\ 0 & 0 \end{bmatrix}$. Then:

$$
\begin{aligned}
\mathbf{e}_1 &= \mathbf{b}_1 = \begin{bmatrix} 1 & 1 \\ 0 & 1 \end{bmatrix} \\
\mathbf{e}_2 &= \mathbf{b}_2 - \frac{\langle \mathbf{b}_2, \mathbf{e}_1 \rangle}{\|\mathbf{e}_1\|^2} \mathbf{e}_1 \\
&= \begin{bmatrix} 1 & 0 \\ 1 & 1 \end{bmatrix} - \tfrac{2}{3} \begin{bmatrix} 1 & 1 \\ 0 & 1 \end{bmatrix} = \tfrac{1}{3} \begin{bmatrix} 1 & -2 \\ 3 & 1 \end{bmatrix}.
\end{aligned}
$$

For the rest of the algorithm, use $\mathbf{e}_2 = \begin{bmatrix} 1 & -2 \\ 3 & 1 \end{bmatrix}$, the result is the same.

$$
\begin{aligned}
\mathbf{e}_3 &= \mathbf{b}_3 - \frac{\langle \mathbf{b}_3, \mathbf{e}_1 \rangle}{\|\mathbf{e}_1\|^2} \mathbf{e}_1 - \frac{\langle \mathbf{b}_3, \mathbf{e}_2 \rangle}{\|\mathbf{e}_2\|^2} \mathbf{e}_2 \\
&= \begin{bmatrix} 1 & 0 \\ 0 & 1 \end{bmatrix} - \tfrac{2}{3} \begin{bmatrix} 1 & 1 \\ 0 & 1 \end{bmatrix} - \tfrac{2}{15} \begin{bmatrix} 1 & -2 \\ 3 & 1 \end{bmatrix} \\
&= \tfrac{1}{5} \begin{bmatrix} 1 & -2 \\ -2 & 1 \end{bmatrix}.
\end{aligned}
$$

Now use $\mathbf{e}_4 = \begin{bmatrix} 1 & -2 \\ -2 & 1 \end{bmatrix}$, the results are unchanged.

$$\mathbf{e}_4 = \mathbf{b}_4 - \frac{\langle \mathbf{b}_4, \mathbf{e}_1 \rangle}{\|\mathbf{e}_1\|^2} \mathbf{e}_1 - \frac{\langle \mathbf{b}_4, \mathbf{e}_2 \rangle}{\|\mathbf{e}_2\|^2} \mathbf{e}_2 - \frac{\langle \mathbf{b}_4, \mathbf{e}_3 \rangle}{\|\mathbf{e}_3\|^2} \mathbf{e}_3$$

$$= \begin{bmatrix} 1 & 0 \\ 0 & 0 \end{bmatrix} - \tfrac{1}{3} \begin{bmatrix} 1 & 1 \\ 0 & 1 \end{bmatrix} - \tfrac{1}{15} \begin{bmatrix} 1 & -2 \\ 3 & 1 \end{bmatrix} - \tfrac{1}{10} \begin{bmatrix} 1 & -2 \\ -2 & 1 \end{bmatrix}$$

$$= \tfrac{1}{2} \begin{bmatrix} 1 & 0 \\ 0 & -1 \end{bmatrix} .$$

Use $\mathbf{e}_4 = \begin{bmatrix} 1 & 0 \\ 0 & -1 \end{bmatrix}$ for convenience. Hence, finally, the Gram-Schmidt algorithm gives the

orthogonal basis $\left\{ \begin{bmatrix} 1 & 1 \\ 0 & 1 \end{bmatrix}, \begin{bmatrix} 1 & -2 \\ 3 & 1 \end{bmatrix}, \begin{bmatrix} 1 & -2 \\ -2 & 1 \end{bmatrix}, \begin{bmatrix} 1 & 0 \\ 0 & -1 \end{bmatrix} \right\} .$

4(b) $\mathbf{e}_1 = 1$

$\mathbf{e}_2 = x - \dfrac{\langle x, \mathbf{e}_1 \rangle}{\|\mathbf{e}_1\|^2} \mathbf{e}_1 = x - \tfrac{2}{2} \cdot 1 = x - 1$

$\mathbf{e}_3 = x^2 - \dfrac{\langle x^2, \mathbf{e}_1 \rangle}{\|\mathbf{e}_1\|^2} \mathbf{e}_1 - \dfrac{\langle x^2, \mathbf{e}_2 \rangle}{\|\mathbf{e}_2\|^2} \mathbf{e}_2 = x^2 - \tfrac{8/3}{2} \cdot 1 - \tfrac{4/3}{2/3} \cdot (x - 1) = x^2 - 2x + \tfrac{2}{3}.$

6(b) $[x \quad y \quad z \quad w]$ is in $U^\perp$ if and only if

$$x + y = [x \quad y \quad z \quad w] \cdot [1 \quad 1 \quad 0 \quad 0] = 0.$$

Thus $y = -x$ and

$$U^\perp = \{[x \quad -x \quad z \quad w] \mid x, z, w \text{ in } \mathbb{R}\}$$
$$= \text{span} \{[1 \quad -1 \quad 0 \quad 0], [0 \quad 0 \quad 1 \quad 0], [0 \quad 0 \quad 0 \quad 1]\} .$$

Hence $\dim U^\perp = 3$ and $U = 1$.

(d) If $p(x) = a + bx + cx^2$, p is in $U^\perp$ if and only if

$$0 = \langle p, x \rangle = \int_0^1 (a + bx + cx^2) x \, dx = \tfrac{a}{2} + \tfrac{b}{3} + \tfrac{c}{4}.$$

Thus $s = 2s + t$, $b = -3s$, $c = -2t$ where s and t are in $\mathbb{R}$, so $p(x) = (2s + t) - 3sx - 2tx^2$. Hence, $U^\perp = \text{span}\{2 - 3x, 1 - 2x^2\}$ and $\dim U^\perp = 2$, $\dim U = 1$.

(f) $\begin{bmatrix} a & b \\ c & d \end{bmatrix}$ is in U if and ony if

$$0 = \left\langle \begin{bmatrix} a & b \\ c & d \end{bmatrix}, \begin{bmatrix} 1 & 1 \\ 0 & 0 \end{bmatrix} \right\rangle = a + b$$

$$0 = \left\langle \begin{bmatrix} a & b \\ c & d \end{bmatrix}, \begin{bmatrix} 1 & 0 \\ 1 & 0 \end{bmatrix} \right\rangle = a + c$$

$$0 = \left\langle \begin{bmatrix} a & b \\ c & d \end{bmatrix}, \begin{bmatrix} 1 & 0 \\ 1 & 1 \end{bmatrix} \right\rangle = a + c + d.$$

The solution $d = 0$, $b = c = -a$, so $U^\perp = \left\{ \begin{bmatrix} a & -a \\ -a & 0 \end{bmatrix} \mid a \text{ in } \mathbb{R} \right\} = \text{span} \left\{ \begin{bmatrix} 1 & -1 \\ -1 & 0 \end{bmatrix} \right\}$.
Thus $\dim U^\perp = 1$ and $\dim U = 3$.

7(b) Write $B_1 = \begin{bmatrix} 1 & 0 \\ 0 & 1 \end{bmatrix}$, $B_2 = \begin{bmatrix} 1 & 1 \\ 1 & -1 \end{bmatrix}$, and $B_3 = \begin{bmatrix} 1 & 1 \\ 0 & 0 \end{bmatrix}$. Then $\{B_1, B_2, B_3\}$ is independent but not orthogonal. The Gram-Schmidt algorithm gives

$$E_1 = B_1 = \begin{bmatrix} 1 & 0 \\ 0 & 1 \end{bmatrix}$$

$$E_2 = B_2 - \frac{\langle B_2, E_1 \rangle}{\|E_1\|^2} E_1 = \begin{bmatrix} 1 & 1 \\ 1 & -1 \end{bmatrix} - \frac{0}{2} \begin{bmatrix} 1 & 0 \\ 0 & 1 \end{bmatrix} = \begin{bmatrix} 1 & 1 \\ 1 & -1 \end{bmatrix}$$

$$E_3 = B_3 - \frac{\langle B_3, E_1 \rangle}{\|E_1\|^2} E_1 - \frac{\langle B_3, E_2 \rangle}{\|E_2\|^2} E_2$$

$$= \begin{bmatrix} 1 & 1 \\ 0 & 0 \end{bmatrix} - \frac{1}{2} \begin{bmatrix} 1 & 0 \\ 0 & 1 \end{bmatrix} - \frac{2}{4} \begin{bmatrix} 1 & 1 \\ 1 & -1 \end{bmatrix}$$

$$= \frac{1}{2} \begin{bmatrix} 0 & 1 \\ -1 & 0 \end{bmatrix}.$$

If $E_3' = \begin{bmatrix} 0 & 1 \\ -1 & 1 \end{bmatrix}$ then $\{E_1, E_2, E_3'\}$ is an orthogonal basis of U. If $A = \begin{bmatrix} 2 & 1 \\ 3 & 2 \end{bmatrix}$ then

$$\text{proj}_U(A) = \frac{\langle A, E_1 \rangle}{\|E_1\|^2} E_1 + \frac{\langle A, E_2 \rangle}{\|E_2\|^2} E_2 + \frac{\langle A, E_3' \rangle}{\|E_3'\|^2} E_3'$$

$$= \frac{4}{2} \begin{bmatrix} 1 & 0 \\ 0 & 1 \end{bmatrix} + \frac{4}{4} \begin{bmatrix} 1 & 1 \\ 1 & -1 \end{bmatrix} + \frac{-2}{2} \begin{bmatrix} 0 & 1 \\ -1 & 0 \end{bmatrix}$$

$$= \begin{bmatrix} 3 & 0 \\ 2 & 1 \end{bmatrix}$$

is the vector in U closest to A.

9(b) $\{1, 2x - 1\}$ is an orthogonal basis of U because $\langle 1, 2x - 1 \rangle = f_0^1(2x - 1)\, dx = 0$. Thus

$$\text{proj}_U(x^2 + 1) = \frac{\langle x^2 + 1, 1 \rangle}{\|1\|^2} 1 + \frac{\langle x^2 + 1, 2x - 1 \rangle}{\|2x - 1\|^2} (2x - 1)$$

$$= \frac{\frac{4}{3}}{1} 1 + \frac{\frac{1}{6}}{\frac{1}{3}} (2x - 1)$$

$$= x + \tfrac{5}{6}.$$

Hence, $x^2 + 1 = (x + \tfrac{5}{6}) + (x^2 - x + \tfrac{1}{6})$ is the required decomposition. Check: $x^2 - x + \tfrac{1}{6}$ is in $U^\perp$ because

$$\left\langle x^2 - x + \tfrac{1}{6}, 1 \right\rangle = \int_0^1 \left(x^2 - x + \tfrac{1}{6}\right) dx = 0$$

$$\left\langle x^2 - x + \tfrac{1}{6}, 2x - 1 \right\rangle = \int_0^1 \left(x^2 - x + \tfrac{1}{6}\right)(2x - 1)dx = 0.$$

11(b) We have $\langle \mathbf{v}+\mathbf{w}, \mathbf{v}-\mathbf{w} \rangle = \langle \mathbf{v}, \mathbf{v} \rangle - \langle \mathbf{v}, \mathbf{w} \rangle + \langle \mathbf{w}, \mathbf{v} \rangle - \langle \mathbf{u}, \mathbf{u} \rangle = \|\mathbf{v}\|^2 - \|\mathbf{w}\|^2$. Hence $\langle \mathbf{v}+\mathbf{w}, \mathbf{v}-\mathbf{u} \rangle = 0$ if and only if $\|\mathbf{v}\| = \|\mathbf{w}\|$. This is what we wanted.

14(b) If $\mathbf{v}$ is in $U^\perp$ then $\langle \mathbf{v}, \mathbf{w} \rangle = 0$ for all $\mathbf{u}$ in U. In particular, $\langle \mathbf{v}, \mathbf{u}_i \rangle = 0$ for $1 \leq i \leq n$, so $\mathbf{v}$ is in $\{\mathbf{u}_1, \ldots, \mathbf{u}_m\}^\perp$. This shows that $U^\perp \subseteq \{\mathbf{u}_1, \ldots, \mathbf{u}_m\}^\perp$. Conversely, if $\mathbf{v}$ is in $\{\mathbf{u}_1, \ldots, \mathbf{u}_m\}^\perp$ then $\langle \mathbf{v}, \mathbf{u}_i \rangle = 0$ for each i. If $\mathbf{u}$ is in U, write $\mathbf{u} = r_1\mathbf{u}_1 + \cdots + r_m\mathbf{u}_m$, r_i in $\mathbb{R}$. Then

$$\langle \mathbf{v}, \mathbf{u} \rangle = \langle \mathbf{v}, r_1\mathbf{u}_1 + \cdots + r_m\mathbf{u}_m \rangle$$

$$= r_1\langle \mathbf{v}, \mathbf{u}_1 \rangle + \cdots + r_m\langle \mathbf{v}, \mathbf{u}_m \rangle$$

$$= r_1 \cdot 0 + \cdots + r_m \cdot 0$$

$$= 0.$$

As $\mathbf{u}$ was arbitrary in U, this shows that a v is in $U^\perp$; that is $\{\mathbf{u}_1, \ldots, \mathbf{u}_m\}^\perp \subseteq U^\perp$.

18(b) Write $\mathbf{e}_1 = (3, -2, 5)$ and $\mathbf{e}_2 = (-1, 1, 1)$, write $B = \{\mathbf{e}_1, \mathbf{e}_2\}$, and write $U = \text{span } B$. Then

B is orthogonal and so is an orthogonal basis of U. Thus if $\mathbf{v} = (-5, 4, -3)$ then

$$
\begin{aligned}
\text{proj}_U(\mathbf{v}) &= \frac{\mathbf{v} \cdot \mathbf{e}_1}{\|\mathbf{e}_1\|^2}\mathbf{e}_1 + \frac{\mathbf{v} \cdot \mathbf{e}_2}{\|\mathbf{e}_2\|^2}\mathbf{e}_2 \\
&= \tfrac{-38}{38}(3, -2, 5) + \tfrac{6}{5}(-1, 1, 1) \\
&= (-5, 4, -3) \\
&= \mathbf{v}.
\end{aligned}
$$

Thus, $\mathbf{v}$ is in U. However, if $\mathbf{v}_1 = (-1, 0, 2)$ then

$$
\begin{aligned}
\text{proj}_U(\mathbf{v}_1) &= \frac{\mathbf{v}_1 \cdot \mathbf{e}_1}{\|\mathbf{e}_2\|^2}\mathbf{e}_1 + \frac{\mathbf{v} \cdot \mathbf{e}_2}{\|\mathbf{e}_2\|^2}\mathbf{e}_2 \\
&= \tfrac{7}{38}(3, -2, 5) + \tfrac{3}{3}(-1, 1, 1) \\
&= \tfrac{1}{38}(-17, 24, 73).
\end{aligned}
$$

As $\mathbf{v}_1 \neq \text{proj}_U(\mathbf{v}_1)$, $\mathbf{v}_1$ is not in U by (a).

Exercises 9.3   Orthogonal Diagonalization

1(b) If $B = \{E_1, E_2, E_3, E_4\}$ where $E_1 = \begin{bmatrix} 1 & 0 \\ 0 & 0 \end{bmatrix}$, $E_2 = \begin{bmatrix} 0 & 1 \\ 0 & 0 \end{bmatrix}$, $E_3 = \begin{bmatrix} 0 & 0 \\ 1 & 0 \end{bmatrix}$ and $E_4 = \begin{bmatrix} 0 & 0 \\ 0 & 1 \end{bmatrix}$, then

$$
T(E_1) = \begin{bmatrix} -1 & 0 \\ 1 & 0 \end{bmatrix} = -E_1 + E_3
$$

$$
T(E_2) = \begin{bmatrix} 0 & -1 \\ 0 & 1 \end{bmatrix} = -E_2 + E_4
$$

$$
T(E_3) = \begin{bmatrix} 1 & 0 \\ 2 & 0 \end{bmatrix} = E_1 + 2E_3
$$

$$
T(E_4) = \begin{bmatrix} 0 & 1 \\ 0 & 2 \end{bmatrix} = E_2 + 2E_4.
$$

Hence,

$$
\begin{aligned}
M_B(T) &= [C_B[T(E_1)] \quad C_B[T(E_2)] \quad C_B[T(E_3)] \quad C_B[T(E_4)]] \\
&= \begin{bmatrix} -1 & 0 & 1 & 0 \\ 0 & -1 & 0 & 1 \\ 1 & 0 & 2 & 0 \\ 0 & 1 & 0 & 2 \end{bmatrix}
\end{aligned}
$$

As $M_B(T)$ is symmetric, T is a symmetric operator.

4(b) If T is symmetric then $\langle \mathbf{v}, T(\mathbf{w}) \rangle = \langle T(\mathbf{v}), \mathbf{w} \rangle$ holds for all $\mathbf{v}$ and $\mathbf{w}$ in V. Given r in $\mathbb{R}$:

$$\langle \mathbf{v}, (rT)(\mathbf{w}) \rangle = \langle \mathbf{v}, rT(\mathbf{w}) \rangle = r\langle \mathbf{v}, T(\mathbf{w}) \rangle = r\langle T(\mathbf{v}), \mathbf{w} \rangle = \langle rT(\mathbf{v}), \mathbf{w} \rangle = \langle (rT)(\mathbf{v}), \mathbf{w} \rangle$$

for all $\mathbf{v}$ and $\mathbf{w}$ in V. This shows that rT is symmetric.

(d) Given $\mathbf{v}$ and $\mathbf{w}$, write $T^{-1}(\mathbf{v}) = \mathbf{v}_1$ and $T^{-1}(\mathbf{w}) = \mathbf{w}_1$. Then

$$\langle T^{-1}(\mathbf{v}), \mathbf{w} \rangle = \langle \mathbf{v}_1, T(\mathbf{w}_1) \rangle = \langle T(\mathbf{v}_1), \mathbf{w}_1 \rangle = \langle \mathbf{v}, T^{-1}(\mathbf{w}) \rangle.$$

5(b) If $B_0 = \{(1,0,0), (0,1,0), (0,0,1)\}$ is the standard basis of $\mathbb{R}^3$:

$$\begin{aligned} M_{B_0}(T) &= [C_{B_0}[T(1,0,0)] \quad C_{B_0}[T(0,1,0)] \quad C_{B_0}[T(0,0,1)]] \\ &= [C_{B_0}(7,-1,0) \quad C_{B_0}(-1,7,0) \quad C_{B_0}(0,0,2)] \\ &= \begin{bmatrix} 7 & -1 & 0 \\ -1 & 7 & 0 \\ 0 & 0 & 2 \end{bmatrix}. \end{aligned}$$

Thus, $c_T(x) = \begin{vmatrix} x-7 & 1 & 0 \\ 1 & x-7 & 0 \\ 0 & 0 & x-2 \end{vmatrix} = (x-6)(x-8)(x-2)$ so the eigenvalues are $\lambda_1 = 6$,

$\lambda_2 = 8$, and $\lambda_3 = 2$, (real as $M_{B_0}(T)$ is symmetric). Corresponding (orthogonal) eigenvectors

are $X_1 = \begin{bmatrix} 1 \\ 1 \\ 0 \end{bmatrix}$, $X_2 = \begin{bmatrix} 1 \\ -1 \\ 0 \end{bmatrix}$, and $X_3 = \begin{bmatrix} 0 \\ 0 \\ 1 \end{bmatrix}$, so

$$\left\{ \frac{1}{\sqrt{2}} \begin{bmatrix} 1 \\ 1 \\ 0 \end{bmatrix}, \frac{1}{\sqrt{2}} \begin{bmatrix} 1 \\ -1 \\ 0 \end{bmatrix}, \begin{bmatrix} 0 \\ 0 \\ 1 \end{bmatrix} \right\}$$

is an orthonormal basis of eigenvectors of $M_{B_0(T)}$. These vectors are equal to $C_{B_0} \left[\frac{1}{\sqrt{2}}(1,1,0) \right]$,

$C_{B_0} \left[\frac{1}{\sqrt{2}}(1,-1,0) \right]$, and $C_{B_0}[(0,0,1)]$ respectively, so

$$\left\{ \frac{1}{\sqrt{2}}(1,1,0), \frac{1}{\sqrt{2}}(1,-1,0), (0,0,1) \right\}$$

is an orthonormal basis of eigenvectors of T.

(d) If $B_0 = \{1, x, x^2\}$ then

$$\begin{aligned} M_{B_0}(T) &= \begin{bmatrix} C_{B_0}[T(1)] & C_{B_0}[T(x)] & C_{B_0}[T(x^2)] \end{bmatrix} \\ &= \begin{bmatrix} C_{B_0}(-1+x^2) & C_{B_0}(3x) & C_{B_0}(1-x^2) \end{bmatrix} \\ &= \begin{bmatrix} -1 & 0 & 1 \\ 0 & 3 & 0 \\ 1 & 0 & -1 \end{bmatrix}. \end{aligned}$$

Hence, $c_T(x) = \begin{vmatrix} x+1 & 0 & -1 \\ 0 & x-3 & 0 \\ -1 & 0 & x+1 \end{vmatrix} = (x-3)x(x+2)$ so the (real) eigenvalues are $\lambda_1 = 3$,

$\lambda_2 = 0$, $\lambda_3 = -2$. Corresponding (orthogonal) eigenvectors are $X_1 = \begin{bmatrix} 0 \\ 1 \\ 0 \end{bmatrix}$, $X_2 = \begin{bmatrix} 1 \\ 0 \\ 1 \end{bmatrix}$,

$X_3 = \begin{bmatrix} 1 \\ 0 \\ -1 \end{bmatrix}$, so

$$\left\{ \begin{bmatrix} 0 \\ 1 \\ 0 \end{bmatrix}, \frac{1}{\sqrt{2}} \begin{bmatrix} 1 \\ 0 \\ 1 \end{bmatrix}, \frac{1}{\sqrt{2}} \begin{bmatrix} 1 \\ 0 \\ -1 \end{bmatrix} \right\}$$

is an orthonormal basis of eigenvectors of $M_{B_0}(T)$. These have the form $C_{B_0}(x)$, $C_{B_0}\left[\frac{1}{\sqrt{2}}(1+x^2)\right]$, and $C_{B_0}\left[\frac{1}{\sqrt{2}}(1-x^2)\right]$, respectively, so

$$\left\{ x, \tfrac{1}{\sqrt{2}}(1+x^2), \tfrac{1}{\sqrt{2}}(1-x^2) \right\}$$

is an orthonormal basis of eigenvectors of T.

7(b) Write $A = \begin{bmatrix} a & b \\ c & d \end{bmatrix}$ and compute:

$$M_B(T) = \left[C_B \left(T \begin{bmatrix} 1 & 0 \\ 0 & 0 \end{bmatrix} \right) \quad C_B \left(T \begin{bmatrix} 0 & 0 \\ 1 & 0 \end{bmatrix} \right) \quad C_B \left(T \begin{bmatrix} 0 & 1 \\ 0 & 0 \end{bmatrix} \right) \quad C_B \left(T \begin{bmatrix} 0 & 0 \\ 0 & 1 \end{bmatrix} \right) \right]$$

$$= \left[C_B \begin{bmatrix} a & 0 \\ c & 0 \end{bmatrix} \quad C_B \begin{bmatrix} b & 0 \\ d & 0 \end{bmatrix} \quad C_B \begin{bmatrix} 0 & a \\ 0 & c \end{bmatrix} \quad C_B \begin{bmatrix} 0 & b \\ 0 & d \end{bmatrix} \right]$$

$$= \begin{bmatrix} a & b & 0 & 0 \\ c & d & 0 & 0 \\ 0 & 0 & a & b \\ 0 & 0 & c & d \end{bmatrix}$$

$$= \begin{bmatrix} A & 0 \\ 0 & A \end{bmatrix}.$$

Hence,

$$c_T(x) = \det[xI - M_B(T)] = \det\left\{\begin{bmatrix} xI & 0 \\ 0 & xI \end{bmatrix} - \begin{bmatrix} A & 0 \\ 0 & A \end{bmatrix}\right\}$$

$$= \det\begin{bmatrix} xI - A & 0 \\ 0 & xI - A \end{bmatrix} = \det(xI - A)\cdot\det(xI - A) = [c_A(x)]^2.$$

14(c) We have

$$M_B(T') = \begin{bmatrix} C_B(T'(\mathbf{e}_1)) & C_B(T'(\mathbf{e}_2)) & \cdots C_B(T'(\mathbf{e}_n)) \end{bmatrix}.$$

Hence, column j of $M_B(T')$ is

$$C_B(T'(\mathbf{e}_j)) = \begin{bmatrix} \langle \mathbf{e}_j, T(\mathbf{e}_1)\rangle \\ \langle \mathbf{e}_j, T(\mathbf{e}_2)\rangle \\ \vdots \\ \langle \mathbf{e}_j, T(\mathbf{e}_n)\rangle \end{bmatrix}$$

by the definition of T'. Hence the (i,j)-entry of $M_B(T')$ is $\langle \mathbf{e}_j, T(\mathbf{e}_i)\rangle$. But this is the (j,i)-entry of $M_B(T') = \langle \mathbf{e}_j, T(\mathbf{e}_i)\rangle$. But this is the (j,i)-entry of $M_B(T)$ by Theorem 2. Thus, $M_B(T')$ is the transpose of $M_B(T)$.

Exercises 9.4   Isometries

2(b) If B_0 is the standard basis of $\mathbb{R}^2$, then

$$M_{B_0}(T) = \begin{bmatrix} C_{B_0}\left(T\begin{bmatrix} 1 \\ 0 \end{bmatrix}\right) & C_{B_0}\left(T\begin{bmatrix} 0 \\ 1 \end{bmatrix}\right) \end{bmatrix}$$

$$= \begin{bmatrix} C_{B_0}\begin{bmatrix} 0 \\ -1 \end{bmatrix} & C_{B_0}\begin{bmatrix} -1 \\ 0 \end{bmatrix} \end{bmatrix}$$

$$= \begin{bmatrix} 0 & -1 \\ -1 & 0 \end{bmatrix}$$

This is a rotation through π by (the discussion following) Theorem 3. This can also be seen directly from the diagram.

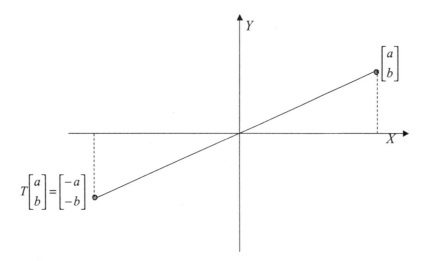

(d) If B_0 is the standard basis of $\mathbb{R}^2$, then

$$M_{B_0}(T) = \left[C_{B_0}\left(T\begin{bmatrix} 1 \\ 0 \end{bmatrix} \right) \quad C_{B_0}\left(\begin{bmatrix} 0 \\ 1 \end{bmatrix} \right) \right]$$

$$= \left[C_{B_0}\begin{bmatrix} 0 \\ -1 \end{bmatrix} \quad C_{B_0}\begin{bmatrix} -1 \\ 0 \end{bmatrix} \right]$$

$$= \begin{bmatrix} 0 & -1 \\ -1 & 0 \end{bmatrix}.$$

Thus $\det T = -1$ so T is a reflection. The vector $\mathbf{d} = \begin{bmatrix} -1 \\ 1 \end{bmatrix}$ is an eigenvector corresonding to the eigenvalue 1, so the fixed line is $\mathbb{R}\mathbf{d}$. This has equation $y = -x$. Again, this can be seen directly from the diagram.

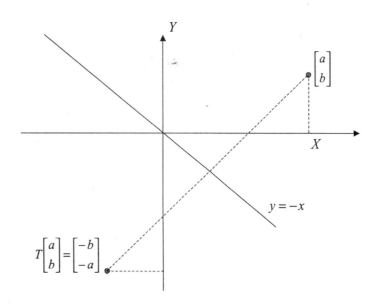

(f) If B_0 is the standard basis of $\mathbb{R}^2$, then

$$M_{B_0}(T) = \left[C_{B_0}\left(T\begin{bmatrix} 1 \\ 0 \end{bmatrix} \right) \quad C_{B_0}\left(T\begin{bmatrix} 0 \\ 1 \end{bmatrix} \right) \right]$$

$$= \left[C_{B_0}\left(\frac{1}{\sqrt{2}}\begin{bmatrix} 1 \\ 1 \end{bmatrix} \right) \quad C_{B_0}\left(\frac{1}{\sqrt{2}}\begin{bmatrix} -1 \\ 1 \end{bmatrix} \right) \right]$$

$$= \frac{1}{\sqrt{2}}\begin{bmatrix} 1 & -1 \\ 1 & 1 \end{bmatrix}.$$

Hence, $\det T = 1$ so T is a rotation. Indeed, (the discussion following) Theorem 3 shows that T is a rotation through an angle θ where $\cos\theta = \frac{1}{\sqrt{2}}$, $\sin\theta = \frac{1}{\sqrt{2}}$; that is $\theta = \frac{\pi}{4}$.

3(b) If B_0 is the standard basis of $\mathbb{R}^3$, then

$$M_{B_0}(T) = \left[C_{B_0}\left(T\begin{bmatrix} 1 \\ 0 \\ 0 \end{bmatrix} \right) \quad C_{B_0}\left(T\begin{bmatrix} 0 \\ 1 \\ 0 \end{bmatrix} \right) \quad C_{B_0}\left(T\begin{bmatrix} 0 \\ 0 \\ 1 \end{bmatrix} \right) \right]$$

$$= \left[C_{B_0}\left(\frac{1}{2}\begin{bmatrix} -1 \\ \sqrt{3} \\ 0 \end{bmatrix} \right) \quad C_{B_0}\left(\frac{1}{2}\begin{bmatrix} 0 \\ 0 \\ 2 \end{bmatrix} \right) \quad C_{B_0}\left(\frac{1}{2}\begin{bmatrix} \sqrt{3} \\ 1 \\ 0 \end{bmatrix} \right) \right]$$

$$= \frac{1}{2}\begin{bmatrix} -1 & 0 & \sqrt{3} \\ \sqrt{3} & 0 & 1 \\ 0 & 2 & 0 \end{bmatrix}.$$

Hence,

$$c_T(x) = \begin{vmatrix} x+\frac{1}{2} & 0 & -\frac{\sqrt{3}}{2} \\ -\frac{\sqrt{3}}{2} & x & -\frac{1}{2} \\ 0 & -1 & x \end{vmatrix} = \begin{vmatrix} x+\frac{1}{2} & 0 & -\frac{\sqrt{3}}{2} \\ -\frac{\sqrt{3}}{2} & 0 & x^2-\frac{1}{2} \\ 0 & -1 & x \end{vmatrix} = \begin{vmatrix} x+\frac{1}{2} & -\frac{\sqrt{3}}{2} \\ -\frac{\sqrt{3}}{2} & x^2-\frac{1}{2} \end{vmatrix} = (x-1)\left(x^2+\tfrac{3}{2}x+1\right).$$

Hence, we are in (1) of Table 8.1 so T is a rotation about the line $\mathbb{R}E$ with direction vector

$$E = \begin{bmatrix} 1 \\ \sqrt{3} \\ \sqrt{3} \end{bmatrix}, \text{ where } E \text{ is an eigenvector corresponding to the eigenvalue 1.}$$

3(d) If B_0 is the standard basis of $\mathbb{R}^3$, then

$$M_{B_0}(T) = \left[C_{B_0}\left(T \begin{bmatrix} 1 \\ 0 \\ 0 \end{bmatrix} \right) \quad C_{B_0}\left(T \begin{bmatrix} 0 \\ 1 \\ 0 \end{bmatrix} \right) \quad C_{B_0}\left(T \begin{bmatrix} 0 \\ 0 \\ 1 \end{bmatrix} \right) \right]$$

$$= \left[C_{B_0} \begin{bmatrix} 1 \\ 0 \\ 0 \end{bmatrix} \quad C_{B_0} \begin{bmatrix} 0 \\ -1 \\ 0 \end{bmatrix} \quad C_{B_0} \begin{bmatrix} 0 \\ 0 \\ -1 \end{bmatrix} \right]$$

$$= \begin{bmatrix} 1 & 0 & 0 \\ 0 & -1 & 0 \\ 0 & 0 & -1 \end{bmatrix}.$$

Hence $c_T(x) = (x-1)(x+1)^2$. So we are in case (4) of Table 8.1. Then $E = \begin{bmatrix} 1 \\ 0 \\ 0 \end{bmatrix}$ is an

eigenvector corresponding to 1, so T is a rotation of π about the line $\mathbb{R}E$ with direction vector E, that is the X-axis.

3(f) If B_0 is the standard basis of $\mathbb{R}^3$, then

$$M_{B_0}(T) = \left[C_{B_0}\left(T \begin{bmatrix} 1 \\ 0 \\ 0 \end{bmatrix} \right) \quad C_{B_0}\left(T \begin{bmatrix} 0 \\ 1 \\ 0 \end{bmatrix} \right) \quad C_{B_0}\left(T \begin{bmatrix} 0 \\ 0 \\ 1 \end{bmatrix} \right) \right]$$

$$= \left[C_{B_0}\left(\frac{1}{\sqrt{2}} \begin{bmatrix} 1 \\ 0 \\ -1 \end{bmatrix} \right) \quad C_{B_0}\left(\frac{1}{\sqrt{2}} \begin{bmatrix} 0 \\ -\sqrt{2} \\ 0 \end{bmatrix} \right) \quad C_{B_0}\left(\frac{1}{\sqrt{2}} \begin{bmatrix} 1 \\ 0 \\ 1 \end{bmatrix} \right) \right]$$

$$= \frac{1}{\sqrt{2}} \begin{bmatrix} 1 & 0 & 1 \\ 0 & -\sqrt{2} & 0 \\ -1 & 0 & 1 \end{bmatrix}.$$

Hence,

$$c_T(x) = \begin{vmatrix} x - \frac{1}{\sqrt{2}} & 0 & -\frac{1}{\sqrt{2}} \\ 0 & x+1 & 0 \\ \frac{1}{\sqrt{2}} & 0 & x - \frac{1}{\sqrt{2}} \end{vmatrix} = (x+1) \begin{vmatrix} x - \frac{1}{\sqrt{2}} & -\frac{1}{\sqrt{2}} \\ \frac{1}{\sqrt{2}} & x - \frac{1}{\sqrt{2}} \end{vmatrix} = (x+1)(x^2 - \sqrt{2}x + 1).$$

Thus we are in case (2) of Table 8.1. Now $E = \begin{bmatrix} 0 \\ 1 \\ 0 \end{bmatrix}$ is an eigenvector corresponding to the eigenvalue -1, so T is rotation $\left(\text{of } \frac{3\pi}{4}\right)$ about the line $\mathbb{R}E$ (the Y-axis) followed by a reflection in the plane $(\mathbb{R}E)^{\perp}$ — the XZ-plane.

6(b) If S and T are isometries, then $\|S(\mathbf{v})\| = \|\mathbf{v}\| = \|T(\mathbf{v})\|$ for all $\mathbf{v}$ in V. Hence,

$$\|ST(\mathbf{v})\| = \|S(T(\mathbf{v}))\| = \|T(\mathbf{v})\| = \|\mathbf{v}\|$$

for all $\mathbf{v}$, whence ST is an isometry.

8. If θ is the angle between $\mathbf{u}$ and $\mathbf{v}$ then $\cos\theta = \dfrac{\langle \mathbf{u}, \mathbf{v} \rangle}{\|\mathbf{u}\| \, \|\mathbf{v}\|}$ by Exercise 32, §9.1. We have $T(\mathbf{u}) \neq \mathbf{0} \neq T(\mathbf{v})$ as T is one-to-one, so the angle φ between $T(\mathbf{u})$ and $T(\mathbf{v})$ is given by

$$\cos\varphi = \frac{\langle T(\mathbf{u}), T(\mathbf{v}) \rangle}{\|T(\mathbf{u})\| \, \|T(\mathbf{v})\|} = \frac{\langle \mathbf{u}, \mathbf{v} \rangle}{\|\mathbf{u}\| \, \|\mathbf{v}\|}$$

using Theorem 1. Hence, $\cos\varphi = \cos\theta$ so $\theta = \varphi$ (as $0 \leq \theta \leq \pi$ and $0 \leq \varphi \leq \pi$).

13(d) As in the hint, use (b) twice:

$$\langle [T(r\mathbf{v}) - rT(\mathbf{v})], T(\mathbf{w}) \rangle = \langle T(r\mathbf{v}), T(\mathbf{w}) \rangle - r\langle T(\mathbf{v}), T(\mathbf{w}) \rangle$$
$$= \langle r\mathbf{v}, \mathbf{w} \rangle - r\langle \mathbf{v}, \mathbf{w} \rangle = 0.$$

Hence, $T(r\mathbf{v}) - rT(\mathbf{v})$ is orthogonal to $T(\mathbf{w})$ for all $\mathbf{w}$. By (c), $T(r\mathbf{v}) - rT(\mathbf{v})$ is orthogonal to each vector in a basis of V. Hence, $T(r\mathbf{v}) - rT(\mathbf{v}) = 0$. Similarly, use (b) thrice to get

$$\langle [T(\mathbf{u} + \mathbf{v}) - T(\mathbf{u}) - T(\mathbf{v})], T(\mathbf{w}) \rangle = \langle T(\mathbf{u} + \mathbf{v}), T(\mathbf{w}) \rangle - \langle T(\mathbf{u}), T(\mathbf{w}) \rangle - \langle T(\mathbf{v}), T(\mathbf{w}) \rangle$$
$$= \langle \mathbf{u} + \mathbf{v}, \mathbf{w} \rangle - \langle \mathbf{u}, \mathbf{w} \rangle - \langle \mathbf{v}, \mathbf{w} \rangle = 0.$$

Hence, $T(\mathbf{u} + \mathbf{v}) - T(\mathbf{u}) - T(\mathbf{v})$ is orthogonal to $T(\mathbf{w})$ for all $\mathbf{w}$; as before, (c) implies that $T(\mathbf{u} + \mathbf{v}) - T(\mathbf{u}) - T(\mathbf{v}) = \mathbf{0}$. Hence T is linear.

Exercises 9.5   Fourier Approximation

The integrations involved in the computation of the Fourier coefficients are omitted in 1(b), 1(d), and 2(b).

1(b) $f_5 = \frac{\pi}{2} - \frac{4}{\pi}\left(\cos x + \frac{\cos 3x}{3^2} + \frac{\cos 5x}{5^2}\right)$

(d) $f_5 = \frac{\pi}{4} + \left(\sin x - \frac{\sin 2x}{2} + \frac{\sin 3x}{3} - \frac{\sin 4x}{4} + \frac{\sin 5x}{5}\right) - \frac{2}{\pi}\left(\cos x + \frac{\cos 3x}{3^2} + \frac{\cos 5x}{5^2}\right)$

2(b) $\frac{2}{\pi} - \frac{8}{\pi}\left(\frac{\cos 2x}{2^2-1} + \frac{\cos 4x}{4^2-1} + \frac{\cos 6x}{6^2-1}\right)$

4. $\int_0^{\pi} (\cos kx \cos lx)\,dx = \frac{1}{2}\left[\dfrac{\sin(k+l)x}{k+l} - \dfrac{\sin(k-l)x}{k-l}\right]_0^{\pi} = 0$ if $k \neq l$.

5(b) By 1(c): $x^2 = \frac{\pi^2}{3} - 4\left[\cos x - \frac{1}{2^2}\cos(2x) + \frac{1}{3^2}(\cos(3x)) - \cdots\right]$. Take $x = 0$.

APPENDIX

Exercises A   Complex Numbers

1(b) $12 + 5i = (2 + xi)(3 - 2i) = (6 + 2x) + (-4 + 3x)i$. Equating real and imaginary parts gives $6 + 2x = 12$, $-4 + 3x = 5$, so $x = 3$.

(d) $5 = (2 + xi)(2 - xi) = (4 + x^2) = 0i$. Hence $4 + x^2 = 5$, so $x = \pm 1$.

2(b) $(3 - 2i)(1 + i) + |3 + 4i| = (5 + i) + \sqrt{9 + 16} = 10 + i$.

(d) $\begin{aligned}
\frac{3 - 2i}{1 - i} - \frac{3 - 7i}{2 - 3i} &= \frac{(3 - 2i)(1 + i)}{(1 - i)(1 + i)} - \frac{(3 - 7i)(2 + 3i)}{(2 - 3i)(2 + 3i)} \\
&= \frac{5 + i}{1 + 1} - \frac{27 - 5i}{4 + 9} \\
&= \tfrac{11}{26} + \tfrac{23}{26}i
\end{aligned}$

(f) $(2 - i)^3 = (2pi)^2(2 - i) = (3 - 4i)(2 - i) = 2 - 11i$

(h) $(1 - i)^2(2 + i)^2 = (-2i)(3 + 4i) = 8 - 6i$

3(b) $iz + 1 = i + z - 6i + 3iz = -5i + (1 + 3i)z$. Hence $1 + 5i = (1 + 2i)z$, so

$$z = \frac{1 + 5i}{1 + 2i} = \frac{(1 + 5i)(1 - 2i)}{(1 + 2i)(1 - 2i)} = \frac{11 + 3i}{1 + 4} = \tfrac{11}{5} + \tfrac{3}{5}i.$$

(d) $z^2 = 3 - 4i$. If $z = a + bi$ the condition is $(a^2 - b^2) + (2ab)i = 3 - 4i$, whence $a^2 - b^2 = 3$ and $ab = -2$. Thus $b = \frac{-2}{9}$, $a^2 - \frac{4}{a^2} = 3$, $a^4 - 3a^2 - 4 = 0$. This factors as $(a^2 - 4)(a^2 + 1) = 0$, so $a = \pm 2$, $b = \mp 1$, $z = \pm(2 - i)$.

(f) Write $z = a + bi$. Then the condition reads

$$(a + bi)(2 - i) = (a - bi + 1)(1 + i)$$
$$(2a + b) + (2b - a)i = (a + 1 + b) + (a + 1 - b)i.$$

Thus $2a + b = a + 1 + b$ and $2b - a = a + 1 - b$; whence $a = 1$, $b = 1$, so $z = 1 + i$.

4(b) $x = \frac{1}{2}\left[-(-1) \pm \sqrt{(-1)^2 - 4}\right] = \frac{1}{2}\left[1 \pm i\sqrt{3}\right]$

(d) $x = \frac{1}{4}\left[-(-5) \pm \sqrt{(-5)^2 - 4 \cdot 2 \cdot 2}\right] = \frac{1}{4}\left[5 \pm \sqrt{9}\right] = 2, \frac{1}{2}$.

5(b) If $x = re^{i\theta}$ then $x^3 = -8$ becomes $r^3 e^{3i\theta} = 8e^{\pi i}$. Thus $r^3 = 8$ (whence $r = 2$) and $3\theta = \pi + 2k\pi$. Hence $\theta = \frac{\pi}{3} + k \cdot \frac{2\pi}{3}$, $k = 0, 1, 2$. The roots are

$$2e^{i\pi/3} = 1 + \sqrt{3}i \quad (k = 0)$$
$$2e^{\pi i} = -2 \qquad\quad (k = 1)$$
$$2e^{5\pi i/3} = 1 - \sqrt{3}i \quad (k = 2).$$

(d) If $x = re^{-i\theta}$ then $x^4 = 64$ becomes $r^4 e^{4i\theta} = 64 e^{i \cdot 0}$. Hence $r^4 = 64$ (whence $r = 2\sqrt{2}$) and $4\theta = 0 + 2k\pi$; $\theta = k\frac{\pi}{2}$, $k = 0, 1, 2, 3$. The roots are

$$2\sqrt{2}e^{0i} = 2\sqrt{2} \quad (k = 0)$$
$$2\sqrt{2}e^{\pi i/2} = 2\sqrt{2}i \quad (k = 1)$$
$$2\sqrt{2}e^{\pi i} = -2\sqrt{2} \quad (k = 2)$$
$$2\sqrt{2}e^{3\pi i/2} = -2\sqrt{2}i \quad (k = 3).$$

6(b) The quadratic is $(x - u)(x - \bar{u}) = x^2 - (u + \bar{u})x + u\bar{u} = x^2 - 4x + 13$. The other root is $\bar{u} = 2 + 3i$.

(d) The quadratic is $(x - u)(x - \bar{u}) = x^2 - (u + \bar{u})x + u\bar{u} = x^2 - 6x + 25$. The other root is $\bar{u} = 3 + 4i$.

8. If $u = 2 - i$, then u is a root of $(x - u)(x - \bar{u}) = x^2 - (u + \bar{u})x + u\bar{u} = x^2 - 4x + 5$. If $v = 3 - 2i$, then v is a root of $(x - v)(x - \bar{v}) = x^2 - (v + \bar{v})x + v\bar{v} = x^2 - 6x + 13$. Hence u and v are roots of

$$(x^2 - 4x + 5)(x^2 - 6x + 13) = x^4 - 10x^3 + 42x^2 - 82x + 65.$$

10(b) Taking $x = u = -2$: $x^2 + ix - (4 - 2i) = 4 - 2i - 4 + 2i = 0$. If v is the other root then $u + v = -i$ (i is the coefficient of x) so $v = -u - i = 2 - i$.

(d) Taking $x = u = -2 + i$: $(-2 + i)^2 \quad +3(1 - i)(-2 + i) - 5i$
$$= (3 - ri) + 3(-1 + 3i) - 5i$$
$$= 0.$$
If v is the other root then $u + v = -3(1 - i)$, so $v = -3(1 - i) - u = -1 + 2i$.

11(b) $x^2 - x + (1 - i) = 0$ gives $x = \frac{1}{2}\left[1 \pm \sqrt{1 - 4(1 - i)}\right] = \frac{1}{2}\left[1 \pm \sqrt{-3 + 4i}\right]$. Write $w = \sqrt{-3 + 4i}$ so $w^2 = -3 + 4i$. If $w = a + bi$ then $w^2 = (a^2 - b^2) + (2ab)i$, so $a^2 - b^2 = -3$, $2ab = 4$. Thus $b = \frac{2}{a}$, $a^2 - \frac{4}{a^2} = -3$, $a^4 + 3a^2 - 4 = 0$, $(a^2 + 4)(a^2 - 1) = 0$, $a = \pm 1$, $b = \pm 2$, $w = \pm(1 + 2i)$. Finally the roots are $\frac{1}{2}\left[1 \pm w\right] = 1 + i, -i$.

(d) $x^2 - 3(1 - i)x - 5i = 0$ gives $x = \frac{1}{2}\left[3(1 - i) \pm \sqrt{9(1 - i)^2 + 20i}\right] = \frac{1}{2}\left[3(1 - i) \pm \sqrt{2i}\right]$. If $w = \sqrt{2i}$ then $w^2 = 2i$. Write $w = a + bi$ so $(a^2 - b^2) + 2abi = 2i$. Hence $a^2 = b^2$ and $ab = 1$; the solution is $a = b = \pm 1$ so $w = \pm(1 + i)$. Thus the roots are $x = \frac{1}{2}(3(1 - i) \pm w) = 2 - i$, $1 - 2i$.

12(b) $|z - 1| = 2$ means that the distance from z to 1 is 2. Thus the graph is the circle, radius 2, center at 1.

(d) If $z = x + yi$, then $z = -\bar{z}$ becomes $x + yi = -x + yi$. This holds if and only if $x = 0$; that is if and only if $z = yi$. Hence the graph is the imaginary axis.

(f) If $z = x + yi$, then im $z = m \cdot$ re z becomes $y = mx$. This is the line through the origin with slope m.

18(b) $-4i = 4e^{3\pi i/2}$.

(d) $\left|-4 + 4\sqrt{3}i\right| = 4\sqrt{1+3} = 8$ and $\cos\varphi = \frac{4}{8} = \frac{1}{2}$. Thus $\varphi = \frac{\pi}{3}$, so $\theta = \frac{2\pi}{3}$. Then $-4 + 4\sqrt{3}i = 8e^{2\pi i/3}$.

(f) $\left|-6 + 6i\right| = 6\sqrt{1+1} = 6\sqrt{2}$ and $\cos\varphi = \frac{6}{6\sqrt{2}} = \frac{1}{\sqrt{2}}$. Thus $\varphi = \frac{\pi}{4}$ so $\theta = \frac{3\pi}{4}$; whence $- + 6i = 6\sqrt{2}e^{-3\pi i/4}$.

19(b) $e^{7\pi i/3} = e^{(\pi/3 + 2\pi)i} = e^{\pi i/3} = \cos\frac{\pi}{3} + i\sin\frac{\pi}{3} = \frac{1}{2} + \frac{\sqrt{3}}{2}i$.

(d) $\sqrt{2}e^{-\pi i/4} = \sqrt{2}\left(\cos\left(\frac{-\pi}{4}\right) + i\sin\left(\frac{-\pi}{4}\right)\right) = \sqrt{2}\left(\frac{1}{\sqrt{2}} - \frac{1}{\sqrt{2}}i\right) = 1 - i$.

(f) $2\sqrt{3}e^{-2\pi i/6} = 2\sqrt{3}\left(\cos\left(\frac{-\pi}{3}\right) + i\sin\left(\frac{-\pi}{3}\right)\right) = 2\sqrt{3}\left(\frac{1}{2} - \frac{\sqrt{3}}{2}i\right) = \sqrt{3} - 3i$.

20(b)
$$
\begin{aligned}
(1 + \sqrt{3}i)^{-4} &= (2e^{\pi i/3})^{-4} = 2^{-4}e^{-4\pi i/3} \\
&= \tfrac{1}{16}\left[\cos(-4\pi/3) + i\sin(-4\pi/3)\right] \\
&= \tfrac{1}{16}\left(-\tfrac{1}{2} + \tfrac{\sqrt{3}}{2}i\right) \\
&= -\tfrac{1}{32} + \tfrac{\sqrt{3}}{32}i.
\end{aligned}
$$

(d)
$$
\begin{aligned}
(1 - i)^{10} &= \left[\sqrt{2}e^{-\pi i/4}\right]^{10} = (\sqrt{2})^{10}e^{-5\pi i/2} = (\sqrt{2})^{10}e^{(-\pi/2 - 2\pi)i} \\
&= (\sqrt{2})^{10}e^{-\pi i/2} = 2^5\left[\cos\left(\frac{-\pi}{2}\right) + i\sin\left(\frac{-\pi}{2}\right)\right] \\
&= 32(0 - i) = -32i.
\end{aligned}
$$

(f)
$$
\begin{aligned}
(\sqrt{3} - i)^9(2 - 2i)^5 &= \left[2e^{-\pi i/6}\right]^9\left[2\sqrt{2}e^{-\pi i/4}\right]^5 \\
&= 2^9 e^{-3\pi i/2}(2\sqrt{2})^5 e^{-5\pi i/4} \\
&= 2^9(i)2^5(\sqrt{2})^4\sqrt{2}\left(-\tfrac{1}{\sqrt{2}} + \tfrac{1}{\sqrt{2}}i\right) \\
&= 2^{16}i(-1 + i) \\
&= -2^{16}(1 + i).
\end{aligned}
$$

23(b) Write $z = re^{i\theta}$. Then $z^4 = 2(\sqrt{3}i - 1)$ becomes $r^4 e^{4i\theta} = 4e^{2\pi i/3}$. Hence $r^4 = 4$, so $r = \sqrt{2}$, and $4\theta = \frac{2\pi}{3} + 2\pi k$; that is

$$
\theta = \tfrac{\pi}{6} + \tfrac{\pi}{2}k \qquad k = 0, 1, 2, 3.
$$

The roots are

$$
\begin{aligned}
\sqrt{2}e^{\pi i/6} &= \sqrt{2}\left(\tfrac{\sqrt{3}}{2} + \tfrac{1}{2}i\right) = \tfrac{\sqrt{2}}{2}\left(\sqrt{3} + i\right) \\
\sqrt{2}e^{4\pi i/9} &= \sqrt{2}\left(-\tfrac{1}{2} + \tfrac{\sqrt{3}}{2}i\right) = \tfrac{\sqrt{2}}{2}\left(-1 + \sqrt{3}i\right) \\
\sqrt{2}e^{7\pi i/6} &= \sqrt{2}\left(-\tfrac{\sqrt{3}}{2} - \tfrac{1}{2}i\right) = -\tfrac{\sqrt{2}}{2}\left(1 + \sqrt{3}i\right) \\
\sqrt{2}e^{10\pi i/6} &= \sqrt{2}\left(\tfrac{1}{2} - \tfrac{\sqrt{3}}{2}i\right) = -\tfrac{\sqrt{2}}{2}\left(-1 + \sqrt{3}i\right)
\end{aligned}
$$

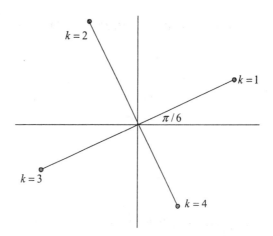

26(b) Each point on the unit circle has polar form $e^{i\theta}$ for some angle θ. As the n points are equally spaced, the angle between consecutive points is $\frac{2\pi}{n}$. Suppose the first one into the first quadrant is $z_0 = e^{\alpha i}$. Write $w = e^{2\pi i/n}$. If the points are labeled $z_1, z_2, z_3, \ldots, z_n$ around the unit circle, they have polar form

$$z_1 = e^{\alpha i}$$

$$z_2 = e^{(\alpha + 2\pi/n)i} = e^{\alpha i}e^{2\pi i/n} = z_1 w$$

$$z_3 = e^{[\alpha + 2(2\pi/n)]i} = e^{\alpha i}e^{4\pi i/n} = z_1 w^2$$

$$z_4 = e^{[\alpha + 3(2\pi/n)]i} = e^{\alpha i}e^{6\pi i/n} = z_1 w^3$$

$$\vdots$$

$$z_n = e^{[\alpha + (n-1)(2\pi/n)]i} = e^{ai}e^{2(n-1)\pi i/n} = z_1 w^{n-1}.$$

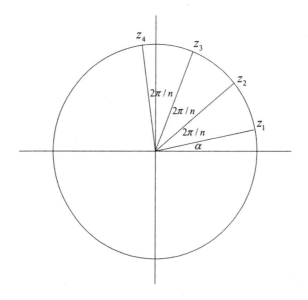

Hence the sum of the roots is

$$z_1 + z_2 + \cdots + z_n = z_1(1 + w + \cdots + w^{n-1}). \tag{*}$$

Now $w^n = \left(e^{2\pi i/n}\right)^n = e^{2\pi i} = 1$ so

$$0 = 1 - w^n = (1 - w)(1 + w + w^2 + \cdots + w^{n-1}).$$

As $w \neq 1$, this gives $1 + w + \cdots + w^{n-1} = 0$. Hence (*) gives

$$z_1 + z_2 + \cdots + z_n = z_1 \cdot 0 = 0.$$

Exercises B   Mathematical Induction

6. Write S_n for the statement

$$\frac{1}{1 \cdot 2} + \frac{1}{2 \cdot 3} + \cdots + \frac{1}{n(n+1)} = \frac{n}{n+1}. \tag{S_n}$$

Then S_1 is true: It reads $\frac{1}{1 \cdot 2} = \frac{1}{1+1}$, which is true. Now <u>assume</u> S_n is true for some $n \geq 1$. We must use S_n to show that S_{n+1} is also true. The statement S_{n+1} reads as follows:

$$\frac{1}{1 \cdot 2} + \frac{1}{2 \cdot 3} + \cdots + \frac{1}{(n+1)(n+2)} = \frac{n+1}{n+2}.$$

The second last term on the left side is $\frac{1}{n(n+1)}$ so we can use S_n:

$$\frac{1}{1 \cdot 2} + \frac{1}{2 \cdot 3} + \cdots + \frac{1}{(n+1)(n+2)} = \left[\frac{1}{1 \cdot 2} + \frac{1}{2 \cdot 3} + \cdots + \frac{1}{n(n+1)} \right] + \frac{1}{(n+1)(n+2)}$$

$$= \frac{n}{n+1} + \frac{1}{(n+1)(n+2)}$$

$$= \frac{n(n+2) + 1}{(n+1)(n+2)}$$

$$= \frac{(n+1)^2}{(n+1)(b+2)}$$

$$= \frac{n+1}{n+2}.$$

Thus S_{n+1} is true and the induction is complete.

14. Write S_n for the statement

$$\frac{1}{1^2} + \frac{1}{2^2} + \cdots + \frac{1}{n^2} \leq 2 - \frac{1}{n}. \tag{S_n}$$

Then S_1 is true as it asserts that $\frac{1}{1^2} \leq 2 - \frac{1}{1}$, which is true. Now assume that S_n is true for some $n \geq 1$. We must use S_n to show that S_{n+1} is also true. The statement S_{n+1} reads as follows:

$$\frac{1}{1^2} + \frac{1}{2^2} + \cdots + \frac{1}{(n+1)^2} = \left[\frac{1}{1^2} + \frac{1}{2^2} + \cdots + \frac{1}{n^2} \right] + \frac{1}{(n+1)^2}$$

$$\leq \left[2 - \frac{1}{n} \right] + \frac{1}{(n+1)^2}$$

$$\leq \left[2 - \frac{1}{n} \right] + \frac{1}{n(n+1)}$$

$$= \left[2 - \frac{1}{n} \right] + \left[\frac{1}{n} - \frac{1}{n+1} \right]$$

$$= 2 - \frac{1}{n+1}.$$

Thus S_{n+1} is true and the induction is complete.